D1109545

Modern Algebra: A First Course

Modern Algebra: a first course

HERBERT A. HOLLISTER

Bowling Green State University

Harper & Row, Publishers
New York / Evanston / San Francisco / London

Modern Algebra: A First Course

Copyright © 1972 by Herbert A. Hollister

Printed in the United States of America. All rights reserved. No part of this book may be used or reproduced in any manner whatsoever without written permission except in the case of brief quotations embodied in critical articles and reviews. For information address Harper & Row, Publishers, Inc., 49 East 33rd Street, New York, N.Y. 10016.

Standard Book Number: 06-042864-3

Library of Congress Catalog Card Number: 74-184932

To my mother and to the memory of my father

Contents

Preface

This text is intended for use in introductory abstract algebra courses taught at the undergraduate level immediately following a sequence of courses in calculus. It is suitable for a two-quarter or two-semester course covering the entire text or a one-semester or one-quarter course covering the essentials of the first six chapters. The sequence of topics was determined by pedagogical reasons rather than a desire for mathematical elegance. An attempt has been made to give the student some feeling for the abstraction process by studying certain specific structures before studying abstract systems. An abundance of examples from the student's previous mathematical experience has been provided, in order for the abstract concepts to have more meaning for him. The student is asked to "dirty his hands" in specific cases, especially on examples of new systems such as factor groups and direct sums. The fundamental homomorphism and isomorphism theorems are delayed until the student, by working with examples, has developed a familiarity with and an appreciation for the concepts involved.

Many of the proofs of elementary propositions and corollaries are left as exercises in order to involve the reader in the development of the theory. The notion of generalizing from specific examples or from one class of systems to another has been emphasized, and the student is frequently asked to adopt or extend a previously given proof to a more general or abstract situation.

The author wishes to thank L. Graue, W. Kirby, and M. Satyanarayana for their help in class-testing most of the material and for their helpful suggestions in preparing the final manuscript. Thanks are also due the typist, Miss Janice Csokmay, and Mrs. Mary Chambers for her stenographic assistance. Finally, I wish to thank my family for their cooperation and tolerance.

HERBERT A. HOLLISTER

Bowling Green, Ohio

Foreword
to the Student

This course may be considerably different from the studies you have made in the past but will be similar to courses that you will take in the future. In elementary courses you frequently skimmed the text and concentrated on the problems, the idea being that the focus of the course was on the ability to do problems. In advanced courses you need to concentrate on the textual material, since the focus of the course is on understanding and mastery of the theory. The problems in this book are designed to help you better understand the text by studying some specific examples and by participating in the development of the theory by proving some of the elementary assertions for yourself. The exercises marked with a double asterisk are intended as challenges for the better students, and those exercises with a triple asterisk are even more challenging.

The text is designed for students who have completed the calculus sequence. It will be helpful if you have been exposed to the rudiments of linear algebra or have had some experience with matrices. If you have not had such exposure, an elementary discussion of 2×2 and 3×3 matrices over the real numbers is provided in Appendix 1. It is assumed that you know the basic terminology of calculus and geometry. Elementary analytic and geometric concepts are frequently used for illustrative purposes. Certain examples and exercises are marked with a single asterisk to indicate that not all readers will have the mathematical background needed to understand the concept under discussion. For familiar-sounding cases, you should consult your textbooks from earlier courses or a mathematical dictionary; ignore the cases not included in your background.

This course will probably introduce you to additional new terminology that has not been included in any of your previous mathematics courses. Comprehension of concepts in the text requires that you develop a good working knowledge of the vocabulary. It is also extremely important that you know the precise meaning of each technical term introduced. Trying to learn the theory without knowing the language is somewhat like trying to read French poetry without knowing any French; you may get the right sounds but you won't understand what it means.

HERBERT A. HOLLISTER

Modern Algebra:
A First Course

1
SETS, FUNCTIONS, AND RELATIONS

The student who has progressed through the calculus or a modern secondary mathematics program is familiar with the intuitive notion of a set and with some of the language of set theory. The axiomatic development of abstract algebra rests upon a set-theoretic foundation. We shall, however, make no attempt to develop set theory in this chapter, but rather shall merely discuss certain basic concepts and establish some notational conventions. We shall give precise set-theoretic definitions of function, relation, partition, and operation and shall develop some properties of these concepts before going on to the actual study of algebra.

§1. Sets

Intuitively, the word *set* means the same as the words "collection," "class," "aggregate," and "bunch." The "things" in a set are called its *elements*. Thus, for example, *Naive Set Theory* is an element in the set of all books written by Paul Halmos, Bowling Green University is an element of the set of all universities in Ohio, and $\sqrt{2}$ is an element of the set of all real numbers. We use letters to name sets and elements of sets, just as we use letters to name numbers in algebra and points in geometry. Usually we use lowercase letters to indicate elements and uppercase letters to denote sets. Occasionally we wish to discuss a set in which each element is itself a set; in such a case we may speak of a *family* of sets rather than of a set of sets. Families of sets will, in this text, frequently be denoted by script letters.

If A is a set and p is an element of A, we can abbreviate the sentence "p is an element of A" by

$$p \in A$$

The symbol "$\in$" is read "is an element of," "belongs to," or "is in."

If A and B are sets such that every element of A is an element of B,

1

we say that A is a *subset* of B. A is a *proper* subset of B if A is a subset of B but A is not equal to B.

The sentence "A is a subset of B" can be abbreviated by

$$A \subseteq B$$

The symbol "$\subseteq$" is read "is a subset of" or "is contained in." Do not confuse the symbols "$\in$" and "$\subseteq$"; they have different meanings. To negate these symbols we follow the customary practice of drawing a line through the symbol. That is, $p \notin A$ means p is not an element of A, and $A \not\subseteq B$ means A is not a subset of B.

One of the simplest methods of describing a set is just to list its elements within a set of braces. For example, the set of prime numbers less than 10 is given by

$$\{2, 3, 5, 7\}$$

Sets can also be described by means of a set of braces, a variable, and a sentence containing that variable. The set so described is the set of objects that when substituted for that variable yields a true sentence. This notation is called "set-builder notation."

Examples. $\{x : x$ is a prime integer$\}$ is read

"The set of all x such that x is a prime integer."

If a and b are real numbers, we can define the intervals determined by a and b by

$$(a, b) = \{x : a < x < b\} \qquad [a, b) = \{x : a \leq x < b\}$$
$$(a, b] = \{x : a < x \leq b\} \qquad [a, b] = \{x : a \leq x \leq b\}$$

In the above examples we would expect all possible replacements for x to be real numbers. In some cases the possible replacements for x may not be clear from the context, and we specify them by requiring in our notation that x be an element of a specific set.

Example.   Let N denote the set of all positive integers. Then $\{x \in N : x > 5\}$ is the set of all positive integers greater than 5. We read the above expression as "The set of all x in N such that x is greater than 5."

Using set-builder notation or some other descriptive method, one might describe a set that has no elements. For example,

$$\{x \in N : x \neq x\}$$

is a set with no elements, as is

$$\{y \in N : y^2 = 2\}$$

Since two sets are the same if they have the same elements, these two sets are the same, as are all sets with no elements. The set with no elements is called the *empty set*, the *void set*, or the *null set* and is denoted by ϕ.

If A and B are sets, then the *union of A and B* is denoted by $A \cup B$ and is defined by

$$A \cup B = \{x : x \in A \text{ or } x \in B\}$$

If $\mathcal{A}$ is a collection of sets, the *union of $\mathcal{A}$* is denoted by $\cup \mathcal{A}$ and defined by

$$\cup \mathcal{A} = \{x : x \in A \text{ for some } A \text{ in } \mathcal{A}\}$$

Frequently, a collection of sets will be indexed by some set; that is, each set in the collection will be associated with one element in the indexing set. If $\mathcal{A} = \{A_i : i \in I\}$ is such a collection with an indexing set I, then we may denote $\cup \mathcal{A}$ by $\cup A_i$ or $\cup_{i \in I} A_i$. For example, if for each positive integer n we define P_n to be the set of all polynomials over the real numbers of degree less than or equal to n, then $\cup P_n$ is the set of all polynomials over the real numbers. We could also denote this set by $\cup_{n \in \mathbf{N}} P_n$.

The *intersection* of two sets, A and B, is denoted by $A \cap B$ and is defined by

$$A \cap B = \{x : x \in A \text{ and } x \in B\}$$

If $A \cap B = \phi$, then we say that A and B are *disjoint*.

If $\mathcal{A}$ is a nonempty collection of sets, then the *intersection* of $\mathcal{A}$ is denoted by $\cap \mathcal{A}$ and defined by

$$\cap \mathcal{A} = \{x : x \in A \text{ for every } A \text{ in } \mathcal{A}\}$$

If $\mathcal{A} = \{A_i : i \in I\}$ is a nonempty collection of sets with an indexing set I, we frequently choose to denote $\cap \mathcal{A}$ by $\cap A_i$ or $\cap_{i \in I} A_i$. Using the example above, $\cap_{n \in \mathbf{N}} P_n$ is the set of all polynomials of degree less than or equal to 1.

If A and B are sets, then the *difference* of A and B is denoted by $A - B$ and is defined by

$$A - B = \{x : x \in A \text{ and } x \notin B\}$$

Examples. If I is the set of all isosceles triangles in the plane, R is the set of all right triangles in the plane, and E is the set of all equilateral triangles in the plane, then $I \cup R$ is the set of all triangles that are either isosceles or right, and $I \cap R$ is the set of all isosceles right triangles. Furthermore, $I - R$ is the set of all isosceles triangles that are not right triangles, and $R - I$ is the set of all right triangles that are not isosceles. Notice that $E \cup I = I$ and $E \cap I = E$, because $E \subseteq I$. Note also that E and R are disjoint.

If $\mathcal{B} = \{[0, r) : r$ is a positive real number$\}$, then $\bigcup \mathcal{B}$ is the set of all nonnegative real numbers and $\bigcap \mathcal{B} = \{0\}$.

In analytic geometry we name points by using ordered pairs of numbers such as $(2, 7)$ and $(-3, 5)$, and the idea of an ordered pair of numbers generalizes easily to the idea of an ordered pair in which the elements come from arbitrary sets.[1]

If A and B are sets, the *cartesian product* of A and B is denoted by $A \times B$ and defined by

$$A \times B = \{(x, y) : x \in A \text{ and } y \in B\}$$

Just as we can study ordered triples of numbers or ordered n-tuples of numbers, we can consider ordered n-tuples where the elements come from arbitrary sets. Thus, if $A_1, A_2, \ldots, A_n$ are sets, we denote the cartesian product of $A_1, A_2, \ldots, A_n$ by $A_1 \times A_2 \times \cdots \times A_n$ and define it by

$$A_1 \times A_2 \times \cdots \times A_n = \{(a_1, a_2, \ldots, a_n) : a_1 \in A_1, a_2 \in A_2, \ldots, a_n \in A_n\}$$

Following are some very useful elementary properties of sets.

(a) If A and B are sets such that $A \subseteq B$ and $B \subseteq A$, then $A = B$.

(b) If A is a set, then $\emptyset \subseteq A$.

(c) If A and B are sets, then $A \cup B = B \cup A$, and $A \cap B = B \cap A$.

(d) If A, B, and C are sets, then

$$A \cup (B \cup C) = (A \cup B) \cup C \quad \text{and}$$
$$A \cap (B \cap C) = (A \cap B) \cap C$$

(e) If A, B, and C are sets, then

$$A \cup (B \cap C) = (A \cup B) \cap (A \cup C) \quad \text{and}$$
$$A \cap (B \cup C) = (A \cap B) \cup (A \cap C)$$

(f) If A, B, and C are sets such that $A \subseteq B$ and $B \subseteq C$, then $A \subseteq C$.

(g) If A and B are sets, then

$$(A \cap B) \subseteq A, \quad (A \cap B) \subseteq B, \quad A \subseteq (A \cup B), \quad \text{and} \quad B \subseteq (A \cup B)$$

Statement (a) is one of the most commonly used tools from set theory. To show that two sets are the same the easiest approach in many cases is to show that each is a subset of the other. In a formal development of set theory this statement is one of the fundamental axioms, called the *Axiom of Extension*. The other assertions above can be proved quite easily using this statement and the definitions. We leave it to the reader

---

[1]One can define ordered pairs precisely and formally using set theory, but such a formulation is not very useful. It is possible to define the ordered pair (a, b) by $(a, b) = \{\{a\}, \{a, b\}\}$ and then prove that $(a, b) = (x, y)$ iff $a = x$ and $b = y$. Such proofs, however, rightly belong to set theory and, although they are interesting, we shall not concern ourselves with them, primarily because so doing would delay our study of algebraic structures. The intuitive notion of an ordered list will suffice for our purposes.

to convince himself of their validity and shall use them without further justification. This course is a course in algebra, and our concern with set theory is primarily as a technical language that we shall use in our development. The study of set theory for its own sake is usually undertaken in a course in the foundations of mathematics.

Venn diagrams frequently are used to describe relationships among sets both intuitively and geometrically. We usually draw a rectangle to represent some set containing all the sets under discussion and call this the universal set or universe of discourse. We then draw all sets under consideration as sets of points inside circles or other closed curves. For example, the union and intersection of sets A and B could be represented by the drawings in Figures 1-1 and 1-2. The union is shaded in the first drawing and the intersection is shaded in the second.

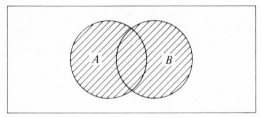

Figure 1-1

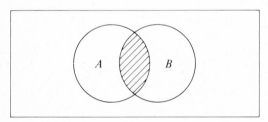

Figure 1-2

The drawing in Figure 1-3 indicates that A and B are disjoint, and Figure 1-4 indicates that $A \subseteq B$.

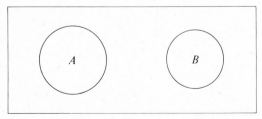

Figure 1-3

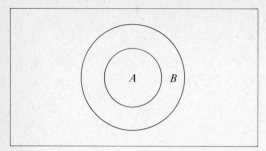

Figure 1-4

We can use Venn diagrams to help our intuition but we must realize that such drawings do not constitute proofs. By shading $A \cup (B \cap C)$ in Figure 1-5 and $(A \cup B) \cap (A \cup C)$ in Figure 1-6 we might begin to expect that these sets were the same, but we could not claim to have a proof that such an equation would always hold.

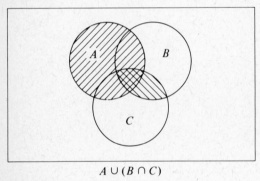

$$A \cup (B \cap C)$$

Figure 1-5

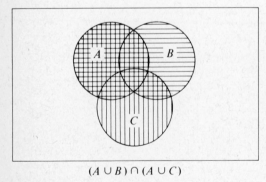

$$(A \cup B) \cap (A \cup C)$$

Figure 1-6

In Exercise 1 of Set A you are asked to decide whether certain set-theoretic statements are true or false. You may use Venn diagrams to help your intuition, but a proof is needed for a final decision. The diagram, however, may help you to construct the needed logical argument.

EXERCISE SET A

1. Let A, B, C, and D be sets. Decide whether the following are true or false. Give reasons for your answers.
   (a) $A \cap (A \cup B) = A$.
   (b) $A \cup (A \cap B) = A$.
   (c) If $A \subseteq B$ and $B \cap C = \emptyset$, then $A \cap C = \emptyset$.
   (d) If $A \subseteq B$ and $A \subseteq C$, then $A \subseteq B \cap C$.
   (e) If $A \cup B \subseteq A \cap B$, then $A = B$.
   (f) $(A - B) \cap C = (A \cap C) - B$.
   (g) If $A \cup B = A \cup C$, then $B = C$.
   (h) If $A \subseteq B$ and $C \subseteq D$, then $A \cap C \subseteq B \cap D$.
   (i) $(A - B) \cap B = \emptyset$.
   (j) If $A \subseteq B$ and $C \subseteq D$, then $A \times C \subseteq B \times D$.
   (k) $A \times \emptyset = \emptyset$.
   (l) $A \times B = B \times A$.
   (m) If $A \times B \subseteq B \times B$, then $A \subseteq B$.
   (n) $A - (B \cap C) = (A - B) \cap (A - C)$.
   (o) $A - (B \cap C) = (A - B) \cup (A - C)$.
   (p) $A - (B \cup C) = (A - B) \cap (A - C)$.

§2. Functions

The word *function* is not new to the student of mathematics. Many times in both mathematical and nonmathematical situations we assign objects of one set to objects of another set. The intuitive idea of a function is some kind of assigning process whereby each object in one set is assigned to some particular element in (possibly) another set. Such an intuitive description is not, however, precise enough for mathematics. The idea of "assignment" is a rather loose intuitive notion, and it is this that we need to make precise. If each element of the set A is assigned by some function to exactly one element of the set B, then this association could be determined just by examining some of the ordered pairs in $A \times B$, in particular by examining those ordered pairs in which the first element is assigned to the second. If we know all of these pairs we know the association or function, and therefore this set of ordered pairs determines the assignment process. For this reason we use ordered pairs to define the word "function," and our intuitive idea is made precise by the following definition.

If A and B are sets and $f \subseteq A \times B$, we say that f is a *function* from

A into B if for each element a in A there is exactly one element b in B such that $(a,b) \in f$.

An equivalent phrasing of the above is as follows:

A function from A into B is defined to be a set of ordered pairs contained in $A \times B$ with the property that each element of A is the first element of exactly one of the pairs in the set.

The notation $(a, b) \in f$ is not very useful. In calculus we usually write $f(a) = b$ instead of $(a, b) \in f$, and in this text we shall write $(a)f = b$ or just $af = b$ instead of $(a, b) \in f$.

It may seem strange to write functions on the right instead of on the left. Realize, however, that you have been writing many functions on the right all along, without thinking of them that way. For example, exponents and factorials are written on the right,

$$2^3 \quad \text{and} \quad 4!$$

and if f is the cubing function and g is the factorial function we have $xf = x^3$ and $ng = n!$

The expression "$af = b$" is read "a acted on by f is b" or just af equals b and is indicative of our intuitive notion that the function f "does something" to a. Since each element of A appears only once as a first element in one of the pairs in f, $(a)f$ is a uniquely determined element of B. Drawings such as Figure 1-7 sometimes help one to get a better intuitive grasp of the concept of a function.

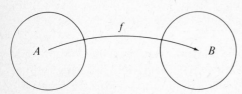

Figure 1-7

If f is a function from A into B, we call A the *domain* of f and B the *range* of f. If d is an element of A, we call df the *image of d under f*. We frequently call df the *value of f at d* even though df may not be a number. If S is a subset of A,

$$Sf = \{af : a \in S\}$$

is called the *image of S under f*.

If b is in B, we define bf^{-1} by

$$bf^{-1} = \{x \in A : xf = b\}$$

The set bf^{-1} is called the *inverse image of b under f*. If H is a subset of B, we define Hf^{-1} by

$$Hf^{-1} = \{x \in A : xf \in H\}$$

The set Hf^{-1} is called the *inverse image of H under f.* The use of the symbol f^{-1} in this definition does not mean that f^{-1} is a function.

The word *mapping* or *map* is used interchangeably with the word function in many parts of mathematics. We shall consider the word mapping to mean the same as the word function, but the intuitive notion of a function "doing something" is emphasized by the word mapping. We shall say that the function f maps a to b if $af = b$, that is, if b is the image of a under f. Similarly, we say that f maps the subset M of A into the subset N of B if $Mf \subseteq N$, that is, if the image of M under f is contained in N and that f maps M *onto* N if $Mf = N$.

If g is a function from the set A into the set B, and M is a subset of A, then for each element p in M, since $p \in A$, there is exactly one element q in B such that $(p, q) \in g$. Thus, the set of pairs in g whose first elements are in M is, in fact, a function from M into B. This function is called the *restriction of g to M* or *g restricted to M.* This function is frequently denoted by $g \mid_M$. Using set-builder notation we have

$$g \mid_M = \{(a,b) \in g : a \in M\}$$

The concept is illustrated in Figure 1–8.

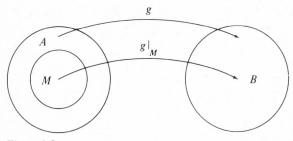

Figure 1-8

Functions from one set of numbers to another are often described by algebraic rules. For example, the function that assigns each integer to its square is given by $xg = x^2$, and we mean that

$$g = \{(x, x^2) : x \text{ is an integer}\}$$

When we describe functions by means of a rule we shall not use the ordered-pairs notation but shall just state what the image of an arbitrary element is under the function. In doing so, however, *we must be sure that the rule or description given does define a function.*

Examples. Let f be the function from $\mathbf{R}$ to $\mathbf{R}$ defined by $xf = x^2 - 2x - 3$; that is,

$$f = \{(x, x^2 - 2x - 3) : x \in \mathbf{R}\}$$

Then $0f = -3$, $5f = 12$, and $(-2)f = 5$, so the image of 0 is -3, the image of 5 is 12, and the image of -2 is 5. We would say that the value of f at 5 is 12 and so on. By graphing this function using calculus or just high school algebra we see that the image of $[1, 2]$ under f is the closed interval $[-4, -3]$, and the image of $[0, 1]$ is also $[-4, -3]$. By computing we see that $0f^{-1} = \{-1, 3\}$ and $5f^{-1} = \{-2, 4\}$. In this case, for example, to find $5f^{-1}$ we need only solve $x^2 - 2x - 3 = 5$.

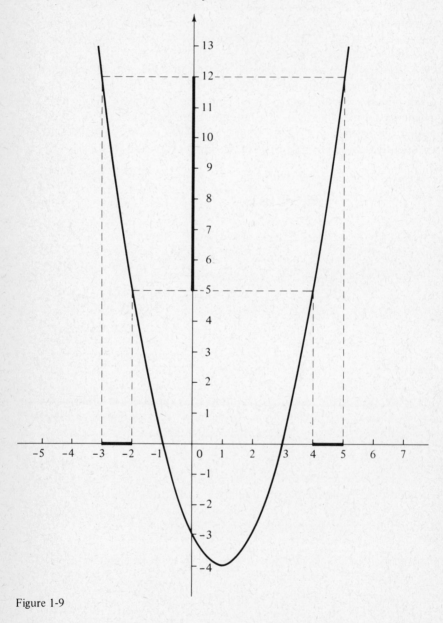

Figure 1-9

We can also determine, by solving the inequalities

$$5 \leq x^2 - 2x - 3 \quad \text{and} \quad x^2 - 2x - 3 \leq 12$$

that $[5, 12]f^{-1} = \{x : x \leq -2 \text{ or } x \geq 4\} \cap [-3, 5] = [-3, -2] \cup [4, 5]$. Figure 1-9 shows this rather simply.

A function in which no two distinct elements have the same image is called a *one-to-one* function. Thus, if f is a function from A into B we say that f is one-to-one if, whenever u and v are distinct elements of A, uf and vf are distinct elements of B. Equivalently, f is one-to-one if, whenever $uf = vf$, $u = v$; to show that a function f is one-to-one it suffices to show that whenever $uf = vf$ it follows that $u = v$.

Examples. Let $\mathbf{R}$ denote the set of real numbers and let g, h, and p be functions from $\mathbf{R}$ into $\mathbf{R}$ defined by

$$xg = 3x, \quad xh = \tfrac{1}{2}x + 1, \quad \text{and} \quad xp = x^3$$

If $u \neq v$, then $3u \neq 3v$, $\tfrac{1}{2}u + 1 \neq \tfrac{1}{2}v + 1$, and $u^3 \neq v^3$. Thus, g, h, and p are all one-to-one functions.

The squaring function s, defined by $xs = x^2$ from $\mathbf{R}$ into $\mathbf{R}$, is an example of a function that is not one-to-one.

If g is a function from A into B, then g is said to be a *constant function* if (as one would expect) there is an element b in B such that $xg = b$ for every element x in A. If A has more than one element, then a constant function is certainly not one-to-one.

If f is a function from A into B, we say that f is *onto* B if, for each element b in B, there is at least one element a in A such that $af = b$. Equivalently, f is onto if the image of A under f is the entire set B.

Examples. Let p be the function defined by $xp = x^3 - x$, where the domain and range are both $\mathbf{R}$, the set of real numbers. Let q be the linear function from $\mathbf{R}$ into $\mathbf{R}$ defined by $xq = 5x + 3$. The functions p and q are both onto; the function p is not one-to-one, whereas q is one-to-one. The function $xg = e^x$ is a one-to-one function from $\mathbf{R}$ to $\mathbf{R}$ that is not onto.

The reader should *not* expect a function to be onto and he should not expect it to be one-to-one. Neither should he expect to find both conditions satisfied at the same time; that is, he should not expect one-to-one functions to be onto or onto functions to be one-to-one.

If f is a function from A into B and g is a function from B into C, there is, as illustrated in Figure 1-10, a natural way of using f and g to form a function from A into C. If a is an element of A, first "apply f to a" and obtain af. Since af is an element of B, it is meaningful to "apply g to af" and obtain $(af)g$, which is an element of C. Thus, we have a new

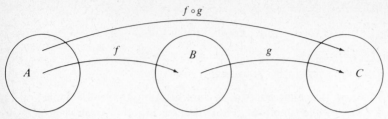

Figure 1-10

function from A into C. We make this notion precise by the following definition.

If f is a function from A into B, and g is a function from B into C, the *composition* of *f and g* or *f composed with g* is denoted by $f \circ g$ and defined by

$$f \circ g = \{(u, v) \in A \times C : (uf)g = v\}$$

Before proceeding further we need to be sure that the description just given does, in fact, define a function.

Proposition 1

If f is a function from A into B, and g is a function from B into C, then $f \circ g$ is a function from A into C.

Proof:　By definition, $f \circ g \subseteq A \times C$. We must show that for each element m in A there is exactly one element n in C such that $(m, n) \in f \circ g$. Thus, let m be any element of A. Then $mf \in B$ and, since g is a function from B into C, $(mf)g$ is an element of C. Thus, for each m in A, there is at least one element n in C such that $(m, n) \in f \circ g$. We need to show that there is only one such element. Now suppose $(m, n) \in f \circ g$ and $(m, p) \in f \circ g$. Then $(mf)g = n$ and $(mf)g = p$, and therefore $n = p$. Since for every element m in A there is exactly one element n in C such that $(m, n) \in f \circ g$, $f \circ g$ is a function.　□

The definition of composition given above illustrates part of the motivation for writing functions on the right. This choice of orientation enables us to compose functions f and g by writing $f \circ g$ and having this mean first apply f and then g. Our order of application is also our order of writing.

When A, B, and C are all sets of numbers, and f and g are given by algebraic rules, it is often possible to express $f \circ g$ also rather simply as a rule.

Example.　Let p and q be functions from $\mathbf{R}$ into $\mathbf{R}$ defined by

$$xp = 2x + 1 \quad \text{and} \quad xq = x^2 - 4$$

Then

$$(x)(q \circ p) = (xq)p = (x^2 - 4)p = 2(x^2 - 4) + 1 = 2x^2 - 7$$

and

$$(x)(p \circ q) = (xp)q = (2x + 1)q = (2x + 1)^2 - 4 = 4x^2 + 4x - 3$$

Notice that in this example $p \circ q \neq q \circ p$. In most cases, if f and g are functions, $f \circ g$ will not be the same as $g \circ f$.

Before proceeding to the proof of Proposition 2, notice that if f and g are functions from A into B such that $xf = xg$ for each element x in A, then

$$f = \{(x, xf): x \in A\} = \{(x, xg): x \in A\} = g$$

Thus, to show that $f = g$ it suffices to show that $xf = xg$ for each x in A.

Proposition 2

If f is a function from A into B, g is a function from B into C, and h is a function from C into D, then $f \circ (g \circ h) = (f \circ g) \circ h$.

Proof: Let x be an arbitrary element of A. Then

$$(x)[f \circ (g \circ h)] = (xf)(g \circ h) = [(xf)g]h$$

and

$$(x)[(f \circ g) \circ h] = [(x)(f \circ g)]h = [(xf)g]h$$

Since $(x)[f \circ (g \circ h)] = (x)[(f \circ g) \circ h]$ for every element x in A, it follows that

$$f \circ (g \circ h) = (f \circ g) \circ h \quad \square$$

Among the nicer characteristics of one-to-one functions is that the composition of two such functions is again a one-to-one function.   A similar statement also holds for onto functions.   We state this more formally in the next assertion.

Proposition 3

If f is a function from A into B and g is a function from B into C, then

(i) If f and g are one-to-one, then $f \circ g$ is one-to-one.
(ii) If f and g are onto, then $f \circ g$ is onto.

Proof: Exercise. $\square$

EXERCISE SET B

1. Which of the following are not functions from **R**, the set of real numbers into **R**? If not, why not?

(a) $\{(x, y): y = x^3 + 3x\}$
(b) $\{(x, y): y = 1/x\}$
(c) $\{(x, y): y = x^2\}$
(d) $\{(x, y): y^2 = x^2\}$
(e) $\{(x, y): y = \tan x\}$
(f) $\{(x, y): \tan y = x\}$

(g) $\{(x, y): y = \sin x\}$
(h) $\{(x, y): x = \sin y\}$
(i) $\{(x, y): y = \cos x\}$
(j) $\{(x, y): y = \sec x\}$
(k) $\{(x, y): y = \tan (x/\sqrt{1 + x^2})\}$
(l) $\{(x, y): y = \sec (x/(1 + |x|))\}$

2. Define the function p from $\mathbf{R}$ into $\mathbf{R}$ by $xp = x^3 + 8$. Determine the following sets. ($[a, b]$ means $\{x \in \mathbf{R}: a \le x \le b\}$).

(a) $3p$
(b) $0p$
(c) $[-2, 3]\, p$

(d) $0p^{-1}$
(e) $35p^{-1}$
(f) $[0, 35]\, p^{-1}$

3. Let c be the cosine function, that is, $xc = \cos x$ for each real number x. Determine the following sets.

(a) πc
(b) $[0, \pi/2]\, c$
(c) $1c^{-1}$

(d) $4c^{-1}$
(e) $[0, 1]\, c^{-1}$
(f) $[-1, 1]\, c^{-1}$

*4. Let $\mathbf{R}^3$ denote the set of all ordered triples of real numbers or Euclidean three-dimensional space. Define T from $\mathbf{R}^3$ into $\mathbf{R}$ by $(x, y, z)\, T = (x, y, z) \cdot (2, 3, 4) = 2x + 3y + 4z$, where $\cdot$ denotes the dot product. Determine the following sets. (Feel free to use geometric descriptions.)

(a) $(5, 2, 1)T$
(b) $0T^{-1}$

(c) $4T^{-1}$
(d) $[0, 4]T^{-1}$

5. Of the examples in Exercise 1 that are functions, which ones are one-to-one? Which ones are onto?

6. Let $xf = \sin x$ and $xg = \cos x$, and let A be the interval $[0, 1]$, B be the interval $[0, \frac{1}{2}]$, C be the interval $[-\frac{1}{2}, \frac{1}{2}]$, and D be the interval $[-\frac{1}{2}, 0]$. Which of the following functions are one-to-one? $f|_A$, $f|_B$, $f|_C$, $f|_D$, $g|_A$, $g|_B$, $g|_C$, $g|_D$.

*7. Prove: If g is a differentiable function from $\mathbf{R}$ into $\mathbf{R}$ such that g' is a positive function, then g is one-to-one.

8. Let f, g, h, and k be functions from $\mathbf{R}$ to $\mathbf{R}$ defined by $xf = x^2 + 1$, $xg = 2x - 3$, $xh = x^3 + 27$, and $xk = -4x + 1$. Compute $f \circ g$, $g \circ f$, $f \circ h$, $g \circ h$, $g \circ k$.

9. Prove Proposition 3.

§3. Relations

In our intuitive idea of a relation we think of two people, two numbers, two triangles, two books, or perhaps two sets as being, in some sense, "related." As with functions, we are thinking essentially about ordered pairs of objects. If our set is the set of integers and the relation is "less than," then we are concerned with, for example, the ordered pairs $(4, 8)$, $(6, 20)$, $(1, 7)$, $(-7, 0)$, and $(-47, 94)$, because in each case the first element is "less than" the second element. Since our intuitive notion of a relation is always concerned with ordered pairs of objects, we make the following definition.

If A and B are sets, we say that R is a *relation* from A to B if $R \subseteq A \times B$. A relation from the set S to S is called a relation on S.

Examples

(i) Every function is a relation.

(ii) If P is the set of all people, and B is the set of all books, the relation "is the author of" would be the set of ordered pairs in which the first element is someone who has written a book and the second element is one of the books he has written.

(iii) If, again, P is the set of all people, the relation "is the father of" would be the set of all ordered pairs in which the first element of each pair is a man and the second element is one of his children.

(iv) If $\mathbf{R}[x]$ is the set of all polynomials over $\mathbf{R}$, the set of real numbers, the relation "is a root of" is the set of ordered pairs in which each first element is a number and the second element is a polynomial having the first element as a root.

(v) If S is any set, the relation "equality" or "equals" on S is the set $\{(x, x):x \in S\}$.

If R is a relation from A into B, and M is a subset of A, then the set of all pairs in R whose first elements are in M is called the *restriction* of R to M and is denoted by $R\mid_M$. Using set-builder notation we have, as with functions,

$$R\mid_M = \{(a, b) \in R : a \in M\}$$

If R is a relation we shall usually not write $(a,b) \in R$, but shall, rather, use a more common notation. When we are discussing the relation "$<$," we would not write $(2, 5) \in$ "$<$" but rather would write $2 < 5$. Also, we would not write $(7, 7) \in$ "$=$" but would write $7 = 7$. In most cases, if R is a relation and (a, b) is an element of R, we shall write aRb.

The relation "equality" on any set S satisfies three important conditions:

(a) If x is in S, then $x = x$.

(b) If x and y are in S and $x = y$, then $y = x$.

(c) If $x, y,$ and z are in S and $x = y$ and $y = z$, then $x = z$.

If S is the set of triangles in the plane and "$\cong$" is the relation of congruence, we can replace "$=$" in the statements above by "$\cong$" and the conditions would all hold. Also, if S is the set of all triangles in the plane and we consider the relation "$\sim$" of similarity, we could replace the symbol "$=$" by "$\sim$" in each of the above statements and they would all hold.

If R is a relation on the set S, then we say that

(a) R is *reflexive* if xRx for every x in S.

(b) R is *symmetric* if, for each x and y in S, if xRy, then yRx.

(c) R is *transitive* if, for each $x, y,$ and z in S, if xRy and yRz, then xRz.

We say that a relation on a set S is an *equivalence relation* if R is reflexive, symmetric, and transitive.

Examples

(a) Let B be the set $\{a, b, c, d\}$, and let the relation D on B be the set $\{(a, a), (b, b), (c, c), (d, d), (a, b), (b, a), (c, d), (d, c)\}$.
(b) Let P be the set of all people, and let F be the relation "having the same father as."
(c) Let Z be the set of integers. Define "$\sim$" on Z by $x \sim y$ if $x - y$ is even.
(d) Again, let P be the set of all people, and let I be the relation "having the same amount of annual income."
(e) Let S be the set of all triangles in the plane, and define the relation A by tAu if t and u are triangles with the same area.

There are many equivalence relations that arise in mathematics. In fact, we often are content if our information is complete "up to some equivalence relation." For example, in geometry we "know" a triangle if we know its three sides. We do not actually know the location or the orientation of the triangle, but we know much of what can be said about it, its area, its angles, the lengths of its medians, and so on. For many purposes this information is all we need, and therefore we are satisfied to know the triangle "up to congruence."

The set of triangles congruent to a given triangle would be a subset of the set of all triangles and would have the property that any two triangles in the set are congruent and that none of them is congruent to any triangle outside the set. The set of all triangles is, then, "partitioned" by congruence into disjoint subsets. Each of these subsets could be called a congruence class, and we can decide whether or not two classes are the same by examining a pair of triangles, one from each class. If the triangles are congruent then the classes are the same. Consider also the relation "having the same last name" defined on the set of residents of Ohio. The relation "partitions" the set of Ohio residents into classes according to their last names. No two distinct classes have any members in common. The concept is not limited to these particular relations.

If S is a set, R is an equivalence relation on S, and a is an element of S, the *equivalence class for a* (or the class for a with respect to the relation R or just the class for a) is denoted by $\bar{a}$ and defined to be all the elements of S that bear the relation R to a. Thus,

$$\bar{a} = \{x \in S : xRa\}$$

The symmetry of an equivalence relation enables us to write

$$\bar{a} = \{x \in S : aRx\}$$

In Example (a) above,

$$\bar{a} = \{a, b\}, \qquad \bar{b} = \{a, b\}, \qquad \bar{c} = \{c, d\}, \qquad \bar{d} = \{c, d\}$$

and therefore the relation D "partitions" the set B into the two classes:

$$\bar{a} = \bar{b} = \{a, b\} \quad \text{and} \quad \bar{c} = \bar{d} = \{c, d\}$$

In Example (b) above, the classes are of the form

"all the children of _____"

in which the blank is filled with some man's name.

In Example (c) above, the classes would be the set of even integers and the set of odd integers, because, if x and y are even, $x - y$ is even and $x \sim y$. Similarly, if a and b are odd, $a - b$ is even and $a \sim b$. Thus, all evens are in a class and all odds are in a class. If p is even and q is odd, then $p - q$ is odd so $p \not\sim q$. Thus, in this case we have just the two classes that again are disjoint.

In each of the examples above, we see that two classes that are different are disjoint. That this is, in general, the case is established by the following theorem.

Theorem 4

If R is an equivalence relation on the set S, and a and b are elements of S such that $\bar{a} \neq \bar{b}$, then $\bar{a} \cap \bar{b} = \phi$.

Proof: We proceed by way of contradiction, in that we shall assume that $\bar{a} \cap \bar{b} \neq \phi$ and show that $\bar{a} = \bar{b}$.

If $\bar{a} \cap \bar{b} \neq \phi$, then for some element c in S, $c \in \bar{a}$, and $c \in \bar{b}$. Then cRa and cRb. By symmetry of R, we now have aRc and cRb, and therefore aRb by the transitivity of R. Also by symmetry then, bRa. Therefore, $a \in \bar{b}$ and $b \in \bar{a}$.

If $x \in \bar{a}$, then xRa and since aRb, by transitivity xRb, and therefore $x \in \bar{b}$. Thus, we have shown that $\bar{a} \subseteq \bar{b}$. On the other hand, if $y \in \bar{b}$, then yRb and since bRa, by transitivity it follows that yRa, and therefore $y \in \bar{a}$. Thus, we see that $\bar{b} \subseteq \bar{a}$. Since $\bar{a} \subseteq \bar{b}$ and $\bar{b} \subseteq \bar{a}$, $\bar{a} = \bar{b}$. But this contradicts the hypothesis that $\bar{a} \neq \bar{b}$, and therefore $\bar{a} \cap \bar{b} = \phi$. □

EXERCISE SET C

1. Decide whether or not the following relations are equivalence relations on the set of all people. If not, why not?
   (a) "is the cousin of"   (d) "is an ancestor of"
   (b) "is a brother of"   (e) "has the same father as"
   (c) "is the father of"   (f) "has a parent in common with"
2. Give examples of relations that are
   (a) reflexive but not symmetric or transitive
   (b) symmetric but not reflexive or transitive

(c) transitive but not reflexive or symmetric

(d) reflexive and symmetric but not transitive

(e) reflexive and transitive but not symmetric

(f) symmetric and transitive but not reflexive

(g) not reflexive, symmetric, or transitive

3. (a) Let $xf = x^2 + 2x - 3$ be a function from the set of real numbers, $\mathbf{R}$, into $\mathbf{R}$. In $\mathbf{R}$ define a relation, $\sim_f$, by $a \sim_f b$ iff $af = bf$. Prove that $\sim_f$ is an equivalence relation.

(b) Let g be the absolute value function from $\mathbf{R}$ into $\mathbf{R}$. As before, define $\sim_g$ by $a \sim_g b$ iff $ag = bg$. Prove that $\sim_g$ is an equivalence relation on $\mathbf{R}$.

4. The results from Exercise 3 can be generalized. Let f be a function from A into B. In A we define the relation "$\sim_f$" by $a \sim_f b$ iff $af = bf$. Prove that $\sim_f$ is an equivalence relation.

5. What are the classes for equality on a set?

6. (a) If f is the function from $\mathbf{R}$ into $\mathbf{R}$ defined by $xf = x^2 + 4$, describe $\bar{2}$, the class for 2 under $\sim_f$; that is, find $\{x:xf = 2f\}$. Compute $\bar{3}$ and $\bar{4}$.

(b) If g is a function from $\mathbf{R}$ into $\mathbf{R}$ and is defined by $xg = x^2 + 2x - 3$, describe $\bar{2}$, the class for 2 under $\sim_g$; also determine $\bar{3}, \bar{4}$, and $\overline{-3}$.

(c) For any number b, determine $\bar{b}$ with respect to $\sim_f$ from part (a); that is, determine $\{x:x \sim_f b\} = \{x:xf = bf\}$.

(d) For any number c, determine $\bar{c}$ with respect to $\sim_g$ from part (b); that is, determine $\{x:x \sim_g c\} = \{x:xg = cg\}$.

7. Prove: If f is a function from A into B and $a \in A$, then the class for a under $\sim_f$ is the set $(af)f^{-1}$, that is, $\bar{a} = (af)f^{-1}$. (Remember that f^{-1} does not necessarily denote a function. Hint: Write down all definitions.)

8. Let G be the greatest integer function defined on $\mathbf{R}$, that is, $xG = [x]$, the greatest integer less than or equal to x. What is the class for $4\frac{1}{2}$ under $\sim_G$? In general, if r is a real number, what is the class for r with respect to $\sim_G$?

9. Let $S = \{a, b, c, d, e, f, g\}$, and let $X = \{a, b\}$, $V = \{c, d, e\}$, and $W = \{f, g\}$. Define a relation R in S by uRv iff both u and v are in X, both u and v are in V, or both u and v are in W. Prove that R is an equivalence relation. What are the classes for R?

§4. Partitions

The relation of congruence on the set of all triangles in the plane separates the set of triangles into disjoint congruence classes. Similarly, the set of even integers and the set of odd integers separate the set of integers into two disjoint sets. In Theorem 4 we saw that, in general, if R is any equivalence relation on a set S and if $\bar{S}$ denotes the set of all equivalence classes under R, then the sets in $\bar{S}$ are disjoint and the set $\bar{S}$ subdivides S into disjoint subsets. This notion can be generalized into the formal concept of a partition.

If S is a set and $\mathcal{S}$ is a family of subsets of S, we say that $\mathcal{S}$ is a *partition* of S if

(a) $\emptyset \notin \mathcal{S}$.

(b) $\cup \mathcal{S} = S$.

(c) If $A \in \mathcal{S}$ and $B \in \mathcal{S}$ and $A \neq B$, then $A \cap B = \emptyset$.

Examples. If R is an equivalence relation on S and $\bar{S}$ is the set of all equivalence classes with respect to R, then, by Theorem 4, $\bar{S}$ is a partition of the set S.

Let S be the set of all points in the plane, let $\mathcal{V}$ be the set of all vertical lines in the plane, let $\mathcal{H}$ be the set of all horizontal lines in the plane, let $\mathcal{F}$ be the set of all lines in the plane with slope 4, and let $\mathcal{C}$ be the set of all circles with center at the origin. Then each of $\mathcal{V}$, $\mathcal{H}$, $\mathcal{F}$, and $\mathcal{C}$ is a partition of S.

If, for each real number r we define $T_r = \{x \in \mathbf{R} : x^2 = r\}$, then $\mathcal{J} = \{T_r : r \in \mathbf{R} \text{ and } r \geq 0\}$ is a partition of $\mathbf{R}$.

If, for each integer a, we define $B_a = \{r \in \mathbf{R} : a \leq r < a + 1\}$ and let $\mathcal{B} = \{B_a : a \text{ is an integer}\}$, then the family $\mathcal{B}$ is also a partition of $\mathbf{R}$.

With each of the partitions above, there is a natural way of defining an associated relation. With the partition $\mathcal{V}$ we could define $\sim$ by $x \sim y$ iff x and y are on the same vertical line. The relation $\sim$ is easily shown to be an equivalence relation. Suppose p is any point in the plane. Then p is on some vertical line, and thus p is on the same vertical line as itself, and therefore $p \sim p$. Now if p and q are points and $p \sim q$, then p and q are on the same vertical line, and therefore $q \sim p$. Furthermore, if p, q, and w are points such that $p \sim q$ and $q \sim w$, then p and q are on the same vertical line and q and w are on the same vertical line. Since q is on one and only one vertical line, p and w must both be on this line, and therefore $p \sim w$. The set $\mathcal{V}$ then becomes the set of equivalence classes for $\sim$ since, if p is a point in the plane, $\bar{p} = \{q : p \sim q\}$ is just the vertical line through p.

With the partition $\mathcal{F}$, we could define a relation by $u \sim v$ iff u and v are on the same line with slope 4. In a manner similar to that above, this relation can be shown to be an equivalence relation, and the set of classes is just $\mathcal{F}$. The next theorem establishes that this process can be generalized for use with any partition.

Theorem 5

If S is a nonempty set and $\mathcal{S}$ is a partition of S, then the relation $\sim$ defined on S by

$$x \sim y \text{ iff } x \in A \text{ and } y \in A \text{ for some set } A \text{ in } \mathcal{S}$$

is an equivalence relation, and the set of all classes with respect to the relation $\sim$ is precisely $\mathcal{S}$.

Proof: If x is any element of S, then since $\cup \mathcal{S} = S$, x must be an element of some set A in $\mathcal{S}$, and therefore $x \sim x$. Thus, the relation is reflexive.

If x and y are elements of S such that $x \sim y$, then for some set A in

S, $x \in A$ and $y \in A$. This implies $y \in A$ and $x \in A$, and therefore $y \sim x$. Hence, the relation is symmetric.

We have yet to show that the relation is transitive. If x, y, and z are elements of S such that $x \sim y$ and $y \sim z$, then for some set A in S, $x \in A$ and $y \in A$. Also, for some set B in S, $y \in B$ and $z \in B$. Since $y \in A$ and $y \in B$, $A \cap B \neq \emptyset$ and, since S is a partition, condition (c) in the definition implies $A = B$. Then, since $x \in A$ and $z \in A$, it follows that $x \sim z$, and therefore the relation is transitive.

From the definition of the relation $\sim$, it follows that if $x \in S$, $\bar{x}$ is the element of S that contains x, and therefore each class is one of the sets in S. However, if $A \in S$, then $A \neq \emptyset$ and $\bar{a} = A$ for some element a in A; that is, A is one of the classes. Thus, the classes for $\sim$ are just the sets in S. □

By the above theorem we see that with every partition there is associated a unique equivalence relation, the one that has for its classes just the elements of the partition. Thus, there is a one-to-one correspondence between equivalence relations on a set and partitions of the set.

EXERCISE SET D

1. Let S be the set of all points in the plane, and let $\mathcal{C}$ be the set of all circles with center at the origin. Under the relation $\sim$ determined by $\mathcal{C}$, is $(2, 4) \sim (4, -2)$? Is $(3, 4) \sim (0, -5)$? Is $(2, 3) \sim (1, 4)$? Is $(7, 9) \sim (6, 8)$? Is $(-2, 3) \sim (2, 1)$? Under what arithmetic conditions is $(a, b) \sim (c, d)$? Can you describe the relation $\sim$ in a familiar geometric manner?

2. Let S again denote the set of points in the plane, and let $\mathcal{L}$ be the set of all lines with slope 2. Then $\mathcal{L}$ is a partition of S. Under the relation $\sim$ determined by $\mathcal{L}$, is $(1, 2) \sim (3, 6)$? Is $(-1, 4) \sim (3, 9)$? Is $(7, 11) \sim (13, 23)$? Under what arithmetic conditions is $(a, b) \sim (c, d)$?

3. Let $\mathbf{R}[x]$ denote the set of all polynomials[2] over $\mathbf{R}$. For each real number a, define $W_a = \{f \in \mathbf{R}[x]: f(0) = a\}$ and let $\mathcal{W} = \{W_z: z \in \mathbf{R}\}$. List three distinct elements of W_5. Prove that $\mathcal{W}$ is a partition of $\mathbf{R}[x]$. Can you give a simpler and more familiar description of the relation $\sim$ determined by $\mathcal{W}$?

4. Let $\mathcal{C}$ denote the set of all continuous functions from $[0, 1]$ into the real numbers. For each real number t, define $\bar{t} = \left\{ f: \int_0^1 f = t \right\}$. For example,
$$\bar{3} = \left\{ f: \int_0^1 f = 3 \right\}.$$
   (a) Determine three distinct elements of $\bar{1}$.
   (b) Can you give a simpler and more familiar description of the equivalence relation on $\mathcal{C}$ determined by $\mathfrak{M}$ if $\mathfrak{M} = \{\bar{t}: t \in \mathbf{R}\}$?

5. Give a geometric description of some interesting partition of the plane. Now give a geometric description of the equivalence relation determined by this partition.

[2] A review of the arithmetic of polynomials is provided in Appendix 2.

6. Give an interesting partition of the set of all real numbers. Can you give an arithmetic, geometric, or analytic description of the corresponding equivalence relation?
7. Determine a nontrivial equivalence relation $\sim$ on $\mathbf{R}[x]$ such that if f, g, and p are polynomials such that $f \sim g$, then $f + p \sim g + p$. Describe the set of all polynomials p such that $p \sim 0$.

§5. Operations

After learning to count, the first mathematical concepts to which a student is introduced are the operations of addition and multiplication. Early mathematical training is, in fact, primarily the study of these operations. The intuitive idea of an operation is that of a process of "combining" two elements of a set to get a third element of the set. The idea of combining is, however, ambiguous and we need, as we did with "function" and "relation," to formulate a precise definition.

An *operation* on a set S is a function from $S \times S$ into S.

Intuitively, an operation assigns each ordered pair of elements of the set S to an element of S.

Examples

(a) Addition and multiplication on the set of positive integers.
(b) Subtraction on the set of integers.
(c) Division on the set of all positive real numbers.
(d) Union on the set of all subsets of the plane.
(e) Intersection on the set of open sets of the real line.

Notice that subtraction is not an operation on the set of positive integers, because $5 - 6$ is not defined on this set. Similarly, division is not an operation on the set of all real numbers, because $1 \div 0$ has no meaning.

Instead of following functional notation such as $(a, b)f$ to denote operations, we shall employ a more common method. We do not write $(3, 4) +$ or $(3, 4) \cdot$, but rather express the image of the pair $(3, 4)$ under addition and multiplication by

$$3 + 4 \quad \text{and} \quad 3 \cdot 4$$

Most operations that will concern us can be expressed in a similar fashion, and we shall continue to do so.

We shall often use the asterisk $*$ for an unspecified or general operation, and we shall use it as we do $+$, $-$, $\cdot$, $\cap$, and $\cup$. Thus, if we "let $*$ be an operation of S," by $a * b$ we mean the image of the ordered pair (a, b) under the operation $*$.

If $*$ is an operation on the set S and T is a subset of S, the function $*$ restricted to $T \times T$ determines a function from $T \times T$ into S. Instead of calling this function $*$ restricted to $T \times T$, it is customary just to call it

$*$ restricted to T. If the operation $*$ restricted to T maps $T \times T$ into T, then we say that T is *closed* under $*$; that is, if T is a subset of S, we say that T is *closed* under the operation $*$ if, whenever a and b are elements of T, $a * b$ is also in T. For example, the set of even integers is closed under both addition and multiplication. The odd integers is closed under multiplication but not under addition. If $*$ is an operation on S, S is, of course, closed under $*$ according to the definition of operation.

A remark on terminology is appropriate at this point. Many texts define a function from $S \times S$ to S to be a *binary* operation on S, a function from S to S to be a *unary* operation on S, a function from $S \times S \times S$ to S to be a *ternary* operation on S, and so on. If we were concerned with such functions, then a distinction would be necessary. In our case, however, all operations considered will be of the type that others call binary, and therefore we shall just refer to them as operations. Some authors define a binary operation on S to be a function with domain $S \times S$ and make no restrictions on the range. Then, to get the conditions included in our definition they will require that their operation be closed, that is, that the range be contained in S. As always in mathematics, one's choices of definitions are arbitrary and the ones we have given suffice for our purposes.

If $*$ is an operation on the set S and θ is an element of S such that

$$x * \theta = x$$

for each element x in S, then θ is said to be a *right identity* element with respect to the operation $*$ in S. Similarly, if e is an element of S such that

$$e * x = x$$

for each element x in S, then e is said to be a *left identity* element with respect to the operation $*$ in S. An element that is both a right and left identity is called a *two-sided identity* or just an *identity*.

Examples. The number 0 is a right identity with respect to the operation of subtraction on the real numbers but is not a left identity.

The number 1 is an identity element with respect to multiplication of the set of integers.

As an example of a useful assertion that can be proved about very general structures, we establish the following statement.

Proposition 6

If $*$ is an operation on the set S, θ is a right identity with respect to $*$, and e is a left identity with respect to $*$, then $\theta = e$.

Proof: Since θ is a right identity,

$$e * \theta = e$$

Since e is a left identity,

$e * \theta = \theta$

Therefore,

$\theta = e$ □

From the above result we see that if a system has a right identity and a left identity it has, in fact, just one identity and this is two-sided. Also, if we know that a particular one-sided identity is not two-sided, then there is no two-sided identity and, in fact, if we have a right (left) identity that is not a left (right) identity, then there is no left (right) identity. For example, because 0 is a right identity with respect to subtraction of real numbers, and because 0 is not a left identity, there can be no left identity.

Two of the conditions that make addition and multiplication of numbers easier to work with are the Associative laws:

For any numbers a, b, and c,

$a + (b + c) = (a + b) + c$

and

$a \cdot (b \cdot c) = (a \cdot b) \cdot c$

and the Commutative laws:

For any numbers a and b,

$a + b = b + a$

and

$a \cdot b = b \cdot a$

If $*$ is an operation on the set S, we say that $*$ satisfies the *Associative law* or is *associative* if, for any elements x, y, and z in S, $x * (y * z) = (x * y) * z$. We say that $*$ satisfies the *Commutative law* or is *commutative* if, for any elements x and y in S, $x * y = y * x$.

Examples. Addition and multiplication, as noted above, are associative and commutative on the set of integers, but subtraction is neither associative nor commutative on that set. Composition of functions is associative but not commutative. Multiplication of matrices is also associative but not commutative. (This should not be surprising; matrix multiplication is essentially composition of functions.) Averaging on the set of positive rational numbers is an example of an operation that is commutative but not associative.

Most of the operations we study in mathematics satisfy one or the other of these laws. The student should not, however, expect either law to hold for an operation unless the particular law has been axiomatized or proved.

Notice that, if $*$ is an associative operation on the set S with a, b, and c elements of S, we can write $a * b * c$ with no ambiguity in meaning, because no matter where we insert parentheses, the results are the same. Similarly, if a, b, c, and d are in S, $a * b * c * d$ now has meaning because, although we originally had meaning only for two elements under an operation, no matter where we insert the parentheses, the result is the same. In fact, if $*$ is an associative operation on the set S any finite "product" will yield the same result no matter where we insert the parentheses, as long as we keep the same order of symbols. If $*$ is also commutative, then we can also rearrange the elements and still obtain the same result.

EXERCISE SET E

1. Let F be the set of all functions from $\mathbf{R}$ into $\mathbf{R}$. Which of the following subsets of F are closed under the operation of composition of functions?
   (a) the one-to-one functions
   (b) the onto functions
   (c) the multiplicative functions
      [f is multiplicative if $(xy)f = (xf)(yf)$ for all x and y]
   (d) the additive functions
   (e) the continuous functions
   (f) the differentiable functions
   (g) the discontinuous functions
   *(h) the increasing functions
   *(i) the monotone functions
   *(j) the bounded functions

*2. In the usual three-dimensional Euclidean space, is the vector cross product associative? Is it commutative? Is there a left or a right identity?

3. Is the set of matrices[3] below closed under matrix addition? Under multiplication?

$$\left\{ \begin{bmatrix} 1 & 0 \\ 0 & 1 \end{bmatrix}, \begin{bmatrix} -1 & 0 \\ 0 & -1 \end{bmatrix}, \begin{bmatrix} 0 & 1 \\ 1 & 0 \end{bmatrix}, \begin{bmatrix} 0 & -1 \\ -1 & 0 \end{bmatrix} \right\}$$

4. Using the definition of associativity only, prove: If $*$ is an associative operation on the set S and a, b, c, and d are elements of S, then

$$a * [(b * c) * d] = (a * b) * (c * d) = [a * (b * c)] * d$$

5. Using only the definitions, prove: If $*$ is an associative and commutative operation on the set S and a, b, and c are elements of S, then

$$(a * b) * c = (c * b) * a$$

*6. Let S be a set with an associative operation $*$. Let a be an element of S and define $C(a)$ by

$$C(a) = \{x \in S : x * a = a * x\}$$

Prove that $C(a)$ is closed under $*$.

[3]The arithmetic of such matrices is discussed in Appendix 1.

2
THE INTEGERS

To construct a rigorous mathematical theory or any logical discourse, one must start with certain undefined terms and some assumptions. In mathematics the basic assumptions are usually called *axioms* or *postulates*, and the method of choosing the undefined terms and a system of axioms or postulates varies with the purpose of the development. The first attempt at a rigorous mathematical exposition was made by Euclid in the *Elements* about 300 B.C. Euclid chose his axioms to describe a particular system that he had in mind in that the axioms he chose were intended to describe space as he saw it. This explains, to some extent, the idea of viewing axioms as "self-evident truths." Modern mathematicians realize that Euclid's axioms are just assumptions and to consider them as "true" in a philosophical sense is to step beyond the scope of mathematics into physics or metaphysics. The mathematician works with a system in which certain assertions (the axioms) are assumed to hold and proceeds to prove that if these axioms hold, then other assertions (called theorems, propositions, lemmas, and corollaries) must hold.

In what might be called the Euclidean approach to a mathematical theory, one first has an intuitive notion of a structure and isolates certain elementary properties of this system. He attempts to choose as axioms sufficiently many of these properties so that he has completely described this intuitive structure but usually wishes to assume no more than necessary. He then uses these assumptions to prove more of the expected properties and attempts to prove or disprove other assertions that he thinks might hold in the system. His intuition, then, is responsible for the fundamental assumptions of the system and for the conjectures that he thinks he may be able to prove. His criteria for acceptance of an assertion is its provability on the basis of his fundamental assumptions.

In this chapter we undertake a rigorous study of the *integers* or whole numbers. The axioms that we choose are elementary properties that we feel the system of whole numbers should satisfy. Most of the assertions established in the first few sections will not be new to you but the proofs of these assertions will be. The primary purpose of these sections is that you may see, on familiar ground, how a theory can be developed from a relatively small set of basic assumptions. In the latter sections we apply some of this theory to other structures.

§1. The Axioms and Elementary Arithmetic Properties

Axioms for the Integers

The integers, denoted by $\mathbf{Z}$, is a set with two operations, called addition and multiplication and denoted by $+$ and $\cdot$, respectively, that satisfies the following axioms.

1. For any integers a, b, and c, $a + (b + c) = (a + b) + c$.
2. For any integers a and b, $a + b = b + a$.
3. There is an integer 0 such that $a + 0 = a$ for each integer a.
4. For each integer a, there is an integer $-a$, called the *negative* of a, such that $a + (-a) = 0$.
5. For any integers a, b, and c, $a \cdot (b \cdot c) = (a \cdot b) \cdot c$.
6. For any integers a, b, and c,

   $$a \cdot (b + c) = a \cdot b + a \cdot c \quad \text{and} \quad (a + b) \cdot c = a \cdot c + b \cdot c$$

7. For any integers a and b, $a \cdot b = b \cdot a$.
8. There is an integer 1 such that $1 \neq 0$ and $a \cdot 1 = a$ for each integer a.
9. There is a relation "$<$" (which is read "less than") defined on $\mathbf{Z}$ such that
   (i) "$<$" is transitive.
   (ii) For any integers a and b, exactly one of the following holds:

   $$a < b, \quad a = b, \quad \text{or} \quad b < a$$

   We say $x \leq y$ if either $x < y$ or $x = y$. The expression $x \leq y$ is read "x is less than or equal to y."
   We define the relation "$>$" on $\mathbf{Z}$ by $a > b$ iff $b < a$. The relation "$\geq$" is defined by $a \geq b$ iff $b \leq a$. The expression "$x > y$" is read "x is greater than y," and the expression "$x \geq y$" is read "x is greater than or equal to y."
10. If a, b, and c are integers such that $a < b$, then $a + c < b + c$.
11. If a and b are integers such that $0 < a$ and $0 < b$, then $0 < a \cdot b$.
    An integer x is said to be *positive* if $0 < x$ and *negative* if $x < 0$. The set of all positive integers is denoted by $\mathbf{N}$ and is frequently called the set of *natural numbers*.
    We say that an element b in a set W of integers is the *smallest* element of the set if b is less than or equal to every element in the set W.
12. If $S \subseteq \mathbf{N}$ and $S \neq \emptyset$, then S contains a smallest element.

Axioms 1 and 5 are called the *Associative laws*, Axioms 2 and 7 are called the *Commutative laws*, Axiom 6 is the *Distributive law*, and Axiom 12 is called the *Well-ordering Principle.*

Now that we have chosen the axioms we proceed to prove the basic assertions that we would expect to hold for all integers. In order to involve the reader in the development, several of the proofs are left as exercises.

Proposition 1

If a, b, and c are integers such that $a + c = b + c$, then $a = b$.

Proof: By Axiom 4, there is an integer $(-c)$ such that $c + (-c) = 0$. Then, by substitution,

$$(a + c) + (-c) = (b + c) + (-c)$$

Axiom 1 implies that

$$a + [c + (-c)] = b + [c + (-c)]$$

Thus,

$$a + 0 = b + 0 \qquad \text{by Axiom 4}$$

and therefore

$$a = b \qquad \text{by Axiom 3} \quad \square$$

Corollary 2

If x is an integer such that $x + x = x$, then $x = 0$.

Proof: Exercise.

Corollary 3

If u and v are integers such that $u + v = 0$, then $u = -v$.

Proof

$$\begin{aligned}
u &= u + 0 \\
&= u + [v + (-v)] \\
&= (u + v) + (-v) \\
&= 0 + (-v) \\
&= -v \quad \square
\end{aligned}$$

Corollary 3 shows that the negative of an integer is unique. Since $0 + 0 = 0$, $0 = -0$ and, furthermore, since $a + (-a) = 0$, $a = -(-a)$ for every integer a. By applying Corollary 3 one can easily prove the following assertion.

Corollary 4

If h and k are integers, then

$$(-h) + (-k) = -(h + k)$$

Proof: Exercise.

We expect the product of 0 with any integer to be 0, but we did not assume this in the axiom system, and therefore we must prove this assertion if we are to use it.

Our defining property for 0 is simply that it is an identity element with respect to the addition operation. The statement that the product of 0 with any integer is 0 is, in fact, a multiplicative property of 0. Since the Distributive law is the only axiom that connects addition and multiplication, we can expect to use this axiom in the proof.

Proposition 5

If a is an integer, then $a \cdot 0 = 0$.

Proof: We know that $0 + 0 = 0$. Then

$$a \cdot (0 + 0) = a \cdot 0$$

and hence

$$a \cdot 0 + a \cdot 0 = a \cdot 0 \qquad \text{by Axiom 6}$$

Then, by Corollary 2,

$$a \cdot 0 = 0 \quad \square$$

When a student is first introduced to "negative numbers," the fact that $(-3)(-5) = 15$ often bothers him and he does not want to believe it. The "Law of Signs" is, in high school mathematics, frequently "proved by intimidation" or just taken as a hard-to-believe axiom. In the following theorem we see that it is a necessary consequence of our axiom system. This theorem is, in fact, an assertion about the multiplicative properties of the negatives of integers. Negatives are defined using only addition and, therefore, as in the proof of Proposition 5, we should expect to use the Distributive laws because they are the only connections we have between addition and multiplication.

Theorem 6 (The Law of Signs)

If a and b are integers, then $a \cdot (-b) = -(a \cdot b) = (-a) \cdot b$ and $(-a) \cdot (-b) = a \cdot b$.

Proof: Since $a \cdot 0 = 0$ and $b + (-b) = 0$,

$$a \cdot [b + (-b)] = 0$$

Then

$$a \cdot b + a \cdot (-b) = 0$$

and therefore

$$-(a \cdot b) = a \cdot (-b) \qquad \text{by Corollary 3}$$

Similarly,

$$0 = 0 \cdot b = [a + (-a)] \cdot b = a \cdot b + (-a) \cdot b$$

and therefore

$$-(a \cdot b) = (-a) \cdot b \qquad \text{by Corollary 3}$$

Now, using these two equations we have

$$(-a) \cdot (-b) = -[a \cdot (-b)] = -[-(a \cdot b)] = a \cdot b \quad \square$$

Notice that this theorem is proved without referring to "positive numbers," "negative numbers," or the concept of "less than."

Our axioms for Z give us only two operations for the structure, but we expect to have a third operation, namely subtraction, defined on this system. Instead of axiomatizing the properties of this operation, however, we *define* subtraction in terms of addition and negatives and then prove that it has the desired properties.

If x and y are integers, then $x - y = x + (-y)$.

As expected, $x - y$ is read "x minus y" and the operation so defined is called *subtraction*. The following assertion shows that multiplication is distributive over subtraction. Its proof requires the use of the definition, Axiom 6, and the Law of Signs. We leave the proof as an exercise.

Proposition 7

If a, b, and c are integers, then

$$a \cdot (b - c) = a \cdot b - a \cdot c$$

Proof: Exercise.

EXERCISE SET A

1. Prove Corollary 2.
2. Prove Corollary 4.
3. Prove Proposition 7.
4. The intuitive idea of subtraction is "$x - y$ is the number whose sum with y is x." Prove that $x - y$, as we defined it, has this property.
5. Let S be a set, and let $\mathcal{S}$ be the set of all subsets of S. If we substitute $\mathcal{S}$ for Z, $\cup$ for $+$, and $\cap$ for $\cdot$, which of Axioms 1–8 will hold?

6. If we let $\mathbf{R}[x]$ denote the set of all polynomials in one variable x with real numbers as coefficients and define "$<$" on $\mathbf{R}[x]$ by $p(x) < q(x)$ if and only if the leading coefficient of $q(x) - p(x)$ is positive, which of Axioms 1–12 will be satisfied?

*7. Let M be the set of all 2×2 matrices over the real numbers. If we substitute M for $\mathbf{Z}$ and let addition and multiplication be the usual addition and multiplication of matrices, which of Axioms 1–8 will hold?

§2. The Order Properties and Divisibility

The assertions proved thus far depended only upon Axioms 1–8, and, in fact, we did not need to use all of those. These assertions are purely arithmetic in character in that they are concerned only with the operations of the system. In the study of $\mathbf{Z}$, the ordering relation is of equal importance, and we now proceed to develop this portion of the theory. As a notational convenience, we will frequently denote products just using juxtaposition rather than the dot symbol.

Proposition 8

If x and y are integers such that $x < y$, then $-y < -x$.

Proof: Exercise. (Hint: Apply Axiom 10.)

The following is an immediate consequence of Proposition 8.

Corollary 9

If z is an integer and $z \neq 0$, then either $0 < z$ or $0 < -z$.

Proof: If $0 \neq z$ and $0 \not< z$, then $z < 0$, by Axiom 9.

Then $-0 < -z$ by Proposition 8, and therefore $0 < -z$. □

Corollary 10

If x is an integer and $x \neq 0$, then $0 < x \cdot x$.

Proof: If $x \neq 0$, then by Corollary 9, either $0 < x$ or $0 < -x$.

If $0 < x$, then

$0 < x \cdot x$ by Axiom 11

If $0 < -x$, then

$0 < (-x)(-x)$ by Axiom 11

and therefore

$0 < x \cdot x$ □

Since $0 < 1 \cdot 1$ and $1 = 1 \cdot 1$, $0 < 1$ and 1, as expected, is positive.

Axiom 10 states that, by adding the same integer to both sides of an inequality, we obtain an inequality in the same direction. That the analogous assertion does not hold for multiplication is easily seen by means of some examples. The next two assertions describe the relationship between multiplication and inequalities.

Proposition 11

If x, y, and z are integers such that $x < y$ and $0 < z$, then $xz < yz$.

Proof: If $x < y$, then $x + (-x) < y + (-x)$ and therefore

$$0 < y - x$$

Then, since $0 < z$,

$$0 < (y - x)z \qquad \text{by Axiom 11}$$

Hence,

$$0 < yz - xz$$

and

$$0 + xz < (yz - xz) + xz$$

Therefore,

$$xz < yz \qquad \square$$

Corollary 12

If x, y, and z are integers such that $x < y$ and $z < 0$, then $yz < xz$.

Proof: Exercise. (Hint: If $z < 0$, then $0 < -z$; now use Proposition 11.)

We expect the product of nonzero numbers to be nonzero, but we have not assumed it and therefore, if we are to use this assertion, we must prove it.

Theorem 13

If a and b are integers such that $a \neq 0$ and $b \neq 0$, then $a \cdot b \neq 0$.

Proof

CASE 1. If $0 < a$ and $0 < b$, then $0 < a \cdot b$ by Axiom 11, and therefore $ab \neq 0$.

CASE 2. If $a < 0$ and $b < 0$, then $0 \cdot b < a \cdot b$ by Corollary 12. Therefore, $0 < a \cdot b$ and $a \cdot b \neq 0$.

CASE 3. If $a < 0$ and $0 < b$, then $a \cdot b < 0 \cdot b$ by Corollary 12. Therefore, $a \cdot b < 0$ and $a \cdot b \neq 0$.

CASE 4. If $0 < a$ and $b < 0$, then $a \cdot b < 0$ again and $a \cdot b \neq 0$. $\square$

Proposition 14

If x, y, and z are integers such that $x \neq 0$ and $xy = xz$, then $y = z$.

Proof: If $xy = xz$, then

$$xy - xz = 0$$

which implies that

$$x(y - z) = 0$$

Since $x \neq 0$, by Theorem 13

$$y - z = 0$$

and therefore,

$$y = z \quad \square$$

Proposition 14 is known as the *Cancellation law*. The proof of this assertion depends directly on Theorem 13 and it can be shown that if one assumes Proposition 14, he can prove Theorem 13 (see Exercise 8 in Set B). Since these statements imply each other, they are logically equivalent.

In elementary arithmetic one would say that 3 divides evenly into 12, because $3 \cdot 4 = 12$, that is, because 3 times some integer is 12. The words "evenly into" are redundant and we shall reduce the phrase to just the word "divides."

If a and b are integers, we say that *a divides b* if $a \cdot n = b$ for some integer n.

It is customary to use the vertical line "$|$" as an abbreviation for the word divides. The expression

$$a \mid b$$

is read "a divides b." Remember that $a \mid b$ is not an integer; it is a sentence.

Theorem 15

If a, b, c, x, and y are integers such that $a \mid b$ and $a \mid c$, then $a \mid (bx + cy)$.

Proof: If $a \mid b$ and $a \mid c$, then

$$an = b \quad \text{and} \quad am = c \quad \text{for some integers } n \text{ and } m$$

Then,

$$anx = bx \quad \text{and} \quad amy = cy$$

Thus,

$$anx + amy = bx + cy;$$

$$a(nx + my) = bx + cy$$

and therefore, by definition,

$$a \mid (bx + cy) \quad \square$$

Proposition 16

If a, b, and c are integers such that $a \mid b$ and $b \mid c$, then $a \mid c$.

Proof: Exercise.

EXERCISE SET B

1. Prove that N is closed under addition. Generalize your technique to show that if p is a positive integer, then $\{x:p < x\}$ is closed under addition.
2. Prove Proposition 8.
3. Prove Corollary 12.
4. Prove: If a and b are integers and $ab > 0$ and $a > 0$, then $b > 0$.
5. Prove: If $a \in Z$ and $b \in N$, then $a - b < a$.
6. Prove Proposition 16.
7. Prove: If a and b are integers and $1 \leq b$, then $0 \leq a - bx$, for some x. (Hint: Either $x = 0$ or $x = a$ will suffice, depending on whether or not a is positive.)
**8. Show that if one assumes Proposition 14 to be true, Proposition 13 can be proved without using the ordering relation.

§3. The Division Algorithm and the Greatest Common Divisor

At this point in the development it is worthwhile to note that none of the assertions thus far established has depended upon Axiom 12. Therefore, the proofs given would be valid in any system satisfying Axioms 1–11. From this point on the development depends quite heavily on the Well-ordering Principle. The reader should note that it is this principle that distinguishes Z from other familiar systems such as R, the real numbers, and Q, the rational numbers. In particular, assertions 17–20 fail in Q and R because they depend on Axiom 12.

Proposition 17

The integer 1 is the smallest positive integer.

Proof: Let $K = \{x \in N : 0 < x < 1\}$.

If we can show that $K = \phi$, the assertion will be proved. Suppose $K \neq \phi$. Then, by Axiom 12, K must have a smallest element. Call this element p. Since $0 < p < 1$,

$$0 \cdot p < p \cdot p < 1 \cdot p$$

Then $0 < p \cdot p < p$. But $p < 1$, and therefore

$$0 < p \cdot p < p < 1$$

Thus, $p \cdot p$ is an element of K smaller than p. But this is impossible, because p is supposedly the smallest element of K. Since the assumption $K \neq \phi$ leads to a contradiction, $K = \phi$ and the proposition is proved. $\square$

An equivalent way of stating this proposition is as follows:

If a is a positive integer, then $1 \leq a$.

Proposition 18

If x is an integer and y is a positive integer such that $x \mid y$, then $x \leq y$.

Proof: Exercise. (Hint: You will need Exercise 4 of Set B and Proposition 17.)

Corollary 19

If x and y are positive integers such that $x \mid y$ and $y \mid x$, then $x = y$.

Proof: Exercise.

In computing with the integers, one of the fundamental properties we expect is to be able to divide one integer by a nonzero integer and obtain either an exact quotient or a quotient along with a nonnegative remainder less than the divisor. That this is indeed the case is established by the following theorem. Note that the technique involved in the proof is actually the technique we usually use for division; that is, we find a suitable remainder by trying quotients.

Theorem 20

If a and b are integers and b is positive, then there exist unique integers q and r such that $a = bq + r$ and $0 \leq r < b$.

Proof: Let $M = \{a - bx : x \in \mathbf{Z}\}$

Let r be the smallest nonnegative element of M. (M contains nonnegative elements by Exercise 7, Set B.) Then for some integer q,

$$r = a - bq \quad \text{or} \quad a = bq + r$$

We yet need to show that $0 \leq r < b$. Since r is nonnegative, we know $0 \leq r$.

Suppose $b \leq r$. Then $0 \leq r - b$ and $r - b < r$. Since $r = a - bq$, $r - b = (a - bq) - b = a - b(q + 1)$, and therefore $r - b \in M$.

Thus, if $b \leq r$, $r - b$ is a nonnegative element of M less than r and this is impossible. Therefore, $r < b$.

Thus, we have shown that there is at least one pair of integers q and r such that

$$a = bq + r \quad \text{and} \quad 0 \leq r < b$$

We need yet to show that q and r are unique. Suppose

$$a = bq + r, \quad a = bp + t, \quad 0 \le r < b, \quad \text{and} \quad 0 \le t < b$$

If $r \ne t$, without loss of generality we may assume that $r < t$ and therefore, $0 < t - r$.

Since

$$bq + r = bp + t$$
$$bq - bp = t - r$$
$$b(q - p) = t - r$$

and therefore

$$b \mid (t - r)$$

Then, since

$$0 < t - r, \quad b \le (t - r)$$

But also, since

$$0 \le r < t, \quad t - r \le t$$

and therefore $b < t$. But we also know that $t < b$ and therefore $b < b$, which is absurd. Thus, the assumption that $r \ne t$ leads to a contradiction, and hence $r = t$.

Then $bq + r = bp + r$, which implies that $bq = bp$. Then, since $b \ne 0$, $q = p$. $\square$

Intuitively, the greatest common divisor of two integers is the largest integer that divides both of them. This notion is made precise by the following definition.

If a and b are integers, we say that the positive integer d is the *greatest common divisor* of a and b if

(i) $d \mid a$ and $d \mid b$.

(ii) If c is an integer such that $c \mid a$ and $c \mid b$, then $c \mid d$.

We shall abbreviate the words "greatest common divisor" by "GCD."

If a and b are integers and p and q are positive integers satisfying conditions (i) and (ii) in the definition of GCD, then $p \mid q$ and $q \mid p$, and p must be the same as q by Corollary 19. Thus, the use of the word "the" in the definition of GCD is justified; that is, if there is a number satisfying (i) and (ii), there is only one such number.

Notice that if d is the GCD of a and b, then, by Proposition 18, d is greater than all other divisors of a and b and is, therefore, the greatest of all the divisors.

The existence of the GCD in the case where at least one of the integers is not zero is assured by the following theorem. The first proof given is an example of an existence proof in that we show that the GCD exists but give no technique for finding it. Notice that the second part of the theorem follows trivially from the technique used in the proof. A technique for computing GCD's is provided by the alternative proof.

Theorem 21

If a and b are integers such that $a \neq 0$ or $b \neq 0$, then the GCD of a and b exists. Furthermore, if d is the GCD of a and b, then there exist integers u and v such that $d = au + bv$.

Proof: Let $M = \{ax + by : x \in \mathbf{Z} \text{ and } y \in \mathbf{Z}\}$.

Since $a, -a, b$, and $-b$ are all elements of M, M contains positive elements. Let d be the smallest positive element of M. Then for some integers u and v, $d = au + bv$. We now proceed to show that d is, in fact, the GCD of a and b.

There exist integers q and r such that $a = dq + r$, and $0 \leq r < d$. Then $a - dq = r$ and, since $d = au + bv$, we can substitute in this equation and obtain $a - (au + bv)q = r$, which implies that

$$a(1 - uq) + b(-vq) = r$$

and therefore r is an element of M.

Since $0 \leq r < d$ and d is the smallest positive element of M, it follows that r must be 0. Then $a = dq + 0$; that is, $a = dq$ and therefore $d \mid a$.

Similarly, we can show that $d \mid b$. Therefore, condition (i) is satisfied by d.

Now suppose c is an integer such that $c \mid a$ and $c \mid b$. Then, by Theorem 15, $c \mid (au + bv)$ and therefore $c \mid d$. Therefore, condition (ii) is satisfied and d is the GCD of a and b. The latter part of the theorem follows from our choice of d. $\square$

Alternative Proof: In this proof we shall assume that b is positive and leave the generalization to the student. The technique uses Theorem 20 many times.

By Theorem 20, there exist integers q_1 and r_1 such that

$$a = bq_1 + r_1 \quad \text{and} \quad 0 \leq r_1 < b$$

If $r_1 = 0$, then the GCD of a and b is b.

If $r_1 \neq 0$, then there exist integers q_2 and r_2 such that

$$b = r_1 q_2 + r_2 \quad \text{and} \quad 0 \leq r_2 < r_1$$

If $r_2 = 0$, then $r_1 \mid b$, because $b = r_1 q_2$.

Since $r_1 \mid b, r_1 \mid (bq_1 + r_1)$, and therefore $r_1 \mid a$.

Thus, we have $r_1 \mid a$ and $r_1 \mid b$.

If c is an integer such that $c \mid a$ and $c \mid b$, then $c \mid [a + b(-q_1)]$.

Since $c \mid (a - bq_1), c \mid r_1$.

Since r_1 satisfies conditions (i) and (ii), r_1 is the GCD of a and b. Thus, we have shown that if $r_2 = 0$, then r_1 is the GCD of a and b. If $r_2 \neq 0$, then there exist integers q_3 and r_3 such that

$$r_1 = r_2 q_3 + r_3 \quad \text{and} \quad 0 \leq r_3 < r_2$$

If $r_3 = 0$, the process stops. If $r_3 \neq 0$, then there exist integers q_4 and r_4 such that

$$r_2 = r_3 q_4 + r_4 \quad \text{and} \quad 0 \leq r_4 < r_3$$

We can continue in this manner obtaining quotients and remainders such that

$$
\begin{aligned}
a &= bq_1 + r_1 & &\text{and } 0 \leq r_1 < b \\
b &= r_1 q_2 + r_2 & &\text{and } 0 \leq r_2 < r_1 \\
r_1 &= r_2 q_3 + r_3 & &\text{and } 0 \leq r_3 < r_2 \\
r_2 &= r_3 q_4 + r_4 & &\text{and } 0 \leq r_4 < r_3 \\
r_3 &= r_4 q_5 + r_5 & &\text{and } 0 \leq r_5 < r_4 \\
r_4 &= r_5 q_6 + r_6 & &\text{and } 0 \leq r_6 < r_5 \\
&\;\;\vdots & &\quad\;\;\vdots \\
r_{k-3} &= r_{k-2} q_{k-1} + r_{k-1} & &\text{and } 0 \leq r_{k-1} < r_{k-2} \\
r_{k-2} &= r_{k-1} q_k + r_k & &\text{and } 0 \leq r_k < r_{k-1} \\
r_{k-1} &= r_k q_{k+1} + r_{k+1} & &\text{and } 0 \leq r_{k+1} < r_k
\end{aligned}
$$

The integers r_j are such that

$$r_1 > r_2 > r_3 > r_4 > r_5 > r_6 > \cdots > r_j \geq 0$$

By Axiom 12, one of the r_i's must be the smallest. If we *denote the smallest nonzero remainder by* r_k, then $r_{k+1} = 0$, the process must stop, and $r_{k-1} = r_k q_{k+1}$. We now claim that r_k, *the last nonzero remainder, is the GCD of a and b.*

Since $r_{k-1} = r_k q_{k+1}, r_k \mid r_{k-1}$.

Since $r_k \mid r_{k-1}$ and $r_k \mid r_k$, it follows that $r_k \mid (r_{k-1} q_k + r_k)$, and therefore $r_k \mid r_{k-2}$.

Since $r_k \mid r_{k-1}$ and $r_k \mid r_{k-2}$, it follows that $r_k \mid (r_{k-2} q_{k-1} + r_{k-1})$, and therefore $r_k \mid r_{k-3}$.

Since $r_k \mid r_{k-2}$ and $r_k \mid r_{k-3}$, it follows that $r_k \mid (r_{k-3} q_{k-2} + r_{k-2})$, and therefore $r_k \mid r_{k-4}$.

We can continue in this manner to show that $r_k \mid r_2$ and $r_k \mid r_1$.

Since $r_k \mid r_2$ and $r_k \mid r_1$, it follows that $r_k \mid (r_1 q_2 + r_2)$, and therefore $r_k \mid b$.

Since $r_k \mid r_1$ and $r_k \mid b$, it follows that $r_k \mid (bq_1 + r_1)$, and therefore $r_k \mid a$.

Since $r_k \mid a$ and $r_k \mid b$, condition (i) is satisfied by r_k.

If c is an integer such that $c \mid a$ and $c \mid b$, then $c \mid r_1$ as we showed before.

If $c \mid b$ and $c \mid r_1$, then $c \mid (b - r_1 q_2)$, and therefore $c \mid r_2$.

If $c \mid r_1$ and $c \mid r_2$, then $c \mid (r_1 - r_2 q_3)$, and therefore $c \mid r_3$.

Continuing in this manner we see that c divides every remainder and, in particular, $c \mid r_k$. Therefore, condition (ii) is satisfied by r_k and r_k is the GCD of a and b.   □

The following examples illustrate the above argument.
(a) To compute the GCD of 798 and 546:

$$798 = 546 \cdot 1 + 252$$
$$546 = 252 \cdot 2 + 42$$
$$252 = 42 \cdot 6 + 0$$

Since 42 is the last nonzero remainder, the GCD of 798 and 546 is 42.
(b) To compute the GCD of 7657 and 2821:

$$7657 = 2821 \cdot 2 + 2015$$
$$2821 = 2015 \cdot 1 + 806$$
$$2015 = 806 \cdot 2 + 403$$
$$806 = 403 \cdot 2 + 0$$

Thus, the last nonzero remainder, 403, is the GCD.

In any study of integers and divisibility, certain positive integers are of special interest in that they are divisible by no positive integers other than themselves and 1. Examples, from our experience, of such integers are 2, 3, 5, 7, and 19, although at this stage in the development we have not proved that these integers have this property.

An integer is said to be *prime* if it is greater than 1 and the only positive integers that divide it are itself and 1.

Closely related to the idea of a prime integer is the idea of a relatively prime pair of integers.

The integers a and b are said to be *relatively prime* if the GCD of a and b is 1.

Proposition 22

If p is prime and w is an integer and p does not divide w, then p and w are relatively prime.

Proof: Exercise.

Proposition 23

If p is prime and a and b are integers such that $p \mid ab$, then $p \mid a$ or $p \mid b$.

Proof: We shall show that if p does not divide a, then p must divide b. If p does not divide a, then p and a are relatively prime. Thus, for some integers u and v, $au + pv = 1$. Hence, $(au + pv)b = 1 \cdot b$, and therefore $abu + pvb = b$. Since $p \mid ab$ and $p \mid p$, $p \mid (abu + pvb)$, and therefore, $p \mid b$.   □

EXERCISE SET C

1. Prove Proposition 18.
2. Prove Corollary 19.
3. Compute the GCD's of the pairs of integers below.
   (a) 1435, 615   (c) 928, 2832   (e) 2059, 2627
   (b) 728, 1470   (d) 1610, 1386   (f) 1024, 197
4. The concept of a GCD can be extended to three integers and, in fact, to any finite collection. Make a suitable definition and prove that the GCD of three nonzero integers exists.
5. Compute the GCD's of the following triples of integers:
   (a) 1435, 615, 125   (c) 625, 256, 80
   (b) 728, 1470, 210   (d) 1044, 1116, 1470
6. Define 2 by $2 = 1 + 1$. Show that 2 is prime.
7. Make a suitable definition of 3 and show that 3 is prime.
8. Prove Proposition 22.
9. Prove the following generalization of Proposition 23: If x, y, and z are integers such that $x \mid yz$ and x and y are relatively prime, then $x \mid z$.
10. Prove: If a, b, c, and d are primes and $ab = cd$, then $a = c$ and $b = d$, or $a = d$ and $b = c$. (Hint: Use Proposition 23.)
**11. For any two nonzero integers m and n, we define the *least common multiple* (LCM) of m and n to be the positive integer k satisfying the following conditions:
   (a) $m \mid k$ and $n \mid k$.
   (b) If c is any integer such that $m \mid c$ and $n \mid c$, then $k \mid c$. Prove that the LCM of two nonzero integers exists.
**12. Prove: If a and b are positive integers, d is the GCD of a and b, and m is the LCM of a and b, then $ab = dm$.

§4. Mathematical Induction and the Fundamental Theorem of Arithmetic

The Principle (or principles) of Mathematical Induction is a formalization of a common intuitive idea. Both forms are statements about sets of positive integers and say, essentially, that if a set of positive integers satisfies certain conditions, then it is the entire set of positive integers. The course-of-values form of the principle states that if a set S of positive integers satisfies the following condition, then it must be the set of all positive integers.

For every positive integer m, if $\{x \in \mathbf{N} : x < m\}$ is a subset of S then $m \in S$.

The standard form of the principle states that if S is a subset of $\mathbf{N}$ satisfying the following two conditions, then it must be the set of all positive integers:

(i) $1 \in S$.
(ii) Whenever $t \in S$, $t + 1 \in S$.

Both of the principles are consequences of the Well-ordering Principle and, in fact, are equivalent to it. We could replace Axiom 12 by either Theorem 24 or Theorem 25 and then prove the other theorem and the Well-ordering Principle.

Theorem 24   (Course of Values Form of the Principle of Mathematical Induction)

Let S be a subset of $\mathbf{N}$ such that for each m in $\mathbf{N}$, whenever $\{x \in \mathbf{N}: x < m\} \subseteq S$, then $m \in S$. Then $S = \mathbf{N}$.

Proof: If $S \neq \mathbf{N}$, then there is a smallest element t in $\mathbf{N}$ that is not in S. Since t is the smallest element of $\mathbf{N}$ that is not in S, every element of $\mathbf{N}$ less than t is in S; that is, $\{x \in \mathbf{N}:x < t\} \subseteq S$. But then, by our hypothesis on S, $t \in S$. Now we have $t \notin S$ and $t \in S$, which is impossible. The assumption $S \neq \mathbf{N}$ leads to a contradiction, and therefore $S = \mathbf{N}$. □

Theorem 25   (Standard Form of the Principle of Mathematical Induction)

Let S be a subset of $\mathbf{N}$ such that

(i) $1 \in S$
(ii) If $t \in S$, then $t + 1 \in S$.

Then $S = \mathbf{N}$.

Proof: Let $t \in \mathbf{N}$ and assume that $\{x \in \mathbf{N}:x < t\} \subseteq S$. If $t = 1$, then $t \in S$ by (i). If $t \neq 1$, then $1 < t$. Therefore, $0 < t - 1$ and $t - 1 \in \mathbf{N}$. Also, $t - 1 < t$, and therefore $t - 1 \in S$. But, since $t - 1 \in S$, by (ii), $(t - 1) + 1 \in S$; that is, $t \in S$. Thus, whenever $\{x \in \mathbf{N}:x < t\} \subseteq S$, $t \in S$. By Theorem 24, $S = \mathbf{N}$. □

The Principle of Mathematical Induction can be used to define functions with domain $\mathbf{N}$ by

(i) Defining $(1)f$.
(ii) Defining $(n + 1)f$ in terms of $(n)f$ and $(1)f$ or in terms of $\{(1)f, \ldots, (n)f\}$.

Such definitions are called *recursive* definitions. For example, we can define positive integral exponents in this manner:

For each a in $\mathbf{Z}$, define a^k for positive integers k by

$$a^1 = a$$
$$a^{n+1} = a^n \cdot a$$

In this case $(1)f = a$ and $(n + 1)f = (nf)(1f)$.

The Principle of Mathematical Induction is a very powerful tool in all branches of mathematics. We shall use it now, along with many of the assertions proved earlier, to establish the following important theorem.

Theorem 26 (The Fundamental Theorem of Arithmetic)

Every integer greater than 1 is either prime or can be expressed uniquely as a product of primes.

Proof: Let $S = \{x \in N : x = 1$, or x is prime, or x can be expressed uniquely as a product of primes$\}$.

We wish to show that $S = N$.

Let $t \in N$ and suppose $\{x \in N : x < t\} \subseteq S$.

If $t = 1$, then $t \in S$.

Now assume $1 < t$.

If t is prime, then $t \in S$.

If t is not prime, then there exist integers a and b such that $1 < a < t$, $1 < b < t$, and $ab = t$. Since $1 < a < t$ and $1 < b < t$, $a \in S$, $b \in S$, and both a and b are either prime or can be expressed as the product of primes. Therefore, ab is the product of primes.

But $ab = t$ and therefore t can be expressed as the product of primes.

Now suppose $t = p_1 p_2 \cdots p_n$ and $t = q_1 q_2 \cdots q_m$, where each p_i and q_i is prime.

Then $p_1 p_2 \cdots p_n = q_1 q_2 \cdots q_m$ and $p_1 \mid q_1(q_2 \cdots q_m)$.

By Proposition 23, either $p_1 \mid q_1$ or $p_1 \mid q_2 \cdots q_m$.

If $p_1 \mid q_1$, then $p_1 = q_1$.

Since $q_2 \cdots q_m < t$, $q_2 \cdots q_m \in S$, and therefore the prime factorization of $q_2 \cdots q_m$ is unique.

Therefore, if $p_1 \mid q_2 \cdots q_m$, $p_1 = q_j$ for some j, $2 \leq j \leq m$.

Thus, $p_1 = q_j$ for some j, $1 \leq j \leq m$.

We may as well assume $p_1 = q_1$. Then $p_1 p_2 \cdots p_n = p_1 q_2 \cdots q_m$, and therefore $p_2 \cdots p_n = q_2 \cdots q_m$. Then, since $q_2 \cdots q_m < t$, the prime factorization is unique and each p_i is some q_j. Thus, the prime factorization of $t = p_1 p_2 \cdots p_n$ is unique and $t \in S$.

We have shown that if t is a positive integer such that $\{x \in N : x < t\} \subseteq S$, then $t \in S$.

By Theorem 24, $S = N$. $\square$

EXERCISE SET D

1. Prove, using the Principle of Mathematical Induction, that if a is an integer and m and n are positive integers, then $a^m \cdot a^n = a^{m+n}$ and $(a^m)^n = a^{mn}$.
   Hint: Let m be fixed.
   Let $S = \{x \in N : a^m \cdot a^x = a^{m+x}\}$, and show that $S = N$.
   Let $T = \{x \in N : (a^m)^x = a^{mx}\}$, and show that $T = N$.
2. Using the Principle of Mathematical Induction prove:
   (a) If n is a positive integer, then $1 + 2 + \cdots + n = n(n + 1)/2$.
   (b) If n is a positive integer, then $n < 2^n$.

(c) $1 + 4 + 9 + \cdots + n^2 = n(n + 1)(2n + 1)/6$.
(d) $1 + 8 + 27 + \cdots + n^3 = n^2(n + 1)^2/4$.

3. Let A be a set and let f be a function from A into A. Recursively define f^n for each positive integer n. If we define the function g from $\mathbf{Z}$ to $\mathbf{Z}$ by $xg = x^2 + 1$, what is g^2? What is g^3?

4. Use the Principle of Mathematical Induction to prove: If q is a prime, x is an integer, and n is a positive integer such that $q \mid x^n$, then $q \mid x$.

5. Prove: There are no integers x and y such that $x^2 = 2y^2$.
(Hint: Use Theorem 26 and the fact that 2 is prime.)

6. A rational number is a real number that can be expressed as the quotient of two integers. Prove (using Exercise 5) that there is no rational number whose square is 2.

7. Generalize Exercises 5 and 6 for primes other than 2.

**8. Generalize Exercise 7 for exponents greater than 2.

**9. Prove: If n is a positive integer and x is a rational number such that x^n is an integer, then x is an integer.

§5. Modular Systems

In the study of any structure, certain equivalence relations are of special interest. In this section we study a collection of equivalence relations on $\mathbf{Z}$ and some techniques that can be adapted for use in the more general situations encountered in Chapters 4, 5, and 7.

Let m be an integer. For any two integers x and y, we shall say that x is *congruent* to y modulo m iff $m \mid (x - y)$. That x is congruent to y modulo m is denoted by

$$x \equiv y \, (\mathrm{mod} \, m)$$

The relation is called *congruence modulo m*. The integer m is called the *modulus*.

Examples

$$5 \equiv 17 \quad (\mathrm{mod} \, 6) \quad \text{because} \quad 6 \mid (5 - 17)$$
$$13 \equiv (-2) \, (\mathrm{mod} \, 5) \quad \text{because} \quad 5 \mid [13 - (-2)]$$
$$7 \equiv 0 \quad (\mathrm{mod} \, 7) \quad \text{because} \quad 7 \mid (7 - 0)$$

The fact that the symbol "$\equiv$" looks somewhat like an equal sign and that the relation of congruence of triangles is an equivalence relation might lead one to expect the symbol and name are chosen because the relation is an equivalence relation. Your guess is correct.

Proposition 27

If m is an integer, then congruence modulo m is an equivalence relation on $\mathbf{Z}$.

Proof: Since $m \mid 0$, $m \mid (x - x)$ for every integer x, and therefore

$$x \equiv x \, (\mathrm{mod} \, m)$$

If x and y are integers such that $x \equiv y \pmod{m}$, then $m \mid (x - y)$. But then $m \mid (x - y)(-1)$, or

$$m \mid (y - x)$$

and therefore

$$y \equiv x \pmod{m}$$

Suppose x, y, and z are integers such that

$$x \equiv y \pmod{m} \quad \text{and} \quad y \equiv z \pmod{m}$$

Since $m \mid (x - y)$ and $m \mid (y - z)$,

$$m \mid [(x - y) + (y - z)]$$

or, equivalently,

$$m \mid (x - z)$$

and therefore,

$$x \equiv z \pmod{m} \quad \square$$

A natural question at this point would be "What other properties of the equality relation are also satisfied by congruence modulo m?" Can we add the same number to both sides of a congruence and obtain a congruence? What happens if we multiply both sides by a number? Suppose we divide both sides by the same number? The first two questions are answered by the following proposition.

Proposition 28

If m, a, b, c, and d are integers such that $a \equiv b \pmod{m}$ and $c \equiv d \pmod{m}$, then $a + c \equiv b + d \pmod{m}$ and $ac \equiv bd \pmod{m}$.

Proof: If $a \equiv b \pmod{m}$ and $c \equiv d \pmod{m}$, then

$$m \mid (a - b) \quad \text{and} \quad m \mid (c - d)$$

Therefore,

$$m \mid [(a - b) + (c - d)]$$

or equivalently,

$$m \mid [(a + c) - (b + d)]$$

and therefore

$$a + c \equiv b + d \pmod{m}$$

Also, since

$$m \mid (a - b) \quad \text{and} \quad m \mid (c - d)$$
$$m \mid (a - b)c \quad \text{and} \quad m \mid b(c - d)$$

or, equivalently,

$$m \mid (ac - bc) \quad \text{and} \quad m \mid (bc - bd)$$

Then

$$m \mid [(ac - bc) + (bc - bd)]$$

or, equivalently,

$$m \mid (ac - bd)$$

and therefore

$$ac \equiv bd \,(\text{mod } m) \quad \square$$

Proposition 28 asserts that the relation of congruence modulo m is compatible with the operations of addition and multiplication. It is also compatible with subtraction and we leave the proof as an exercise (Set E, Exercise 1). That congruence modulo m need not be compatible with division is noted by observing that $8 \equiv 2 \,(\text{mod } 6)$ although $4 \not\equiv 1 \,(\text{mod } 6)$.

If we divide both sides of a congruence by a number that is relatively prime to the modulus m, then the resulting quotients will be congruent modulo m. We leave the proof of this assertion to the reader (Set E, Exercise 2).

Since any equivalence relation partitions a set into equivalence classes, congruence modulo m partitions $\mathbf{Z}$ into disjoint classes. The equivalence classes with respect to congruence modulo m are usually called *congruence classes*. If m is the modulus, we let $\mathbf{Z}_m$ denote the set of all congruence classes with respect to congruence modulo m.

Examples. If 5 is the modulus, then

$$\begin{aligned}
\bar{0} &= \{\ldots, -15, -10, -5, 0, 5, 10, 15, 20, \ldots\} \\
\bar{1} &= \{\ldots, -14, -9, -4, 1, 6, 11, 16, 21, \ldots\} \\
\bar{2} &= \{\ldots, -18, -13, -8, -3, 2, 7, 12, 17, 22, \ldots\} \\
\bar{3} &= \{\ldots, -7, -2, 3, 8, 13, 18, \ldots\} \\
\bar{4} &= \{\ldots, -6, -1, 4, 9, 14, 19, 24, \ldots\}
\end{aligned}$$

Notice that $\bar{2} = \overline{-8} = \bar{7} = \overline{22}$. We can represent $\mathbf{Z}_5$ by $\mathbf{Z}_5 = \{\bar{0}, \bar{1}, \bar{2}, \bar{3}, \bar{4}\}$, but this representation is not unique, since

$$\mathbf{Z}_5 = \{\overline{10}, \overline{-4}, \bar{7}, \bar{3}, \bar{9}\}$$

If 7 is the modulus, then

$$\begin{aligned}
\bar{0} &= \{\ldots, -14, -7, 0, 7, 14, 21, \ldots\} \\
\bar{3} &= \{\ldots, -18, -11, -4, 3, 10, 17, 24, 31, \ldots\}
\end{aligned}$$

and

$$\bar{5} = \{\ldots, -23, -16, -9, -2, 5, 12, 19, 26, 33, 40, \ldots\}$$

Notice that for any modulus m,

$$\bar{0} = \{\ldots, -4m, -3m, -2m, -m, 0, m, 2m, 3m, 4m, \ldots\}$$

that is,

$$\bar{0} = \{xm : x \in \mathbf{Z}\} = \{p \in \mathbf{Z} : m \mid p\} \qquad \text{the set of all multiples of } m$$

because

$$y \equiv 0 \, (\text{mod } m) \quad \text{iff} \quad m \mid (y - 0) \quad \text{iff} \quad m \mid y$$

Before proceeding to the generality of Theorem 29, let us examine one specific example. If the modulus is 8, then we can denote the set of congruence classes by $\mathbf{Z_8} = \{\bar{0}, \bar{1}, \bar{2}, \bar{3}, \bar{4}, \bar{5}, \bar{6}, \bar{7}\}$. That all the classes are listed is assured by noticing that every integer is congruent modulo m to the remainder obtained by dividing it by m; that is, if $a = bm + r$ with $0 \le r < m$, then $a \equiv r \, (\text{mod } m)$. Thus, the set of possible remainders after dividing by 8 determines all the congruence classes. In $\mathbf{Z_8}$ we wish to define a new addition and multiplication. These new operations will not be the same as the original addition and multiplication but will behave in much the same manner. The rules we use to define the new operations will use the old operations and the symbols for the operations will be the same.

Consider first the two classes

$$\bar{2} = \{\ldots, -14, -6, 2, 10, 18, \ldots\}$$

and

$$\bar{5} = \{\ldots, -19, -11, -3, 5, 13, 21, \ldots\}$$

If we pick an element in $\bar{2}$ and an element in $\bar{5}$ and add them together, the sum must be in some class. Forming sums of elements in $\bar{2}$ with elements in $\bar{5}$ using the particular elements given above we get

$$\{-33, -25, -17, -9, -1, 7, 15, 23, 31, 39\}$$

These are all in $\bar{7}$.

The example leads us to believe that if we pick a number in $\bar{2}$ and a number in $\bar{5}$, their sum will be in $\bar{7}$. This is, in fact, the case and is not hard to establish.

If $x \in \bar{2}$ and $y \in \bar{5}$, then

$$x \equiv 2 \, (\text{mod } 8) \quad \text{and} \quad y \equiv 5 \, (\text{mod } 8)$$

Then

$$x + y \equiv 2 + 5 \, (\text{mod } 8) \qquad \text{by Proposition 28}$$

that is, $x + y \equiv 7 \, (\text{mod } 8)$, and it follows that

$$x + y \in \bar{7}$$

There is nothing special about $\bar{2}$ and $\bar{5}$ in the above argument, and we can define an addition of congruence classes modulo 8 by the rule $\bar{x} + \bar{y} = \overline{x + y}$. Similarly, if we form the set of products of elements of $\bar{2}$ with those of $\bar{5}$, all the products we obtain are in $\overline{10} = \bar{2}$.

For example,

$$\{266, 66, -6, 10, 50, -110, -30\} \subseteq \bar{2}$$

In general, if

$$x \in \bar{2} \quad \text{and} \quad y \in \bar{5}$$

then

$$x \equiv 2 \,(\text{mod } 8) \quad \text{and} \quad y \equiv 5 \,(\text{mod } 8)$$

By Proposition 28, then

$$xy \equiv 10 \,(\text{mod } 8)$$

which implies

$$xy \equiv 2 \,(\text{mod } 8)$$

and therefore

$$xy \in \bar{2}$$

As before, there was nothing special about $\bar{2}$ and $\bar{5}$ in the above argument, and we can define multiplication of congruence classes modulo 8 by

$$\bar{x} \cdot \bar{y} = \overline{xy}$$

The reader should have noticed that there was nothing special about the modulus 8 in the above arguments. We now proceed with the general case.

If m is an integer, we define an addition and multiplication on the set $\mathbf{Z}_m$ by $\bar{x} + \bar{y} = \overline{x + y}$ and $\bar{x} \cdot \bar{y} = \overline{xy}$ for any integers x and y. Before considering any of the properties of these operations, we must be sure that our rules do, in fact, define operations on $\mathbf{Z}_m$. Our rules say to take an element from the class for x and one from the class for y, add and multiply them together and let the sum of the classes be the class to which this sum belongs, and let the product of the classes be the class to which the product belongs.

How do we know we shall get the same resultant *class* no matter which elements we choose? If a, b, c, and d are integers such that $\bar{a} = \bar{c}$ and $\bar{b} = \bar{d}$, then $a \equiv c \,(\text{mod } m)$ and $b \equiv d \,(\text{mod } m)$. By Proposition 28,

$$a + b \equiv c + d \,(\text{mod } m) \quad \text{and} \quad ac \equiv bd \,(\text{mod } m)$$

and therefore,

$$\overline{a + b} = \overline{c + d} \quad \text{and} \quad \overline{ac} = \overline{bd}$$

Thus, no matter what elements we use from $\overline{a}$ and $\overline{b}$, their sum will be in the same class as any other such sum, and thus the class obtained is uniquely determined. Similarly, if we take the product of an element in $\overline{a}$ and one in $\overline{b}$, the product will be in the same class as any other such product and the class is uniquely determined. Thus, the rules do define operations on $\mathbf{Z}_m$.

Theorem 29

If m is an integer and addition and multiplication in $\mathbf{Z}_m$ are defined by $\overline{x} + \overline{y} = \overline{x + y}$ and $\overline{x} \cdot \overline{y} = \overline{xy}$, then

(i) Addition is associative.
(ii) Addition is commutative.
(iii) For each $\overline{x}$, $\overline{x} + \overline{0} = \overline{x}$.
(iv) For each $\overline{x}$, $\overline{x} + (\overline{-x}) = \overline{0}$.
(v) Multiplication is associative.
(vi) For each $\overline{x}$, $\overline{y}$, $\overline{z}$,

$$\overline{x} \cdot (\overline{y} + \overline{z}) = (\overline{x} \cdot \overline{y}) + (\overline{x} \cdot \overline{z})$$

and

$$(\overline{x} + \overline{y}) \cdot \overline{z} = (\overline{x} \cdot \overline{z}) + (\overline{y} \cdot \overline{z})$$

(vii) Multiplication is commutative.
(viii) For each $\overline{x}$, $\overline{x} \cdot \overline{1} = \overline{x}$.

Proof: We shall prove only parts of the theorem and leave the remaining proofs as exercises.

PROOF OF (i). Let $\overline{x}$, $\overline{y}$, and $\overline{z} \in \mathbf{Z}_m$. Then

$$\begin{aligned}
\overline{x} + (\overline{y} + \overline{z}) &= \overline{x} + (\overline{y + z}) \\
&= \overline{x + (y + z)} \\
&= \overline{(x + y) + z} \\
&= (\overline{x + y}) + \overline{z} \\
&= (\overline{x} + \overline{y}) + \overline{z}
\end{aligned}$$

PROOF OF (iii). Let $x \in \mathbf{Z}_m$. Then $\overline{x} + \overline{0} = \overline{x + 0} = \overline{x}$.

PROOF OF (vi). Let $\overline{x}, \overline{y}, \overline{z} \in \mathbf{Z}_m$.

$$\begin{aligned}
\overline{x} \cdot (\overline{y} + \overline{z}) &= \overline{x} \cdot (\overline{y + z}) \\
&= \overline{x \cdot (y + z)}
\end{aligned}$$

$$= \overline{xy + xz}$$
$$= \overline{xy} + \overline{xz}$$
$$= (\overline{x} \cdot \overline{y}) + (\overline{x} \cdot \overline{z}) \quad \square$$

By Theorem 29 we see that by using the integers and the congruence relations defined on $\mathbf{Z}$ we can construct infinitely many new structures. Furthermore, since Axioms 1–8 for $\mathbf{Z}$ are also satisfied by each $\mathbf{Z}_m$, the assertions about $\mathbf{Z}$ that we proved using only these axioms will also be valid for each of these structures. We leave as an exercise the determination of just which assertions these are.

EXERCISE SET E

1. Prove: If m, a, b, c, and d are integers such that $a \equiv b$ (mod m) and $c \equiv d$ (mod m), then $a - c \equiv b - d$ (mod m).
2. Prove: If m, a, b, and c are integers such that $ac \equiv bc$ (mod m) and c and m are relatively prime; then $a \equiv b$ (mod m). (Hint: See Exercise 9, Set C.)
3. Let m be a positive integer; in $\mathbf{Z}$ define $\tilde{m}$ by: $x\tilde{m}y$ iff x and y leave the same remainder after division by m. Prove that $\tilde{m}$ is an equivalence relation.
4. Describe each of the equivalence classes in $\mathbf{Z}$ with respect to $\tilde{6}$. Do the same for $\tilde{9}$. Do the same for the general relation $\tilde{m}$.
5. Describe the congruence classes modulo 6 and modulo 9; and modulo any m.
6. What is the relation between congruence modulo m and the relation $\tilde{m}$ defined in Exercise 3? Prove your assertion.
7. In $\mathbf{Z}_{12}$, solve the following equations (if they have solutions):
   (a) $\overline{x} + \overline{5} = \overline{8}$ (d) $(\overline{2} \cdot \overline{x}) + \overline{5} = \overline{9}$
   (b) $\overline{5} \cdot \overline{x} = \overline{3}$ (e) $\overline{11} \cdot \overline{x} = \overline{1}$
   (c) $\overline{4} \cdot \overline{x} = \overline{3}$ (f) $\overline{3} \cdot \overline{x} = \overline{1}$
8. Prove the remaining portions of Theorem 29.
9. Which of Propositions 1–7 are valid in $\mathbf{Z}_m$ if we make the natural translations?
10. Does Theorem 13 or Proposition 14 hold, in general, for $\mathbf{Z}_m$? Proof?
11. If m is an integer we define the natural function η from $\mathbf{Z}$ to $\mathbf{Z}_m$ by $x\eta = \overline{x}$. Prove: If x and y are integers, then $(x + y)\eta = x\eta + y\eta$ and $(xy)\eta = (x\eta)(y\eta)$.

3
PERMUTATIONS

The theory of groups had its beginnings in the study of sets of permutations of finite sets. In this chapter we develop some of the elementary properties of permutations and, in doing so, involve the reader with several applications of the Principle of Mathematical Induction.

§1. Inverses of Functions, Permutations

If f is a function from A into B, we define f^{-1} by

$$f^{-1} = \{(p, q):(q, p) \in f\}$$

Thus, f^{-1} is a subset of $B \times A$, and a pair (m, n) is in f^{-1} iff (n, m) is in f. This set is called "f inverse" and the symbol f^{-1} is read "f inverse." *In general f^{-1} is not a function.* Our reason for introducing the notion is, however, that we are interested in the case when f^{-1} is, in fact, a function.

The use of the symbol f^{-1} in more than one way may, at first, seem confusing. If f is a function and b is in the range of f, bf^{-1} is a well-defined set even though f^{-1} may not be a function. In the case where f^{-1} is a function and $af = b$, $bf^{-1} = \{a\}$ according to our original meaning of the symbol, and $bf^{-1} = a$ when f^{-1} is viewed as a function. It is possible to introduce another notation for bf^{-1} in the general sense and clear up the minor confusion that exists. The confusion, however, is only a possibility and in practice hardly ever arises. The context usually makes the meaning clear and most mathematicians use the same symbol for both meanings. With practice the student will have no difficulty either. Remember that, in general, $(af)f^{-1} \neq a$ but $(bf^{-1})f = b$ in all cases in which $bf^{-1} \neq \phi$.

If S is a set, the *identity function on S* is denoted by I_S, or, if no ambiguity can arise, just by I, and is defined by

$$I_S = \{(x, x):x \in S\}$$

Thus, for every element t in S, $tI_S = t$. Clearly I_S is a one-to-one and onto

49

function. Notice that if A and B are sets, and f is a function from A into B, then $f \circ I_B = f$ and $I_A \circ f = f$.

Proposition 1

If A and B are sets and f is a one-to-one function from A onto B, then f^{-1} is a one-to-one function from B onto A. Furthermore, $f \circ f^{-1}$ is the identity function on A and $f^{-1} \circ f$ is the identity function on B.

Proof: Clearly $f^{-1} \subseteq B \times A$.

First we need to show that f^{-1} is a function; that is, we need to show that for every element y in B there is exactly one element x in A such that $(y, x) \in f^{-1}$. If y is an element of B, since f is onto, there is an element x in A such that $xf = y$. Hence, $(x, y) \in f$ and $(y, x) \in f^{-1}$. Thus, for each element y in B there is at least one element x in A such that $(y, x) \in f^{-1}$. Since f is a one-to-one function, there is only one element in A whose image is y; that is, there is only one element, namely x, such that $(x, y) \in f$ and therefore, there is a unique element, namely x, such that $(y, x) \in f^{-1}$. Thus, by definition, f^{-1} is a function from B to A.

To show that f^{-1} is onto, let t be an element of A. Then $(t, tf) \in f$, which implies $(tf, t) \in f^{-1}$, that is, $(tf)f^{-1} = t$ for each t in A, and therefore f^{-1} is onto. To show f^{-1} is one-to-one, suppose $xf^{-1} = yf^{-1}$ for some x and y in B. Then $(x, xf^{-1}) \in f^{-1}$ and $(y, yf^{-1}) \in f^{-1}$, which implies $(xf^{-1}, x) \in f$ and $(yf^{-1}, y) \in f$. Since f is a function and $xf^{-1} = yf^{-1}$, x is the same as y. Thus, f^{-1} is one-to-one.

That $f \circ f^{-1}$ is the identity function on A follows from the argument in which we showed that $t = t(f \circ f^{-1})$ for each t in A. If y is in B, then $(y, yf^{-1}) \in f^{-1}$, and therefore $(yf^{-1}, y) \in f$. Thus, $(yf^{-1})f = y$, that is $(y)(f^{-1} \circ f) = y$, and therefore $(f^{-1} \circ f)$ is the identity function on B. □

Our intuitive notion of the inverse of a function is that of "undoing" what the original function has "done." We would expect the inverse of the cubing function to be the cube root function and the arctan y function to be the inverse of tan x [where the domain of tan x is defined to be $(-\pi/2, \pi/2)$]. In the second part of Proposition 1 we were assured that this is in general the case. The following proposition assures us that this is, in fact, the characteristic property of the inverse of a function.

Proposition 2

If f is a function from A into B and g is a function from B into A such that $f \circ g = I_A$ and $g \circ f = I_B$, then $g = f^{-1}$.

Proof: We shall first show that f^{-1} is a function and then we shall use associativity of composition and properties of identity functions to show that $g = f^{-1}$.

Suppose x and y are elements of A such that $xf = yf$. Then $(xf)g = (yf)g$, and therefore $(x)(f \circ g) = (y)(f \circ g)$. Since $f \circ g = I_A$, $xI_A = yI_A$, and therefore $x = y$. Thus, f is a one-to-one function.

Now let t be an element of B. Then $t = tI_B = t(g \circ f) = (tg)f$. Since $(tg)f = t$, t is the image of tg under f, and therefore f is onto.

Since f is a one-to-one function from A onto B, Proposition 1 implies that f^{-1} is a one-to-one function from B onto A with $f \circ f^{-1} = I_A$ and $f^{-1} \circ f = I_B$.

Since

$$f \circ g = I_A$$
$$f^{-1} \circ (f \circ g) = f^{-1} \circ I_A$$

Then

$$(f^{-1} \circ f) \circ g = f^{-1}$$
$$I_B \circ g = f^{-1}$$

and therefore

$$g = f^{-1} \quad \square$$

Proposition 3

If p is a one-to-one function from A onto B and q is a one-to-one function from B onto C, then $p \circ q$ is a one-to-one function from A onto C, and furthermore, $(p \circ q)^{-1} = q^{-1} \circ p^{-1}$.

Proof: From previous results we know $p \circ q$ is a one-to-one and onto function.

Proof: Since q^{-1} and p^{-1} are functions, we know that $q^{-1} \circ p^{-1}$ is a function. Then

$$
\begin{aligned}
(p \circ q) \circ (q^{-1} \circ p^{-1}) &= [(p \circ q) \circ q^{-1}] \circ p^{-1} \\
&= [p \circ (q \circ q^{-1})] \circ p^{-1} \\
&= [p \circ (I_B)] \circ p^{-1} \\
&= p \circ p^{-1} \\
&= I_A
\end{aligned}
$$

and

$$
\begin{aligned}
(q^{-1} \circ p^{-1}) \circ (p \circ q) &= q^{-1} \circ [p^{-1} \circ (p \circ q)] \\
&= q^{-1} \circ [(p^{-1} \circ p) \circ q] \\
&= q^{-1} \circ [I_B \circ q] \\
&= q^{-1} \circ q \\
&= I_C
\end{aligned}
$$

Thus, we have $(p \circ q) \circ (q^{-1} \circ p^{-1}) = I_A$ and $(q^{-1} \circ p^{-1}) \circ (p \circ q) = I_C$. By Proposition 2, $(p \circ q)^{-1} = q^{-1} \circ p^{-1}$. $\square$

The following examples show that, if p and q are functions such that $p \circ q = I$, we *cannot* infer that $q \circ p = I$. Thus, both of the hypotheses of Proposition 2 are necessary.

Examples. Let Z denote the set of all integers. Define f and g to be functions from Z to Z by $xf = 2x$ for all integers x and $yg = \frac{1}{2}y$ if y is even and $yg = 0$ if y is odd. Then $f \circ g = I$, since

$$(x)(f \circ g) = (xf)g = (2x)g = x$$

But $g \circ f \neq I$, because

$$(3)(g \circ f) = (3g)f = (0)f = 0$$

Let $R[x]$ denote the set of all polynomials over the real numbers, and let D denote the derivative operator; that is, D is the function that maps each polynomial to its derivative. If for each real number m we define S_m by

$$fS_m = \int_m^x f(t) \, dt, \text{ then } S_m \circ D = I; \text{ that is}$$

$$f(S_m \circ D) = f$$

for each polynomial, because

$$f(S_m \circ D) = (fS_m)D = (fS_m)' = \left[\int_m^x f(t) \, dt \right]' = f$$

But

$$D \circ S_m \neq I$$

because

$$4(D \circ S_m) = \int_m^x (4)' \, dt = \int_m^x 0 \, dt = 0$$

Of special interest among one-to-one and onto functions are those that map a set onto itself.

If S is a nonempty set, a one-to-one function from S onto S is called a *permutation* of S. We shall denote the set of all permutations of the set S by $P(S)$.

From our study of one-to-one functions and onto functions, we know that if $f \in P(S)$, and $g \in P(S)$, then $f \circ g \in P(S)$ and also that I_s is in $P(S)$. Furthermore, we have shown that:

(i) If f, g, and h are elements of $P(S)$, then $f \circ (g \circ h) = (f \circ g) \circ h$.

(ii) If $f \in P(S)$, then $f \circ I_s = f = I_s \circ f$.

(iii) If $f \in P(S)$, then $f^{-1} \in P(S)$ and $f \circ f^{-1} = I_s = f^{-1} \circ f$. (Proposition 1).

(iv) If $f \in P(S)$ and $g \in P(S)$ and $f \circ g = I_s$, then $f = g^{-1}$ (Proposition 2).

(v) If $f \in P(S)$ and $g \in P(S)$, then $(f \circ g)^{-1} = g^{-1} \circ f^{-1}$ (Proposition 3).

One way of describing a permutation is to use functional notation. For example, one of the permutations of $\{1, 2, 3, 4\}$ could be described by

$$1f = 2 \qquad 2f = 4 \qquad 3f = 3 \qquad 4f = 1$$

Another commonly used notation for permutations of a finite set is to list the elements of the set on one line with their images directly under them and then to set this off by means of parentheses. Thus, the function f described above could be given by

$$\begin{pmatrix} 1 & 2 & 3 & 4 \\ 2 & 4 & 3 & 1 \end{pmatrix}$$

By

$$\begin{pmatrix} 1 & 2 & 3 & 4 \\ 3 & 1 & 4 & 2 \end{pmatrix}$$

we would mean the function that maps 1 to 3, 2 to 1, 3 to 4, and 4 to 2.

Permutations denoted this way can be composed by using the definition of composition and computing. Thus, for example,

$$\begin{pmatrix} 1 & 2 & 3 & 4 \\ 2 & 3 & 4 & 1 \end{pmatrix} \circ \begin{pmatrix} 1 & 2 & 3 & 4 \\ 3 & 1 & 4 & 2 \end{pmatrix} = \begin{pmatrix} 1 & 2 & 3 & 4 \\ 1 & 4 & 2 & 3 \end{pmatrix}$$

because

$$1 \begin{pmatrix} 1 & 2 & 3 & 4 \\ 2 & 3 & 4 & 1 \end{pmatrix} = 2 \quad \text{and} \quad 2 \begin{pmatrix} 1 & 2 & 3 & 4 \\ 3 & 1 & 4 & 2 \end{pmatrix} = 1$$

$$2 \begin{pmatrix} 1 & 2 & 3 & 4 \\ 2 & 3 & 4 & 1 \end{pmatrix} = 3 \quad \text{and} \quad 3 \begin{pmatrix} 1 & 2 & 3 & 4 \\ 3 & 1 & 4 & 2 \end{pmatrix} = 4$$

$$3 \begin{pmatrix} 1 & 2 & 3 & 4 \\ 2 & 3 & 4 & 1 \end{pmatrix} = 4 \quad \text{and} \quad 4 \begin{pmatrix} 1 & 2 & 3 & 4 \\ 3 & 1 & 4 & 2 \end{pmatrix} = 2$$

$$4 \begin{pmatrix} 1 & 2 & 3 & 4 \\ 2 & 3 & 4 & 1 \end{pmatrix} = 1 \quad \text{and} \quad 1 \begin{pmatrix} 1 & 2 & 3 & 4 \\ 3 & 1 & 4 & 2 \end{pmatrix} = 3$$

EXERCISE SET A

1. Define f, g, and h from $\mathbf{R}$ to $\mathbf{R}$ by $xf = 2x + 1$, $xg = x^3 + 3$, and $xh = \sqrt[5]{x}$. Determine $f \circ g$, $g \circ h$, $f \circ h$. Determine f^{-1}, g^{-1}, and h^{-1}. Now compute $(f \circ g)^{-1}$, $(g \circ h)^{-1}$, and $(f \circ h)^{-1}$. (Hint: Use Proposition 3.)
2. Construct two functions f and g from $\mathbf{R}$ to $\mathbf{R}$ such that $f \circ g = I_{\mathbf{R}}$ and $g \circ f \neq I_{\mathbf{R}}$.
3. Describe three permutations of $\mathbf{R}$ that are not continuous functions. Now describe their inverses.

4. Below is a table listing all of the permutations of $\{1, 2, 3\}$. The entry in the row following f and the column under g should be $f \circ g$. Complete the table.

$\circ$	$\begin{pmatrix} 1 & 2 & 3 \\ 1 & 2 & 3 \end{pmatrix}$	$\begin{pmatrix} 1 & 2 & 3 \\ 3 & 1 & 2 \end{pmatrix}$	$\begin{pmatrix} 1 & 2 & 3 \\ 2 & 3 & 1 \end{pmatrix}$	$\begin{pmatrix} 1 & 2 & 3 \\ 2 & 1 & 3 \end{pmatrix}$	$\begin{pmatrix} 1 & 2 & 3 \\ 1 & 3 & 2 \end{pmatrix}$	$\begin{pmatrix} 1 & 2 & 3 \\ 3 & 2 & 1 \end{pmatrix}$
$\begin{pmatrix} 1 & 2 & 3 \\ 1 & 2 & 3 \end{pmatrix}$						
$\begin{pmatrix} 1 & 2 & 3 \\ 3 & 1 & 2 \end{pmatrix}$						$\begin{pmatrix} 1 & 2 & 3 \\ 1 & 3 & 2 \end{pmatrix}$
$\begin{pmatrix} 1 & 2 & 3 \\ 2 & 3 & 1 \end{pmatrix}$						
$\begin{pmatrix} 1 & 2 & 3 \\ 2 & 1 & 3 \end{pmatrix}$					$\begin{pmatrix} 1 & 2 & 3 \\ 3 & 1 & 2 \end{pmatrix}$	
$\begin{pmatrix} 1 & 2 & 3 \\ 1 & 3 & 2 \end{pmatrix}$						
$\begin{pmatrix} 1 & 2 & 3 \\ 3 & 2 & 1 \end{pmatrix}$						

5. Use the table from Exercise 4 to solve the following equations. X is to be a permutation of $\{1, 2, 3\}$ and $I = \begin{pmatrix} 1 & 2 & 3 \\ 1 & 2 & 3 \end{pmatrix}$.

(a) $\begin{pmatrix} 1 & 2 & 3 \\ 3 & 1 & 2 \end{pmatrix} \circ X = \begin{pmatrix} 1 & 2 & 3 \\ 1 & 3 & 2 \end{pmatrix}$

(b) $X \circ \begin{pmatrix} 1 & 2 & 3 \\ 3 & 1 & 2 \end{pmatrix} = \begin{pmatrix} 1 & 2 & 3 \\ 1 & 3 & 2 \end{pmatrix}$

(c) $X \circ X = \begin{pmatrix} 1 & 2 & 3 \\ 3 & 1 & 2 \end{pmatrix}$

(d) $X \circ X = I$

(e) $X \circ X \circ X = I$

6. Determine the following permutations:

(a) $\begin{pmatrix} 1 & 2 & 3 & 4 \\ 2 & 3 & 4 & 1 \end{pmatrix}^{-1}$

(b) $\begin{pmatrix} 1 & 2 & 3 & 4 \\ 3 & 4 & 1 & 2 \end{pmatrix}^{-1}$

(c) $\begin{pmatrix} 1 & 2 & 3 & 4 \\ 3 & 2 & 1 & 4 \end{pmatrix}^{-1}$

7. Solve the following equations (X is to be a permutation of $\{1, 2, 3, 4\}$).

(a) $\begin{pmatrix} 1 & 2 & 3 & 4 \\ 2 & 3 & 4 & 1 \end{pmatrix} \circ X = \begin{pmatrix} 1 & 2 & 3 & 4 \\ 4 & 2 & 3 & 1 \end{pmatrix}$

(b) $X \circ \begin{pmatrix} 1 & 2 & 3 & 4 \\ 2 & 3 & 4 & 1 \end{pmatrix} = \begin{pmatrix} 1 & 2 & 3 & 4 \\ 4 & 2 & 3 & 1 \end{pmatrix}$

(c) $X \circ \begin{pmatrix} 1 & 2 & 3 & 4 \\ 3 & 4 & 1 & 2 \end{pmatrix} = \begin{pmatrix} 1 & 2 & 3 & 4 \\ 4 & 3 & 2 & 1 \end{pmatrix}$

(d) $\begin{pmatrix} 1 & 2 & 3 & 4 \\ 3 & 4 & 1 & 2 \end{pmatrix} \circ X = \begin{pmatrix} 1 & 2 & 3 & 4 \\ 4 & 3 & 2 & 1 \end{pmatrix}$

8. If S is any nonempty set and f and g are permutations of S, solve the following equations in $P(S)$:

(a) $f \circ X = g$ 　　　　　　　　　　(b) $X \circ f = g$

§2. Exponents

Many of the fundamental theorems concerning permutations, and especially permutations of finite sets, can be proved by using one of the forms of the Principle of Mathematical Induction. Thus, by studying permutations the student gets a greater familiarity with mathematical induction and gets an opportunity to use it several times in proving some of these results.

There is a natural way of defining a meaning for positive integral exponents for any functions from a set onto itself and, in fact, this was done in an exercise in Chapter 2.

If f is a function from the set S into S, we define f^n recursively by

$$f^1 = f$$
$$f^{n+1} = f^n \circ f \qquad \text{for all positive integers } n$$

At this point let us mention that it is common to use a multiplicative or juxtaposition notation for permutations rather than the composition notation. If p and q are permutations, by pq we mean $p \circ q$. As long as there is no confusion as to what is meant, we can use either notation.

The following assertions follow quite easily, using the Principle of Mathematical Induction.

Lemma 4

If f is a function from the set S to S and m is a positive integer, then $f \circ f^m = f^{m+1}$.

Proof: Exercise.

Proposition 5

If p is a permutation of the set S and n is a positive integer, then $(p^n)^{-1} = (p^{-1})^n$.

Proof: Exercise.

For permutations we can extend the definition of positive integral exponents to give a meaning to all integral exponents just as we do for all nonzero real numbers. Recall that $5^0 = 1$, $17^0 = 1$, and $(-3)^0 = 1$ while $2^{-3} = 1/2^3$, $7^{-4} = 1/7^4$, and $(\frac{2}{3})^{-2} = 1/(\frac{2}{3})^2$. In general, for nonzero real numbers r we define $r^0 = 1$ and $r^{-m} = 1/r^m$ if m is a positive integer. This definition generalizes easily for permutations.

If S is a set and p is a permutation of S, p^0 is defined by $p^0 = I$, and if m is a positive integer, p^{-m} is defined by $p^{-m} = (p^m)^{-1}$.

From our definition and Proposition 5, notice that

$$p^{-m} = (p^m)^{-1} = (p^{-1})^m$$

if p is a permutation and m is a positive integer. Clearly

$$p^{-0} = p^0 = I = (p^{-1})^0 = (p^0)^{-1}$$

Also, if n is a negative integer,

$$(p^{-1})^n = [(p^{-1})^{-n}]^{-1} = [(p^{-n})^{-1}]^{-1} = [p^n]^{-1}$$

and

$$(p^{-1})^n = [(p^{-1})^{-n}]^{-1} = [(p^{-1})^{-1}]^{-n} = [p]^{-n} = p^{-n}$$

Thus, we have proved the following assertion.

Proposition 6

If p is a permutation of the set S and d is an integer, then $p^{-d} = (p^d)^{-1} = (p^{-1})^d$.

Proposition 7

If p is a permutation of the set S and m and n are integers, then $p^m p^n = p^{m+n}$.

Proof: First assume m and n are both positive. Let m be fixed but arbitrary, and let

$$S = \{x \in \mathbf{N} : p^m p^x = p^{m+x}\}$$

Since $p^m p^1 = p^m p = p^{m+1}$, 1 is in S.
If $t \in S$, then

$$p^m p^{t+1} = p^m(p^t p) = (p^m p^t)p = (p^{m+t})p = p^{(m+t)+1} = p^{m+(t+1)}$$

and therefore $t + 1 \in S$.

By the Principle of Mathematical Induction, $S = \mathbf{N}$ and the theorem holds if both m and n are positive.

Now, if m and n are both negative,

$$
\begin{aligned}
p^m p^n &= (p^{-m})^{-1}(p^{-n})^{-1} \\
&= [(p^{-n})(p^{-m})]^{-1} \\
&= [p^{(-n)+(-m)}]^{-1} \qquad \text{by the above arguments, since } -m \text{ and} \\
&\qquad\qquad\qquad\qquad\quad -n \text{ are positive} \\
&= [p^{(-m)+(-n)}]^{-1} \\
&= [p^{-(m+n)}]^{-1} \\
&= p^{m+n}
\end{aligned}
$$

Now suppose $m < 0, n > 0$ and $m + n > 0$.

Then

$$p^{-m}p^{m+n} = p^{(-m)+(m+n)}$$
$$= p^n$$

Thus,

$$p^m(p^{-m}p^{m+n}) = p^m p^n$$
$$(p^m p^{-m})p^{m+n} = p^m p^n$$
$$I_S \circ p^{m+n} = p^m p^n$$
$$p^{m+n} = p^m p^n$$

The case where $m > 0$, $n < 0$, and $m + n > 0$ is essentially the same as the above with a "right-handed argument."
If $m < 0, n > 0$, and $m + n < 0$, then

$$p^{m+n}p^{-n} = p^{(m+n)+(-n)} = p^m$$

because $m + n$ and $-n$ are both negative.
Thus,

$$(p^{m+n}p^{-n})p^n = p^m p^n$$

and

$$p^{m+n} = p^m p^n$$

We leave as an exercise the case where $m + n = 0$ and the cases where $m = 0$ or $n = 0$. □

Notice that since addition of integers is commutative, powers of the same permutation must commute.

Proposition 8

If p is a permutation of the set S and m and n are integers, then $(p^m)^n = p^{mn}$.

Proof: If either $m = 0$ or $n = 0$, both terms are equal to I_S and the equation holds.
Now let m be fixed but arbitrary.
Let $T = \{x \in \mathbf{N} : (p^m)^x = p^{mx}\}$.
Since $(p^m)^1 = p^m = p^{m \cdot 1}$, 1 is in T.
If $w \in T$, then

$$(p^m)^{w+1} = (p^m)^w(p^m) \quad \text{by definition}$$
$$= (p^{mw})(p^m) \quad \text{because } w \in T$$
$$= p^{mw+m} \quad \text{by Proposition 7}$$
$$= p^{m(w+1)}$$

Therefore, $w + 1 \in T$.
By the Principle of Mathematical Induction, $T = \mathbf{N}$, and therefore $(p^m)^n = p^{mn}$ for every positive integer n.

If n is negative,

$$\begin{aligned}
(p^m)^n &= [(p^m)^{-n}]^{-1} \\
&= [p^{m(-n)}]^{-1} \qquad \text{by the argument above, because } -n \text{ is positive} \\
&= [p^{-mn}]^{-1} \\
&= p^{mn} \qquad \text{by Proposition 6} \quad \square
\end{aligned}$$

The above assertions show that many of the usual laws for powers of real numbers hold also for permutations. However, not all of them can be extended. For example, if p and q are permutations, $(pq)^2 = p^2q^2$ only if p and q commute. For if $(pq)^2 = p^2q^2$ then $pqpq = ppqq$ and if we compose p^{-1} on the left of both expressions and q^{-1} on the right of both, we obtain $qp = pq$.

If p is a permutation of the set S and, for some positive integer n, $p^n = I$, then the *order of* p is defined to be the smallest positive integer k such that $p^k = I$. If $p^n \neq I$ for all positive integers n, p is said to have *infinite* order.

Proposition 9

If p is a permutation of the set S and the order of p is k and m is an integer such that $p^m = I$, then $k \mid m$.

Proof: There exist integers q and r such that $m = kq + r$ and $0 \leq r < k$.
Thus, $p^m = p^{kq+r} = p^{kq}p^r = (p^k)^q p^r = (I)^q p^r = Ip^r = p^r$.
Since $p^m = I$, $p^r = I$.
Since r is nonnegative and less than k and k is, by definition, the smallest positive integer such that $p^k = I$, r must be 0.
Therefore, $m = kq$, which implies that $k \mid m$. $\square$

EXERCISE SET B

1. Prove Lemma 4.
2. Prove Proposition 5.
3. Complete the proof of Proposition 8 in the case where $m + n = 0$ and the cases where $m = 0$ or $n = 0$.
4. What are the orders of the permutations listed below?

   (a) $\begin{pmatrix} 1 & 2 & 3 & 4 \\ 3 & 4 & 1 & 2 \end{pmatrix}$ (c) $\begin{pmatrix} 1 & 2 & 3 & 4 \\ 2 & 1 & 4 & 3 \end{pmatrix}$

   (b) $\begin{pmatrix} 1 & 2 & 3 & 4 \\ 2 & 3 & 1 & 4 \end{pmatrix}$ (d) $\begin{pmatrix} 1 & 2 & 3 & 4 & 5 & 6 \\ 3 & 2 & 1 & 5 & 6 & 4 \end{pmatrix}$

5. If p is a permutation of the set S, we define a relation $\tilde{\tilde{p}}$ on S by $x \; \tilde{\tilde{p}} \; y$ iff $xp^z = y$ for some integer z. Prove that $\tilde{\tilde{p}}$ is an equivalence relation on S.
6. The equivalence classes under the relation $\tilde{\tilde{p}}$ are called the *orbits* of p. Determine the orbits of the permutations given in Exercise 4.

§3. Cycles; Even and Odd Permutations

Under the permutation

$$p = \begin{pmatrix} 1 & 2 & 3 & 4 & 5 \\ 2 & 4 & 5 & 3 & 1 \end{pmatrix}$$

the image of 1 is 2, the image of 2 is 4, the image of 4 is 3, the image of 3 is 5, and the image of 5 is 1. This permutation could be described by the list (1 2 4 3 5), that is, by following each element by its image, the last element being mapped back to the first. Similarly,

$$q = \begin{pmatrix} 1 & 2 & 3 & 4 \\ 3 & 4 & 2 & 1 \end{pmatrix}$$

can be described by (1 3 2 4), because $1q = 3$, $3q = 2$, $2q = 4$, and $4q = 1$. Notice also that $1q = 3$, $1q^2 = 2$, $1q^3 = 4$, and $1q^4 = 1$. The permutation

$$h = \begin{pmatrix} 1 & 2 & 3 & 4 & 5 & 6 \\ 2 & 3 & 4 & 5 & 6 & 1 \end{pmatrix}$$

follows a similar pattern in that $1h = 2$, $1h^2 = 3$, $1h^3 = 4$, $1h^4 = 5$, $1h^5 = 6$, and $1h^6 = 1$, and we could just write (1 2 3 4 5 6). The permutation

$$f = \begin{pmatrix} 1 & 2 & 3 & 4 & 5 \\ 2 & 3 & 1 & 5 & 4 \end{pmatrix}$$

could not be so described, because $1f = 2$, $2f = 3$, and $3f = 1$, while $4f = 5$ and $5f = 4$.

Permutations such as p, q, and h are of special interest, but before studying them we need some precise definitions.

If p is a permutation of the set S and t is in S, we say p *fixes* t if $tp = t$, and we say that p *moves* t if $tp \neq t$. The *support* of a permutation p is the set of all elements moved by p.

Notice that if t is in the support of the permutation p, then tp is also in the support of p, since the fact that p is one-to-one implies that if $tp \neq t$, then $(tp)p \neq tp$. Notice also that $tp \neq t$ implies $t \neq tp^{-1}$, and therefore tp^{-1} is also in the support of p if t is. Thus, a permutation maps its support onto its support.

If f and g are permutations of the set S such that the support of f and the support of g are disjoint sets, then we say that f and g are *disjoint*.

Proposition 10

If f and g are disjoint permutations of the set S, then $f \circ g = g \circ f$.

Proof: Exercise (Hint: Show that $(x)(f \circ g) = (x)(g \circ f)$ in three cases.)

If p is a permutation of the set S and p has order n, p is said to be a *cycle* if, for each element x in S, either $xp = x$ or the support of p is the set $\{xp^i : 0 \leq i < n\}$. A cycle of order n is said to have *length n*. A cycle of length 2 is called a *transposition*. The identity cycle is said to have length 1.

From the definition it follows immediately that if p is a cycle and x and y are elements in the support of p, then for some positive integers k and h, $xp^k = y$ and $yp^h = x$.

In Exercise 5 of Set B, we defined the relation $\tilde{\tilde{p}}$ determined by the permutation p by

$$x \, \tilde{\tilde{p}} \, y \text{ iff } \tilde{\tilde{x}}p^z = y \text{ for some integer } z$$

You were asked to show that this relation is an equivalence relation. In Exercise 6 of Set B an *orbit* of p was defined to be an equivalence class with respect to the relation p.

From the definition it now follows that a cycle is a permutation with finite support that is either the identity or whose support consists of precisely one of its orbits.

A convenient way to describe a cycle is just to list its support by first listing one element and then its successive images. Thus, by (1 4 7 3 2) we mean the permutation that maps 1 to 4, 4 to 7, 7 to 3, 3 to 2, and 2 to 1.

Our way of listing the elements is not unique, because

$$(1 \quad 4 \quad 7 \quad 3 \quad 2) = (4 \quad 7 \quad 3 \quad 2 \quad 1) = (7 \quad 3 \quad 2 \quad 1 \quad 4)$$
$$= (3 \quad 2 \quad 1 \quad 4 \quad 7) = (2 \quad 1 \quad 4 \quad 7 \quad 3)$$

We can start our list with any element in the support of the cycle.

This method of representation does not immediately provide a way of describing the identity permutation. By a symbol such as (1), (8), or (b) we shall mean the identity permutation.

One of the reasons for studying cycles is for ease in computation. We can compose two or more cycles just by "following the paths" and writing down the successive images. For example,

$$(1 \quad 2 \quad 4 \quad 5) \circ (1 \quad 3 \quad 5 \quad 7) \circ (2 \quad 6 \quad 5 \quad 7 \quad 3)$$
$$= (1 \quad 6 \quad 5 \quad 2 \quad 4 \quad 3 \quad 7)$$

Under the first cycle, 1 is mapped to 2, 2 is fixed by the second cycle, and finally 2 is mapped to 6 by the third cycle. Now we determine the image of 6. The number 6 is fixed by the first cycle, fixed by the second, and mapped to 5 by the third. Now determine the image of 5. The number 5 is mapped to 1 under the first cycle, 1 is mapped to 3 under the second cycle, and 3 is mapped to 2 under the third cycle. We leave the rest of the verification to the reader.

Notice that

$$(a \ b) \circ (a \ b) = I$$

and that for any objects $a_1, a_2, \ldots, a_n$,

$$(a_1 \ a_2 \ \cdots \ a_n) = (a_1 \ a_2 \ \cdots \ a_{n-1}) \circ (a_1 \ a_n)$$

Permutations of finite sets are of special interest. Note, however, that instead of studying the permutations of some finite set $\{a_1, \ldots, a_n\}$ we might as well study the permutations of $\{1, 2, \ldots, n\}$. Any permutation of $\{1, \ldots, n\}$ can, however, be viewed as a permutation of $\mathbf{Z}$ whose support is contained in the set $\{1, \ldots, n\}$. Thus, we make the following definition. For every positive integer n, S_n *is the set of all permutations of $\mathbf{Z}$ whose support is contained in the set* $\{1, \ldots, n\}$. When describing a permutation in S_n, we need only describe what that permutation does to elements in its support. Elements not listed will be assumed to be fixed. Thus, for example, $(1 \ 2 \ 3) \in S_4$, because the support of $(1 \ 2 \ 3)$ is $\{1, 2, 3\}$, which is contained in $\{1, 2, 3, 4\}$. Note that our definition of S_n implies that $S_k \subseteq S_m$ if $k \le m$.

The cycle notation is convenient for permutations with finite support but, as we saw in examining

$$\begin{pmatrix} 1 & 2 & 3 & 4 & 5 \\ 2 & 3 & 1 & 5 & 4 \end{pmatrix}$$

not all such permutations are cycles. However, the permutation above can be expressed as a product of cycles by $(1 \ 2 \ 3)(4 \ 5)$. We now show that this is the case for all permutations in S_n.

Proposition 11

If p is a permutation in S_n, then p can be expressed as a product of cycles.

Proof: We proceed by induction.

If $p \in S_1$, then $p = (1)$, the identity permutation, and therefore p is a cycle. Now assume that every permutation in S_k can be expressed as a product of cycles, and let p be an element of S_{k+1}. If $(k + 1)p = k + 1$, then $p \in S_k$ and p can be expressed as a product of cycles. If $(k + 1)p \ne k + 1$, let $w = (k + 1)p^{-1}$, and let $q = (k + 1 \ \ w) \circ p$. Then

$$(k + 1)q = (k + 1)[(k + 1 \ \ w) \circ p] = [(k + 1)(k + 1 \ \ w)]p$$
$$= (w)p = k + 1$$

Therefore, $q \in S_k$ and q can be expressed as a product of cycles, $C_1, C_2, \ldots, C_t$.
Since $q = C_1 C_2 \cdots C_t$ and $q = (k + 1 \ \ w)p$,

$$C_1 C_2 \cdots C_t = (k + 1 \ \ w)p$$

$$(k + 1 \quad w)C_1 C_2 \cdots C_t = (k + 1 \quad w)(k + 1 \quad w)p = Ip = p$$

and therefore

$$(k + 1 \quad w)C_1 C_2 \cdots C_t = p$$

Thus, every element of S_{k+1} can be expressed as a product of cycles. The theorem now follows by the Principle of Mathematical Induction. □

We shall see in Exercise Set C that every permutation in S_n can be expressed as the product of disjoint cycles.

Proposition 12

If p is a cycle in S_n, then p can be expressed as a product of transpositions of the form $(1 \quad k)$, where k is a positive integer.

Proof: If $p = (1), p = (1 \quad 2)(1 \quad 2)$.
If $p = (a \quad b)$, then

$$p = (1 \quad a)(1 \quad b)(1 \quad a)$$

Since

$$1(1 \quad a)(1 \quad b)(1 \quad a) = 1$$
$$a(1 \quad a)(1 \quad b)(1 \quad a) = b$$
$$b(1 \quad a)(1 \quad b)(1 \quad a) = a$$

Thus, cycle s of length 1 and 2 can be so expressed.
Assume the assertion is true for cyle s of length n and let

$$p = (a_1 \ a_2 \cdots a_n \ a_{n+1})$$

But $(a_1 \ a_2 \cdots a_n \ a_{n+1}) = (a_1 \ a_2 \cdots a_n) \circ (a_1 \ a_{n+1})$.

Since $(a_1 \ a_2 \cdots a_n)$ is of length n and $(a_1 \ a_{n+1})$ is of length 2, both can be expressed as a product of transpositions of the form $(1 \quad k)$, and therefore their product can be so expressed. □

Corollary 13

Every element of S_n can be expressed as a product of transpositions of the form $(1 \quad k)$, where k is a positive integer.

Proof: Immediate from Propositions 11 and 12. □

Since every permutation in S_n can be expressed as a product of transpositions, it seems natural to classify permutations according to whether they can be expressed as a product of an even number of transpositions or as an odd number of transpositions.

A permutation is said to be *even* if it can be expressed as a product of an even number of transpositions and *odd* if it can be expressed as a product of an odd number of transpositions. We denote the set of all even permutations in S_n by A_n.

How do we know that a permutation could not be both even and odd? Might not one way of expressing it use an even number of transpositions and another use an odd number? That this is impossible is seen by the following development.

For each positive integer n we define the polynomial Δ by

$$\Delta = \prod_{1 \leq i < j \leq n} (x_i - x_j)$$

For example

$$\prod_{1 \leq i < j \leq 3} (x_i - x_j) = (x_1 - x_2)(x_1 - x_3)(x_2 - x_3)$$

and

$$\prod_{1 \leq i < j \leq 5} (x_i - x_j) = (x_1 - x_2)(x_1 - x_3)(x_1 - x_4)(x_1 - x_5)(x_2 - x_3)$$
$$\cdot (x_2 - x_4)(x_2 - x_5)(x_3 - x_4)(x_3 - x_5)(x_4 - x_5)$$

If p is a permutation in S_n, we define Δp by

$$\Delta p = \prod_{1 \leq i < j \leq n} (x_{ip} - x_{jp})$$

For example, if $p = (1 \quad 2 \quad 3)$ and

$$\Delta = \prod_{1 \leq i < j \leq 3} (x_i - x_j)$$

then

$$\Delta(1 \quad 2 \quad 3) = \prod_{1 \leq i < j \leq 3} (x_{i(1 \ 2 \ 3)} - x_{j(1 \ 2 \ 3)})$$
$$= (x_{1(1 \ 2 \ 3)} - x_{2(1 \ 2 \ 3)})(x_{1(1 \ 2 \ 3)} - x_{3(1 \ 2 \ 3)})$$
$$\cdot (x_{2(1 \ 2 \ 3)} - x_{3(1 \ 2 \ 3)})$$
$$= (x_2 - x_3)(x_2 - x_1)(x_3 - x_1)$$

Notice that in this case

$$\Delta(1 \quad 2 \quad 3) = (x_2 - x_3)(-1)(x_1 - x_2)(-1)(x_1 - x_3)$$
$$= (x_1 - x_2)(x_1 - x_3)(x_2 - x_3)$$
$$= \Delta$$

In general, if p is a permutation in S_n, what must Δp be?

$$\Delta p = \prod_{1 \leq i < j \leq n} (x_{ip} - x_{jp})$$

Since p is a permutation and $i \neq j$, $ip \neq jp$.

If $ip < jp$, then $(x_{ip} - x_{jp})$ is one of the factors of Δ.

If $jp < ip$, then $(x_{jp} - x_{ip}) = -(x_{ip} - x_{jp})$ is a factor of Δ.

Thus, the factors of Δp are either factors of Δ or are negatives of factors of Δ. Therefore, Δp is either equal to Δ or to $-\Delta$.

Notice that $(-\Delta)p = -(\Delta p)$. Also, if p and q are permutations in S_n, then

$$\Delta(p \circ q) = \prod_{1 \le i < j \le n} (x_{i(p \circ q)} - x_{j(p \circ q)})$$
$$= \prod_{1 \le i < j \le n} (x_{(ip)q} - x_{(jp)q})$$
$$= (\Delta p)q$$

Consider the transposition $(1 \quad 2)$ in S_3.

$$\Delta(1 \quad 2) = \prod_{1 \le i < j \le 3} (x_{i(1 \quad 2)} - x_{j(1 \quad 2)})$$
$$= (x_{1(1 \quad 2)} - x_{2(1 \quad 2)})(x_{1(1 \quad 2)} - x_{3(1 \quad 2)})(x_{2(1 \quad 2)} - x_{3(1 \quad 2)})$$
$$= (x_2 - x_1)(x_2 - x_3)(x_1 - x_3)$$

Every factor of Δ except $(x_1 - x_2)$ appears as a factor of $\Delta(1 \quad 2)$ and $(x_2 - x_1)$ appears instead of $(x_1 - x_2)$. Therefore, $\Delta(1 \quad 2) = -\Delta$ in this case. The general argument follows the same path.

Proposition 14

If $(a \quad b)$ is any transposition in S_n, then $\Delta(a \quad b) = -\Delta$.

Proof: First we shall show that $\Delta(1 \quad 2) = -\Delta$.

If

$$\Delta = \prod_{1 \le i < j \le n} (x_i - x_j)$$

then

$$\Delta(1 \quad 2) = \prod_{1 \le i < j \le n} (x_{i(1 \quad 2)} - x_{j(1 \quad 2)})$$

The only factor of Δ to have both subscripts changed by $(1 \quad 2)$ is $(x_1 - x_2)$ and this becomes $(x_2 - x_1)$. In all other factors of Δ, 2 is less than j, and therefore $j(1 \quad 2) = j$. For $j > 2$, the factor $(x_1 - x_j)$ becomes $(x_{1(1 \quad 2)} - x_{j(1 \quad 2)}) = (x_2 - x_j)$ and similarly, the factor $(x_2 - x_j)$ becomes $(x_1 - x_j)$. Thus, except for $(x_2 - x_1)$, all the factors of $\Delta(1 \quad 2)$ are factors of Δ. Therefore, $\Delta(1 \quad 2) = -\Delta$.

Now consider any transposition of the form $(1 \quad a)$.

$$\Delta(1 \quad a) = \Delta[(2 \quad a)(1 \quad 2)(2 \quad a)]$$
$$= [\Delta(2 \quad a)](1 \quad 2)(2 \quad a)$$

But $[\Delta(2 \quad a)]$ is either Δ or $-\Delta$ and therefore,

$$[\Delta(2 \quad a)](1 \quad 2) = -[\Delta(2 \quad a)]$$

Hence

$$\begin{aligned}
\Delta(1 \quad a) &= (-[\Delta(2 \quad a)])(2 \quad a) \\
&= -[\Delta(2 \quad a)(2 \quad a)] \\
&= -\Delta
\end{aligned}$$

Now consider a transposition of the form $(a \quad b)$.

$$(a \quad b) = (1 \quad a)(1 \quad b)(1 \quad a)$$

Therefore,

$$\begin{aligned}
\Delta(a \quad b) &= \Delta[(1 \quad a)(1 \quad b)(1 \quad a)] \\
&= ([\Delta(1 \quad a)](1 \quad b))(1 \quad a) \\
&= [(-\Delta)(1 \quad b)](1 \quad a) \\
&= [-(-\Delta)](1 \quad a) \\
&= \Delta(1 \quad a) \\
&= -\Delta
\end{aligned}$$

Thus, $\Delta(a \quad b) = -\Delta$ for any transposition $(a \quad b)$. $\square$

Proposition 15

If $T_1, \ldots, T_n$ are transpositions, then $\Delta T_1 T_2 \cdots T_n = \Delta$ if n is even and $\Delta T_1 T_2 \cdots T_n = -\Delta$ if n is odd.

Proof: We proceed by induction.
If $n = 1$, n is odd and $\Delta T_1 = -\Delta$.
Assume the proposition is true for k transpositions.
If k is odd, $k + 1$ is even, and

$$\begin{aligned}
\Delta T_1 T_2 \cdots T_k T_{k+1} &= (\Delta T_1 T_2 \cdots T_k) T_{k+1} \\
&= (-\Delta) T_{k+1} \\
&= -(\Delta T_{k+1}) \\
&= -(-\Delta) \\
&= \Delta
\end{aligned}$$

If k is even, $k + 1$ is odd and

$$\begin{aligned}
\Delta T_1 T_2 \cdots T_k T_{k+1} &= (\Delta T_1 T_2 \cdots T_k) T_{k+1} \\
&= (\Delta) T_{k+1} \\
&= -\Delta \quad \square
\end{aligned}$$

That a permutation cannot be both even and odd now follows immediately, along with the following corollary.

Corollary 16

If $p \in S_n$, then p is even iff $\Delta p = \Delta$ and p is odd iff $\Delta p = -\Delta$.

Corollary 17

A cycle of length n is even if n is odd and is odd if n is even.

Proof: $(a_1 \ a_2 \cdots a_n) = (a_1 \ a_2)(a_1 \ a_3) \cdots (a_1 \ a_n)$.
Therefore a cycle of length n can be expressed as the product of $n - 1$ transpositions. $\square$

EXERCISE SET C

1. Prove Proposition 10.
2. Express the following permutations as products of cycles, using as few cycles as possible.

   (a) $\begin{pmatrix} 1 & 2 & 3 & 4 & 5 & 6 & 7 \\ 2 & 5 & 3 & 6 & 7 & 1 & 4 \end{pmatrix}$
   (d) $\begin{pmatrix} 1 & 2 & 3 & 4 & 5 & 6 & 7 & 8 \\ 8 & 3 & 4 & 2 & 6 & 1 & 7 & 5 \end{pmatrix}$

   (b) $\begin{pmatrix} 1 & 2 & 3 & 4 & 5 & 6 & 7 \\ 2 & 3 & 4 & 1 & 6 & 7 & 5 \end{pmatrix}$
   (e) $\begin{pmatrix} 1 & 2 & 3 & 4 & 5 & 6 & 7 & 8 & 9 \\ 9 & 4 & 5 & 3 & 7 & 1 & 8 & 6 & 2 \end{pmatrix}$

   (c) $\begin{pmatrix} 1 & 2 & 3 & 4 & 5 & 6 & 7 \\ 2 & 4 & 5 & 3 & 7 & 1 & 6 \end{pmatrix}$

3. Express the permutations in Exercise 2 as products of transpositions. Which are even and which are odd?
4. Prove that $(1) \in A_n$.
5. Prove: If $p \in S_n, p \in A_n$ iff $p^{-1} \in A_n$.
6. Prove: If $p \in A_n$ and $q \in A_n$, then $p \circ q \in A_n$.
7. Prove: If $p \in S_n - A_n$ and $q \in S_n - A_n$, then $p \circ q \in A_n$.
8. Prove: If $p \in A_n$ and $q \in S_n - A_n$, then $p \circ q \in S_n - A_n$ and $q \circ p \in S_n - A_n$.
9. Prove: If $p \in S_n$ and $q \in S_n$, then $p \in A_n$ iff $q^{-1} \circ p \circ q \in A_n$.
10. List the elements of A_3; of A_4. Express them as products of cycles.
11. List the following sets.

    (a) $A_4(1 \ 2) = \{p(1 \ 2): p \in A_4\}$
    (b) $A_4(3 \ 4) = \{p(3 \ 4): p \in A_4\}$
    (c) $(1 \ 2)A_4 = \{(1 \ 2)p: p \in A_4\}$

12. Prove: Exactly one-half of the elements in S_n are even.
13. In S_8, what is the order of $(1 \ 2)(3 \ 4 \ 5)$? Of $(1 \ 2 \ 3)(4 \ 5 \ 6 \ 7)$? Of $(1 \ 2 \ 3 \ 4)(5 \ 6)$?
**14. Prove: If p is a permutation and $p = p_1 p_2 \cdots p_k$, where each p_i is a cycle of length m_i and the p_i's are pairwise disjoint, then the order of p is the least common multiple of $m_1, \ldots, m_k$.

15. Let p be a permutation of S, and let A be an orbit of p. Define p_A by $xp_A = x$ for all x not in A and $xp_A = xp$ for x in A. Prove that p_A is a permutation of S.

16. Prove: If p is a permutation with finite support and A is an orbit of p, then p_A is a cycle.

**17. Prove: Every permutation in S_n can be expressed as the product of disjoint cycles. (Hint: A permutation in S_n has finitely many orbits each with finitely many elements.)

4
GROUPS

In Chapter 2 we axiomatized the integers, and from this set of axioms we derived many assertions about the particular system we had in mind. This is what we might call the *first level of abstraction* in that from our intuitive notions we developed a formal abstract system. In calculus and real analysis we axiomatize the real numbers, $\mathbf{R}$, in a similar fashion and then proceed to develop the theory of the real numbers and the theory of functions from $\mathbf{R}$ into $\mathbf{R}$. It is also possible to start with the integers (or the positive integers for that matter) and *construct* a system that conforms to our intuitive notion of what the real numbers should be. All rigorous studies of particular structures follow one of these forms. Either we axiomatize a system, as we did $\mathbf{Z}$, or we construct the system as we did $\mathbf{Z}_m$ for each m. In the case of S_n or, in fact, $P(S)$ for any set S, the structure is dependent only on set-theoretic considerations and the existence of certain kinds of functions.

A *second level of abstraction* takes place when we examine a collection of specific structures, describe certain properties that they have in common, and then study all structures that have these particular properties.

The reader has already seen such a process in Chapter 1 when we discussed equivalence relations. After noting that equality had the three properties we later named reflexive, symmetric, and transitive and noting also that congruence of triangles and similarity of triangles also had these properties, we abstracted the notion of an equivalence relation, that is, a relation that was reflexive, symmetric, and transitive. We then proved some assertions that held for all equivalence relations. We were no longer trying to prove theorems concerning a few specific relations but were proving statements about abstract equivalence relations.

In general, after defining a class or type of abstract systems, one usually sets out to prove theorems about all systems in the class. The notion of what to try to prove will usually come from one's knowledge about some of the particular systems that give rise to the abstraction.

Thus, the idea of a congruence class or a similarity class might give rise to the general idea of an equivalence class, and the fact that distinct congruence classes are disjoint might lead one to try to show that distinct equivalence classes are disjoint. Once some general theory is developed, then when one encounters any particular system in the class, he can apply this theory to the structure at hand. Thus, for example, once we saw that congruence modulo m was an equivalence relation, we knew that the relation partitioned Z into disjoint congruence classes, and we used these classes to construct Z_m.

Abstract algebra is the study of certain classes of structures in which the basic properties of the structure are defined, described, or determined by certain operations and their properties. In this chapter we shall consider groups, while in Chapter 5 we shall study rings and fields. The student who has studied linear algebra has already studied a class of algebraic structures known as vector spaces.

§1. The Axioms for a Group

In Chapter 2 we axiomatized the integers and required the following:

(i) If a, b, and c are integers, then $a + (b + c) = (a + b) + c$.
(ii) $a + 0 = a$ for each integer a.
(iii) For each element b in Z, there exists a unique element $-b$ in Z such that $b + (-b) = 0$.

In the study of Z_m we found that the following hold:

(i) For $\bar{a}$, $\bar{b}$, and $\bar{c}$ in Z_m, $\bar{a} + (\bar{b} + \bar{c}) = (\bar{a} + \bar{b}) + \bar{c}$.
(ii) $\bar{a} + \bar{0} = \bar{a}$ for each $\bar{a}$ in Z_m.
(iii) $\bar{a} + (-\bar{a}) = \bar{0}$ for each $\bar{a}$ in Z_m.

When we studied $P(S)$, the set of permutations of the set S we found that the following hold:

(i) If f, g, and h are permutations of S, then $f \circ (g \circ h) = (f \circ g) \circ h$.
(ii) If $f \in P(S)$ and I is the identity function on S, then $f \circ I = f$.
(iii) For each $f \in P(S)$ there is a permutation f^{-1} in $P(S)$ such that $f \circ f^{-1} = I$.

What do these three structures have in common? If we examine the three sets of assertions above, we see that they look somewhat alike. In each case we have a set, an operation on the set, and three statements that hold for these sets and the corresponding operations. The similarity of these structures and other structures has prompted mathematicians to study systems with these properties.

A set G with an operation (which we denote below by $*$) is a *group* if the following three conditions are satisfied:

(i) If a, b, and c are elements of G, then $a * (b * c) = (a * b) * c$.

(ii) There is an element e in G such that $a * e = a$ for every element a in G. The element e is called a *right identity element*.

(iii) For each element b in G, there is an element c in G such that $b * c = e$.

A group G is said to be *commutative* or *abelian* if the operation is commutative, that is, if $a * b = b * a$ for all a and b in G.

Notice that, because of Axiom (ii), a group must be a nonempty set. The three structures that we mentioned above certainly satisfy the axioms and, therefore, are groups.

Other examples of groups are the following:

(a) The rational numbers with the operation of addition.

(b) The real numbers with the operation of addition.

(c) The complex numbers with the operation of addition.

(d) The positive real numbers with the operation of multiplication.

(e) The set $\{1, -1\}$ with the operation of multiplication.

(f) The set of nonzero complex numbers with the operation of multiplication.

These examples are quite common. Let us now examine some other structures that are also groups, but are not quite so apparent as examples.

(g) The functions from the real numbers into the real numbers with the usual addition of functions as the operation, that is,

$$(x)(f + g) = xf + xg$$

(h) The continuous functions from $[0, 1]$ into the real numbers with the usual addition of functions as the operation.

(i) The set of all rotations of the plane with composition of rotations as the operation.

(j) The set of translations of the plane with composition of translations as the operation.

(k) The set of all two-dimensional vectors with the usual vector addition.

(l) (For students who have studied linear algebra.) The set of nonsingular 3×3 real matrices with the usual matrix multiplication.

(m) The set of all linear functions from the real numbers into the real numbers of the form $xf = mx + b$, where m is a nonzero real number and b is any real number with the operation of composition.

(n) The set of order-preserving permutations of the real numbers with the operation of composition. (A function f is order-preserving if, whenever, $x < y$, then $xf < yf$.)

EXERCISE SET A

1. Which of Examples (a)–(n) above are commutative?
2. Verify that Example (m) above is a group.
3. Verify that Example (n) above is a group.
4. Which of the following are groups? For the structures that you claim are not groups, list the axioms that fail to hold.
   (a) The integers with the operation of subtraction.
   (b) The positive integers with the operation of addition.
   (c) The continuous functions f such that $\int_0^1 f(x)\,dx = 0$ with the operation of addition.
   (d) The continuous functions from $\mathbf{R}$ to $\mathbf{R}$ with the operation of composition.
   (e) The nonzero rational numbers with the operation of division.
   (f) The nonnegative rational numbers with the operation of averaging.
   (g) The subsets of $\{a, b, c, d\}$ with the operation of union.
   (h) All integral multiples of 7 with the operation of addition.
   (i) The set $\{0\}$ with the operation of multiplication.
   (j) The set of all polynomials $f(x)$ such that $f(5) = 0$ with the usual addition of polynomials.
5. List two examples of structures, other than those in Exercise 4, that satisfy Axiom (i) but do not satisfy Axioms (ii) or (iii).
6. List two examples of structures, other than those in Exercise 4, that satisfy Axioms (i) and (ii) but not Axiom (iii).
7. List two examples of structures, other than those in Exercise 4, that satisfy Axioms (ii) and (iii) but not Axiom (i).
**8. Can you give an example of a system satisfying Axioms (i) and (iii) but not Axiom (ii)?
**9. Let T be any set, and let S be the set of all subsets of T. In S define the operation ∇ by $A \,\nabla\, B = \{x : x \in A \text{ or } x \in B \text{ and } x \notin A \cap B\}$, that is, $A \,\nabla\, B = (A - B) \cup (B - A)$. Show that S is a group with the operation ∇.

§2. Elementary Properties of Groups, Alternative Axiom Systems

In our development of $\mathbf{Z}$, after choosing the axioms we proved the rest of the theorems using only these axioms and set theory. In the development of the theory of groups, our proofs will depend only on the axioms and the results that we have already obtained in set theory and the study of integers. In studying $\mathbf{Z}$, however, we had a particular intuitive concept in mind and were guided by our intuition and experience in deciding which assertions to try to prove. In studying groups in general, we often attempt to prove an assertion for all groups once we know that it holds for some particular group or groups.

In any of the examples of groups we have considered, whenever an element, different from the identity, is "combined" with itself, a new element is obtained.

For example,

$5 + 5 \neq 5$ ⠀⠀⠀in $\mathbf{Z}$

$(1 \quad 2 \quad 3) \circ (1 \quad 2 \quad 3) \neq (1 \quad 2 \quad 3)$ ⠀⠀⠀in S_3

$(\frac{3}{4})(\frac{3}{4}) \neq (\frac{3}{4})$ ⠀⠀⠀in positive rationals

and

$$\begin{bmatrix} 2 & 0 \\ 0 & 1 \end{bmatrix} \begin{bmatrix} 2 & 0 \\ 0 & 1 \end{bmatrix} \neq \begin{bmatrix} 2 & 0 \\ 0 & 1 \end{bmatrix}$$

Here, then, is the motivation for the first proposition.

Proposition 1

If G is a group with the operation $*$ and a right identity element e, and t is an element of G such that $t * t = t$, then $t = e$.

Proof: Suppose t is an element of G such that $t * t = t$. By Axiom (iii) there is some element w in G such that $t * w = e$. Then $(t * t) * w = t * w$ and, by associativity, $t * (t * w) = t * w$. But $t * w = e$ and, therefore, $t * e = e$. Hence, by Axiom (ii), $t = e$. ⠀$\square$

In all our examples of groups we have seen that there was only one identity element. That this is the case in all groups is assured by the following result.

Corollary 2

If G is a group with operation $*$ and right identity e, then e is the only right identity element in G. Thus, the right identity element in a group is unique.

Proof: Suppose z is a right identity element in G. Then $a * z = a$ for all elements a in G and hence, in particular, $z * z = z$. Then $z = e$ by Proposition 1. ⠀$\square$

In studying permutations we saw that if $f \circ f^{-1} = I$, then $f^{-1} \circ f = I$ and that if $f \circ g = I$, then $g = f^{-1}$. That this is the case for all groups is shown by the following propositions.

Proposition 3

If G is a group with the operation $*$ and right identity element e, and c and d are elements of G such that $c * d = e$, then $d * c = e$.

Proof: ⠀We shall apply Proposition 1 to the element $d * c$.

$$\begin{aligned}
(d * c) * (d * c) &= [(d * c) * d] * c & &\text{by Axiom (i), associativity} \\
&= [d * (c * d)] * c & &\text{by Axiom (i), again} \\
&= (d * e) * c & &\text{because } c * d = e \\
&= d * c & &\text{because } e \text{ is a right identity}
\end{aligned}$$

Thus, we have shown that

$$(d * c) * (d * c) = d * c$$

and therefore, by Proposition 1,

$$d * c = e \quad \square$$

Proposition 4

If G is a group with operation $*$ and e is the right identity of G, then $e * w = w$ for all w in G; that is, e is also a left identity.

Proof: Let w be an element of G. By Axiom (iii) there is an element u in G such that $w * u = e$. Then

$$
\begin{aligned}
e * w &= (w * u) * w && \text{by substitution, since } e = w * u \\
&= w * (u * w) && \text{by associativity} \\
&= w * e && \text{by Proposition 3} \\
&= w && \text{because } e \text{ is a right identity} \quad \square
\end{aligned}
$$

Proposition 4 shows that the right identity element assured by Axiom (ii) is, in fact, a two-sided identity, and therefore we can drop the "right" designation and just refer to e as the identity element in G.

Proposition 5

If G is a group with operation $*$, and a, b, and c are elements of G such that $b * a = c * a$, then $b = c$.

Proof: Let e denote the identity element of G. By Axiom (iii) there is an element d in G such that $a * d = e$.
Since $b * a = c * a$, $(b * a) * d = (c * a) * d$.
Then $b * (a * d) = c * (a * d)$, by associativity.
Thus, $b * e = c * e$, and therefore $b = c$. $\quad \square$

Corollary 6

If G is a group with operation $*$, and a, b, and c are elements of G such that $a * b = a * c$, then $b = c$.

Proof: Exercise.

Corollary 7

If G is a group with operation $*$ and identity element e, and a, b, and c are elements of G such that $a * b = e$ and $a * c = e$, then $b = c$.

Proof: Exercise.

Corollary 7 states that the element whose existence is assured by Axiom (iii) is unique, that is:
For each element a in G, there is a *unique element b* in G such that

$a * b = e$. We shall call this unique element the *inverse of a* or just "a inverse."

The system of Axioms (i), (ii), and (iii) is not the only set that can be used to define a group. We could also use the following:

(I) For each a, b, and c in G, $(a * b) * c = a * (b * c)$.

(II) There is an element e in G such that $e * a = a$ for each a in G.

(III) For each element a in G, there is an element b in G such that $b * a = e$.

Axioms (II) and (III) are just "left-handed" versions of Axioms (ii) and (iii), and we could prove analogous assertions to 1–7 using these axioms. In fact, the objective of these assertions is to establish the two-sidedness and the uniqueness of the identity and the inverses.

We could also choose the following axioms for a group, which, as we have shown above, turn out to be equivalent to the given set.

(A) For each a, b, and c in G, $(a * b) * c = a * (b * c)$.

(B) There is a unique element e in G such that $a * e = e * a = a$ for each element a in G.

(C) For each element a in G there is a unique element b in G such that $a * b = b * a = e$.

If however, we choose this axiom system, then to show that a particular structure is a group we have to prove the "two-sidedness" of the identity, its uniqueness, the "two-sidedness" of inverses, and their uniqueness. This really requires us to repeat, in a particular case, the proofs of some of the assertions we have established. If we wish to show that a structure is a group, we can show that (i), (ii), and (iii) are satisfied or that (I), (II), and (III) are satisfied; this is less work than establishing (A), (B), and (C) above.

All the assertions proved thus far started with the phrase:

"If G is a group with the operation $*$"

In practice, most groups are denoted either multiplicatively or additively; that is, the operation is just indicated by juxtaposition as we usually do in multiplication or is denoted by a "+" sign. Usually the additive notation is reserved for groups that are commutative. From this point on in this chapter we adopt the following notational conveniences:

(a) Unless otherwise specified, all groups will be denoted multiplicatively. Instead of writing $a * b$ we shall just write ab. Additive notation will be used only for certain commutative groups.

(b) Unless otherwise specified, e will be used to denote the identity element of a group. If additive notation is used, 0 will denote the identity element.

(c) Using multiplicative notation, the inverse of the element a will be denoted by a^{-1}. Using additive notation, the inverse of a will be denoted by $-a$.

Proposition 8

If G is a group and $c \in G$, then $(c^{-1})^{-1} = c$.

Proof: Exercise.

Proposition 9

Let G be a group, and let a and b be elements of G. Then the equations $ax = b$ and $ya = b$ have unique solutions in G.

Proof: Let $x = a^{-1}b$ and $y = ba^{-1}$.
Then

$$ax = a(a^{-1}b) = (aa^{-1})b = eb = b$$

and

$$ya = (ba^{-1})a = b(a^{-1}a) = be = b$$

The uniqueness of these solutions follows from Proposition 5 and Corollary 6. $\square$

Proposition 10

Let G be a group, and let a and b be elements of G. Then $(ab)^{-1} = b^{-1}a^{-1}$.

Proof

$$
\begin{aligned}
(ab)(b^{-1}a^{-1}) &= a[(bb^{-1})a^{-1}] \\
&= a(ea^{-1}) \\
&= aa^{-1} \\
&= e
\end{aligned}
$$

Therefore, $(ab)(b^{-1}a^{-1}) = e$ and $b^{-1}a^{-1} = (ab)^{-1}$ by Corollary 7, that is, because inverses are unique. $\square$

If we consider a finite group or, in fact, any finite set with an operation, the operation frequently can be easily described by means of a table. For example, the addition table for $\mathbf{Z}_4$ is the following.

The entry in the row following $\bar{2}$ in the column under $\bar{3}$ is $\bar{2} + \bar{3}$.

$+$	$\bar{0}$	$\bar{1}$	$\bar{2}$	$\bar{3}$
$\bar{0}$	$\bar{0}$	$\bar{1}$	$\bar{2}$	$\bar{3}$
$\bar{1}$	$\bar{1}$	$\bar{2}$	$\bar{3}$	$\bar{0}$
$\bar{2}$	$\bar{2}$	$\bar{3}$	$\bar{0}$	$\bar{1}$
$\bar{3}$	$\bar{3}$	$\bar{0}$	$\bar{1}$	$\bar{2}$

The multiplication table for $\mathbf{Z}_7$ is below.

$\cdot$	$\bar{0}$	$\bar{1}$	$\bar{2}$	$\bar{3}$	$\bar{4}$	$\bar{5}$	$\bar{6}$
$\bar{0}$	$\bar{0}$	$\bar{0}$	$\bar{0}$	$\bar{0}$	$\bar{0}$	$\bar{0}$	$\bar{0}$
$\bar{1}$	$\bar{0}$	$\bar{1}$	$\bar{2}$	$\bar{3}$	$\bar{4}$	$\bar{5}$	$\bar{6}$
$\bar{2}$	$\bar{0}$	$\bar{2}$	$\bar{4}$	$\bar{6}$	$\bar{1}$	$\bar{3}$	$\bar{5}$
$\bar{3}$	$\bar{0}$	$\bar{3}$	$\bar{6}$	$\bar{2}$	$\bar{5}$	$\bar{1}$	$\bar{4}$
$\bar{4}$	$\bar{0}$	$\bar{4}$	$\bar{1}$	$\bar{5}$	$\bar{2}$	$\bar{6}$	$\bar{3}$
$\bar{5}$	$\bar{0}$	$\bar{5}$	$\bar{3}$	$\bar{1}$	$\bar{6}$	$\bar{4}$	$\bar{2}$
$\bar{6}$	$\bar{0}$	$\bar{6}$	$\bar{5}$	$\bar{4}$	$\bar{3}$	$\bar{2}$	$\bar{1}$

Notice that the set $\mathbf{Z}_7$ is not a group with respect to multiplication modulo 7. (Why?) If we consider the set $\mathbf{Z}_7{}^* = \mathbf{Z}_7 - \{\bar{0}\}$, however, we see that $\mathbf{Z}_7{}^*$ is a group with the operation of multiplication modulo 7. We see from the table that the product of two elements of $\mathbf{Z}_7{}^*$ is an element of $\mathbf{Z}_7{}^*$. From Chapter 2 we know that this multiplication is associative. $\bar{1}$ is easily seen to be the identity element. If we examine the table we find that $\bar{1}$ appears in each row, and therefore each element has an inverse.

Below is the table for S_3 using cyclic notation. The entry in the row following (2 3) in the column under (1 3) is (2 3)$\circ$(1 3).

$\circ$	i	(1 2)	(1 3)	(2 3)	(1 2 3)	(1 3 2)
i	i	(1 2)	(1 3)	(2 3)	(1 2 3)	(1 3 2)
(1 2)	(1 2)	i	(1 2 3)	(1 3 2)	(1 3)	(2 3)
(1 3)	(1 3)	(1 3 2)	i	(1 2 3)	(2 3)	(1 2)
(2 3)	(2 3)	(1 2 3)	(1 3 2)	i	(1 2)	(1 3)
(1 2 3)	(1 2 3)	(2 3)	(1 2)	(1 3)	(1 3 2)	i
(1 3 2)	(1 3 2)	(1 3)	(2 3)	(1 2)	i	(1 2 3)

If we reexamine Proposition 5, Corollary 6, and Proposition 9 as they pertain to the table for a group, we see that every element must appear exactly once in each row and each column. If d were in the row following a under the columns headed by b and by c,

	e	a		b	c
e	e	a		b	c
a	a			d	d
b	b				
c	c				

then $ab = d$ and $ac = d$, that is, $ab = ac$, and Corollary 6 would imply that $b = c$. Similarly, by Proposition 5 it is impossible for an element to appear twice in the same column. If a and b are any two elements of a group G, Proposition 9 implies that b must appear in the row following a (in the column under $a^{-1}b$) and that b must appear in the column under a (in the row following ba^{-1}).

	e	a	b	$a^{-1}b$	ba^{-1}
e	e	a	b	$a^{-1}b$	ba^{-1}
a	a			b	
b	b				
$a^{-1}b$	$a^{-1}b$				
ba^{-1}	ba^{-1}	b			

EXERCISE SET B

1. State assertions 1–10 using additive notation.
2. Prove Corollary 6.
3. Prove Corollary 7.
4. Let ABC be an equilateral triangle with the lines u, v, w as shown.

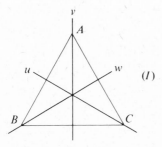

(I)

Let us define certain motions of the triangle as follows: R is a counterclockwise rotation through $120°$; that is, after performing R the triangle is in the position shown below.

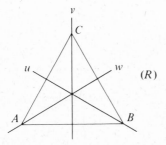

(R)

Notice that the lines are always fixed.

Define T to be a clockwise rotation through 120°; that is, after performing T the triangle is in the position shown below.

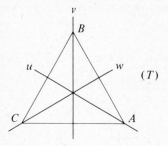

(T)

We can spin the triangle around each line; for example, after spinning the triangle around u it would be in the position shown below. We call this "flipping" movement U.

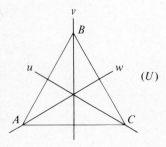

(U)

Spinning around v would yield the position shown below; we call this motion V.

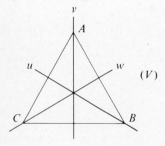

(V)

Also, spinning around w would yield the next position shown below; we call this motion W.

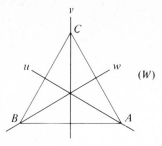

(W)

We could, of course, just leave the triangle alone; we call this "motion" I.

Thus, we have a set M of six "motions" through which we can move the triangle and still have the figure "covering the same space." We now define an operation in M by the following rule: $Y \circ X$ is the result of first performing Y and then performing X. For example, to obtain $U \circ R$, first perform U and then R. Performing U we have

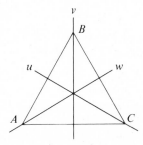

and then performing R we have

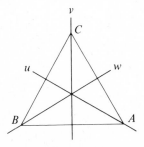

The result of $U \circ R$ is the same as W, and therefore we say that $U \circ R = W$. If we view these motions as functions, the operation is the usual composition. Complete the following table and verify the group axioms.

∘	I	R	T	U	V	W
I	I	R	T	U	V	W
R	R			V		
T	T					
U	U	W				
V	V					
W	W					

The group of motions above is called the group of symmetries of a triangle.

5. Consider a square $HKPQ$ with diagonals a and b and reflecting lines m and n:

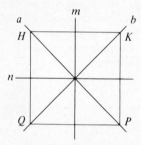

Let R be the motion of rotating the square through a clockwise motion of 90°, S of rotating the square through a clockwise motion of 180°, T of rotating the square through a clockwise motion of 270°, A be a spin about line a, B a spin about line b, M a spin about line m, and N a spin about line n. Thus, the results of applying these motions would be as indicated below. Remember that the lines, as in Exercise 4, are fixed.

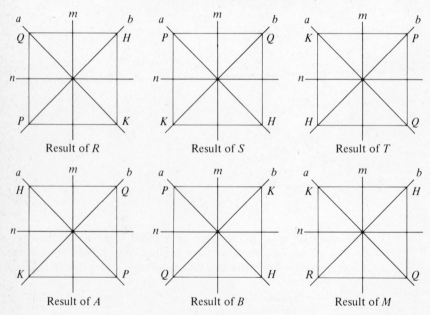

Result of R Result of S Result of T

Result of A Result of B Result of M

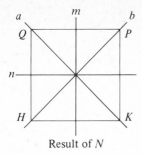

Result of N

If we let I be the identity "motion," that is, the "motion" that fixes every-thing, then the motions $I, R, S, T, A, B, M,$ and N form a collection that, with the operation of composition, forms a group known as the symmetries of a square.

Complete the following table. Remember that $R \circ A$ is first do R, then A. Verify the group axioms.

$\circ$	I	R	S	T	A	B	M	N
I	I	R	S	T	A	B	M	N
R	R	S	T	I	N			
S	S							
T	T							
A	A	M			I			
B	B							
M	M							
N	N							I

6. Consider the rectangle $XYZW$ and the lines m and n shown below. If we rotate the rectangle through an angle of 180° we shall again "take up the same space" and obtain the second figure below.

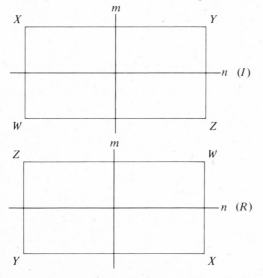

If we spin the rectangle about the line m we get the first figure below, and if we spin about the line n, we get the second figure below. We call these motions I, R, M, and N, where I means to fix the rectangle.

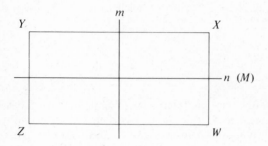

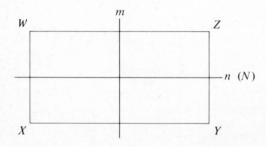

As before, $R \circ M$ means first do R and then M. Make a table for this operation. Verify the group axioms.

7. In S_4 solve the following equations (see Proposition 9):
   (a) $(1 \quad 2 \quad 3) \circ X = (2 \quad 4 \quad 3)$
   (b) $X \circ (1 \quad 2 \quad 3) = (2 \quad 4 \quad 3)$
   (c) $X \circ (1 \quad 2 \quad 4 \quad 3) = (1 \quad 4 \quad 2 \quad 3)$
   (d) $Y \circ (1 \quad 2)(3 \quad 4) = (1 \quad 2 \quad 3 \quad 4)$

8. In S_5 solve the following equations:
   (a) $X \circ (1 \quad 3 \quad 5) = (2 \quad 4 \quad 1)$
   (b) $(1 \quad 4 \quad 3 \quad 5) \circ Y = (1 \quad 5 \quad 3 \quad 2 \quad 4)$

9. Consider the set $S = \{a, b, c, d\}$. In S define the operation $*$ by $x * y = y$ for all x, y in S; that is, $x * y$ is just the element on the right. For example, $a * b = b$ and $b * c = c$. Prove: (i) a is a left identity element; (ii) a is a right inverse for each element. Why is S not a group with the operation $*$?

10. Prove the following generalization of Proposition 10. If $a_1, \ldots, a_n$ are elements of the group G, then $(a_1 a_2 \cdots a_n)^{-1} = a_n^{-1} a_{n-1}^{-1} \cdots a_2^{-1} a_1^{-1}$.

11. Factor $(1 \quad 3 \quad 5 \quad 4 \quad 2)$ into the product of transpositions. Use Exercise 10 to determine the inverse of $(1 \quad 3 \quad 5 \quad 4 \quad 2)$.

**12. Prove: If K is a nonempty set with an associative operation $*$ such that each equation of the form $x * a = b$ or $a * y = b$ has a unique solution in K, then K is a group.

§3. Exponents in Group Theory

The use of multiplicative notation for the operation in a group provides a rather natural meaning for integral exponents. If G is a group and a is an element of G, we define positive integral powers of a recursively by

$$a^1 = a \quad \text{and} \quad a^{k+1} = a^k a \qquad \text{for any positive integer } k$$

If the group is written additively, we have an analogous definition of positive integral multiples of a; that is,

$$1a = a \quad \text{and} \quad (k+1)a = ka + a \qquad \text{for every positive integer } k$$

We give meaning to integral exponents in the same way we defined exponents for nonzero real numbers and for permutations. If the group G is denoted multiplicatively and a is an element of G, $a^0 = e$ and if m is a positive integer, a^{-m} is defined by $a^{-m} = (a^m)^{-1}$.

If a group is denoted additively, we have analogous meanings for integral multiples; that is,

$$0a = 0 \quad \text{and} \quad (-m)a = -(ma)$$

The statements in this section are analogous to assertions that we proved in Chapter 3. We know that $P(S)$ is a group, and if the reader examines the proofs given in Chapter 3 he will find that the basic group properties were all that we needed to make the proofs. Thus, these proofs can be adapted to handle the general case here, and we leave the adaptions and translations as exercises.

Lemma 11

If a is an element of the group G and m is a positive integer, then $aa^m = a^{m+1}$.

Proof: Exercise.

Proposition 12

If b is an element of the group G and n is a positive integer, then $(b^n)^{-1} = (b^{-1})^n$.

Proof: Exercise.

From the definition and Proposition 12, it follows that

$$a^{-m} = (a^m)^{-1} = (a^{-1})^m$$

if a is a group element and m is a positive integer.

Also, if a is in the group G and n is a negative integer, then

$$(a^{-1})^n = [(a^{-1})^{-n}]^{-1} = [(a^{-n})^{-1}]^{-1} = [a^n]^{-1}$$

and

$$(a^{-1})^n = [(a^{-1})^{-n}]^{-1} = [(a^{-1})^{-1}]^{-n} = [a]^{-n} = a^{-n}$$

These statements can be condensed into the following assertion.

Proposition 13

If d is an element of the group G and p is an integer, then $d^{-p} = (d^p)^{-1} = (d^{-1})^p$.

Proposition 14

If b is an element of the group G and p and q are integers, then $b^p b^q = b^{p+q}$.

Proof: Exercise.

From the above assertion we see that, since addition of integers is commutative, powers of the same element commute.

Proposition 15

If d is an element of the group G and m and n are integers, then $(d^m)^n = d^{mn}$.

Proof: Exercise.

We see that we again have the usual laws for exponents, but we cannot expect $(ab)^n = a^n b^n$. That this is, in general, not the case is pointed out by Exercise 7 in Set C.

EXERCISE SET C

1. Translate assertions 11–15 into additive notation.
2. (a) In Z_{12}, list $\{n\bar{2}{:}n \in N\}$.
   (b) In Z_9, list $\{n\bar{2}{:}n \in N\}$.
   (c) In S_5, list $\{[(1\ 2)(3\ 4\ 5)]^n{:}n \in N\}$.
   (d) In S_6, list $\{[(1\ 2)(3\ 4\ 5\ 6)]^n{:}n \in N\}$.
3. Prove Lemma 11.
4. Prove Proposition 12.
5. Prove Proposition 14.
6. Prove Proposition 15.
7. Prove: If G is a group such that $(ab)^2 = a^2 b^2$ for every pair of elements a and b in G, then G is abelian.
8. Prove: If S is a nonempty subset of the group G and S is closed under the operation of G, then if $a \in S$ and n is a positive integer, $a^n \in S$.
9. Prove: If b is an element of the group G and n and m are distinct integers such that $b^n = b^m$, then $b^p = e$ for some integer p.
10. Prove: If G is a finite group and a is an element of G, then $a^q = e$ for some positive integer q.
11. Prove: If x and y are elements of the group G and p is any integer, then $y^{-1}(x^p)y = (y^{-1}xy)^p$.

§4. Subgroups

If G is a group with operation $*$ and H is a subset of G, H is said to be a *subgroup* of G if H is closed under $*$ and Axioms (i), (ii), and (iii) hold in H, that is, *if H is a group under the operation $*$ restricted to H.*

One's experience in mathematics abounds with examples of subgroups. With respect to the operation of addition:

Z is a subgroup of Q, the rational numbers.
Q is a subgroup of R, the real numbers.
R is a subgroup of C, the complex numbers.

If $\mathcal{F}$ denotes the group of functions from the real numbers to the real numbers with the operation of functional addition, then

The group $\mathcal{C}$ of continuous functions is a subgroup of $\mathcal{F}$.
The group $\mathcal{D}$ of differentiable functions is a subgroup of $\mathcal{C}$.
The group $\mathcal{P}$ of polynomial functions is a subgroup of $\mathcal{D}$.

Notice also that S_1 is a subgroup of S_2, S_2 is a subgroup of S_3, S_3 is a subgroup of S_4, and so on. Also, each S_n is a subgroup of $P(Z)$.

As the reader verifies that the above examples are, in fact, subgroups, he may notice that he did not in each case need to verify the associative law. If G is a group, H is a subset of G, and u, v, and w are elements of H, then we know that $(uv)w = u(vw)$, because G satisfies the associative law. The equation must hold for elements of H, because it holds for *all* elements of G and the elements of H are elements of G. We say that the associative property is *hereditary* in that it holds for all subsets of G, because it holds for G. We can argue, similarly, that commutavity is hereditary; that is, if G is commutative and M is a subset of G, then $ab = ba$ for every pair of elements a and b from M. Since we do not have to prove associativity to show that a subset of a group is a subgroup, it seems reasonable to ask just what is necessary.

Theorem 16

Let G be a group and H a nonempty subset of G. The following three statements are equivalent:

(a) H is a subgroup of G.
(b) If $u \in H$ and $v \in H$, then $uv \in H$ and $u^{-1} \in H$.
(c) If $a \in H$ and $b \in H$, then $ab^{-1} \in H$.

Proof: We shall show that (a) implies (b), (b) implies (c), and (c) implies (a).

(i) If H is a subgroup of G and u and v are elements of H, then $uv \in H$, because H is closed under the operation of G. If θ is the identity of H, then $\theta\theta = \theta$, and therefore $\theta = e$. Thus, the identity of H is e. Since

u must have an inverse in H and the inverse of u in G is unique, u^{-1} must be the inverse of u in H, and therefore $u^{-1} \in H$. Thus, (a) implies (b).

(ii) Now assume that (b) holds and let a and b be elements of H. Then, by (b), a and b^{-1} are elements of H and, furthermore, by (b), $ab^{-1} \in H$. Thus, (b) implies (c).

(iii) Now assume that (c) holds. Since $H \neq \phi$, some element of G, say g, is in H. By (c), $gg^{-1} \in H$. Thus, $e \in H$ and Axiom (ii) is satisfied. [Condition (c) does not require a and b to be distinct.] Now let t be any element of H. Since $e \in H$ and $t \in H$, then $et^{-1} \in H$; that is, $t^{-1} \in H$. Thus, Axiom (iii) is satisfied. We yet need to show that H is closed under the operation of G. [Since G is a group and associativity is hereditary, Axiom (i) is satisfied.] Let $x \in H$ and $y \in H$. Then, by what we have shown above, $x \in H$ and $y^{-1} \in H$. Now, by (c), $x(y^{-1})^{-1} \in H$; that is, $xy \in H$. Thus, (c) implies (a) and the proof is complete. $\square$

The equivalence of the three conditions enables us to use conditions (b) or (c) when attempting to show that a certain subset of a group is, in fact, a subgroup.

Proposition 17

Let G be a group, and let A and B be subgroups of G. Then $A \cap B$ is a subgroup of G.

Proof: Since e is an element of every subgroup, $e \in A \cap B$, and therefore $A \cap B \neq \phi$. Let $x \in A \cap B$ and $y \in A \cap B$. Then $x \in A$ and $y \in A$, and therefore $xy^{-1} \in A$. Also, $x \in B$ and $y \in B$ imply $xy^{-1} \in B$. Therefore, $xy^{-1} \in A \cap B$ and condition (c) of Theorem 16 is satisfied by $A \cap B$. By Theorem 16, $A \cap B$ is a subgroup. $\square$

The reasoning used in the above proof generalizes to any collection of subgroups.

Proposition 18

If G is a group and $\mathcal{A}$ is any collection of subgroups of G, then $\cap \mathcal{A}$ is a subgroup of G.

Proof: Since $e \in A$ for each A in $\mathcal{A}$, $e \in \cap \mathcal{A}$, and therefore $\cap \mathcal{A} \neq \phi$. If $x \in \cap \mathcal{A}$ and $y \in \cap \mathcal{A}$, then, for each A in $\mathcal{A}$, $x \in A$ and $y \in A$. Since each A in $\mathcal{A}$ is a subgroup, by condition (c) of Theorem 16, for each A in $\mathcal{A}$, $xy^{-1} \in A$. Since $xy^{-1} \in A$ for each A in $\mathcal{A}$, $xy^{-1} \in \cap \mathcal{A}$. Thus, by Theorem 16, $\cap \mathcal{A}$ is a subgroup. $\square$

Examples

(a) If S denotes the set of all multiples of 6 in the group $\mathbf{Z}$ and F denotes the set of all multiples of 4 in the group $\mathbf{Z}$, then S and F are both subgroups of $\mathbf{Z}$ and $S \cap F$ is a subgroup of $\mathbf{Z}$. $S \cap F$ is, in fact, the set of all multiples of 12.

(b) In the usual three-dimensional Euclidean space, the XY plane is a subgroup and the XZ plane is a subgroup. The intersection of the XY plane and the XZ plane is just the X axis, which is, of course, a subgroup.

(c) In the group of polynomials over **R**, the set of all polynomials having 2 as a root is a subgroup and the set of all polynomials having 9 as a root is also a subgroup. The intersection of these two subgroups is just the set of polynomials having both 2 and 9 as roots.

(d) Again in the group of polynomials over **R**, for each integer q, define P_q to be the set of all polynomials that have q for a root. Each P_q is again a subgroup. If we form $\cap P_q$, the intersection of all these subgroups, we get just the zero polynomial, because no other polynomial can have infinitely many roots.

(e) If we now consider the group $\mathcal{C}$ of all continuous functions from **R** into **R** and for each integer q, define F_q to be the set of all continuous functions that take the value 0 at q, then each F_q is a subgroup of $\mathcal{C}$. If we form $\cap F_q$ we again get a subgroup, but this time it contains elements other than 0. For example, $\cap F_q$ contains $\sin 2\pi x$, because $\sin 2\pi q = 0$ for every integer q. Can you list some of the other elements of $\cap F_q$?

EXERCISE SET D

1. State Theorem 16 in additive notation.
2. Which of the following subsets are subgroups of **Z** with respect to addition? If not, why not?
   (a) **N**
   (b) all even integers
   (c) all odd integers
   (d) $\{x \in \mathbf{Z} : 5 < x\}$
   (e) $\{y \in \mathbf{Z} : 5 \mid y\}$
   (f) $\{w \in \mathbf{Z} : w \mid 8\}$
   (g) $\{p \in \mathbf{Z} : 3 \mid p \text{ and } 8 \mid p\}$
   (h) $\{q \in \mathbf{Z} : q + q = q\}$
3. Let $\mathcal{C}$ denote the set of all continuous functions from **R** into **R**. Which of the following subsets are subgroups of $\mathcal{C}$ with respect to addition?
   (a) the set of all functions $f(x)$ such that $\int_0^1 f(x)\,dx = 0$
   (b) the set of all functions $g(x)$ such that $\int_0^1 g(x)\,dx = 3$
   (c) the set of all functions that are differentiable at least seven times
   (d) the set of increasing functions
4. Which of the following subsets of S_4 are subgroups? If not, why not?
   (a) $\{I, (1 \quad 2 \quad 3), (1 \quad 3 \quad 2)\}$
   (b) $\{I, (1 \quad 2 \quad 3), (1 \quad 3 \quad 2), (1 \quad 3 \quad 4), (1 \quad 4 \quad 3)\}$
   (c) $\{I, (1 \quad 2 \quad 3 \quad 4), (1 \quad 3)(2 \quad 4), (1 \quad 4 \quad 3 \quad 2)\}$
   (d) $\{I, (1 \quad 2), (1 \quad 3), (1 \quad 4)\}$
   (e) $\{I, (1 \quad 2), (3 \quad 4), (1 \quad 2)(3 \quad 4)\}$
   (f) $\{(1 \quad 2 \quad 3 \quad 4), (1 \quad 4 \quad 3 \quad 2), (1 \quad 3 \quad 2 \quad 4), (1 \quad 4 \quad 2 \quad 3)\}$

5. For each integer m, define $m\mathbf{Z}$ to be the set of all multiples of m; that is, $m\mathbf{Z} = \{mx:x \in \mathbf{Z}\} = \{q \in \mathbf{Z}:m \mid q\}$. Prove that, for each integer m, $m\mathbf{Z}$ is a subgroup of $\mathbf{Z}$.

6. Let F be the set of all polynomials of degree 2 or less that have 5 as a root. Let S be the set of all polynomials of degree 2 or less that have 7 as a root. Show that F and S are subgroups of $\mathbf{R}[x]$. List two of the elements of $F \cap S$. Can you simply describe all the elements of $F \cap S$?

7. Let $\mathcal{D}$ be the group of all differentiable functions, and for each integer q define A_q to be the set of all functions with a zero derivative at q, that is, $f \in A_q$ iff $f'(q) = 0$. Prove that each A_q is a subgroup. By Proposition 18, then, $\cap A_q$ is also a subgroup. Can you name three elements in $\cap A_q$? Can you name infinitely many of them?

8. Prove: If G is a group and b is an element of G, then $\{b^n:n \in \mathbf{Z}\}$ is a subgroup of G.

9. Let $\mathcal{W}$ be the set of all subgroups of $\mathbf{Z}$, under addition, that contain 6. Since the set of all even integers contains 6 and is a subgroup of $\mathbf{Z}$, $\mathcal{W} \neq \emptyset$. Therefore, $\cap \mathcal{W}$ is a subgroup. Can you describe $\cap \mathcal{W}$ more simply?

10. Let $\mathbf{R}[x]$ be the additive group of polynomials over the real numbers. Let $S = \{x^n:n \text{ is a nonnegative integer}\}$; that is, $S = \{1, x, x^2,\dots\}$. Let $\mathcal{S}$ be the set of all subgroups containing S. Since $\mathbf{R}[x] \in \mathcal{S}$, $\mathcal{S} \neq \emptyset$, and therefore $\cap \mathcal{S}$ is also a subgroup. Clearly, $x + x^2$ is in $\cap \mathcal{S}$. Also notice that $3x^2$ and $4x^2 - 5x$ would also be in $\cap \mathcal{S}$. Can you more simply describe the group $\cap \mathcal{S}$?

11. Prove: If G is a finite group and H is a nonempty subset of G, then H is a subgroup of G iff H is closed under the operation of G. (Hint: use Exercises 8, 9, and 10 from Exercise Set C.)

12. Let G be a group, and let b and c be elements of G. Define ϕ and λ to be functions from G into G given by $x\phi = b^{-1}xb$ and $x\lambda = xc$ for each x in G. Prove that ϕ and λ are one-to-one functions from G onto G.

13. List the elements of the subgroup A_4 of S_4. For each permutation p in S_4, define $A_4 p = \{q \circ p:q \in A_4\}$.
    (a) List the elements of $A_4(1 \quad 2)$.
    (b) List the elements of $A_4(2 \quad 3)$.
    (c) List the elements of $A_4(1 \quad 2 \quad 3)$.
    (d) Without computing, what do you think are the elements of $A_4(1 \quad 3)$?
    (e) What do you think are the elements of $A_4(2 \quad 3 \quad 4)$?

14. Let H be a subgroup of G. For each element p in G define Hp by $Hp = \{hp:h \in H\}$, and define pH by $pH = \{ph:h \in H\}$. Prove that if $p \in H$, $Hp = H = pH$ and conversely; that is, if $Hp = H$, then $p \in H$.

15. Again let H be a subgroup of the group G and let p be an element of G. Use Exercise 12 to show that there is a one-to-one correspondence between H and Hp. Show also that there is a one-to-one correspondence between pH and H.

**16. Let $P(\mathbf{Z})$ denote the set of all permutations of $\mathbf{Z}$. For each integer w define the function f_w by $xf_w = x + w$ for every integer x. Prove that each $f_w \in P(\mathbf{Z})$ and that $\{f_w:w \in \mathbf{Z}\}$ is a subgroup of $P(\mathbf{Z})$. Now define ϕ from $\mathbf{Z}$ into $P(\mathbf{Z})$ by $w\phi = f_w$. Prove that ϕ is one-to-one and that $(a + b)\phi = f_a \circ f_b$ for every pair of integers a and b.

§5. Sets of Generators

Since the intersection of any collection of subgroups of a group G is a subgroup of G, we can, as in some of the previous examples and exercises, describe a subgroup by describing a collection of subgroups and then forming its intersection. If G is a group and S is a subset of G, then, by Proposition 18 the intersection of the set of all the subgroups containing S is a subgroup of G. (The set of all subgroups of G containing S is not empty, because $S \subseteq G$ and G is a subgroup of itself.)

The intersection of all subgroups of G containing the set S is called the *subgroup of G generated by S* and is denoted by $\langle S \rangle$.

If S is a singleton, say $S = \{a\}$, then it is customary to write $\langle a \rangle$ instead of $\langle \{a\} \rangle$.

If G is a group, H is a subgroup of G, and S is a subset of G such that $\langle S \rangle = H$, then S is said to be a set of *generators* for H or H is *generated* by S. A subgroup H is said to be *cyclic* if H is generated by one element, that is, if $H = \langle a \rangle$ for some a in H. If $G = \langle S \rangle$, then S is a set of generators for G itself.

Proposition 19

Let G be a group, and let a be an element of G. Then
$$\langle a \rangle = \{a^n : n \in \mathbf{Z}\}$$

Proof: If H is any subgroup of G containing a, then $a^{-1} \in H$, because H is a subgroup, and therefore $a^n \in H$ for all integers n, because H is closed under the operation of G. Thus, $\{a^n : n \in \mathbf{Z}\} \subseteq H$. Since $\{a^n : n \in \mathbf{Z}\}$ is a subset of every subgroup containing a, $\{a^n : n \in \mathbf{Z}\}$ is contained in the intersection of all subgroups containing a; that is, $\{a^n : n \in \mathbf{Z}\} \subseteq \langle a \rangle$. But $\{a^n : n \in \mathbf{Z}\}$ is itself a subgroup containing a (Exercise 8 in Set D.) Therefore, $\{a^n : n \in \mathbf{Z}\}$ contains the intersection of all subgroups containing a; that is, $\langle a \rangle \subseteq [a^n : n \in \mathbf{Z}\}$. Thus, $\langle a \rangle = \{a^n : n \in \mathbf{Z}\}$. □

At this point it is worthwhile to look at the technique in the above proof. If we wish to show that a subgroup is the subgroup generated by S, we can first show that the subgroup is contained in every subgroup containing S. This implies that it is contained in the intersection of all subgroups containing S and, therefore, is contained in the subgroup generated by S. On the other hand, if it is a subgroup and contains S, it must contain the intersection of all subgroups containing S; that is, it must contain the subgroup generated by S. We apply this same technique to the proof of the following proposition.

Proposition 20

If G is a group and S is a nonempty subset of G, then $\langle S \rangle$ is the set of all elements in G that can be expressed as products of powers of elements in S; that is, $\langle S \rangle = \{s_1^{x_1} s_2^{x_2} \cdots s_n^{x_n} : n \in \mathbf{N}, s_i \in S, x_i \in \mathbf{Z}\}$.

Proof: Let W be the set of all elements of G that can be expressed as a product of powers of elements in S. If $g = s_1^{x_1} s_2^{x_2} \cdots s_n^{x_n}$ and $h = t_1^{y_1} t_2^{y_2} \cdots t_m^{y_m}$, where each s_i and t_j are in S, then $gh^{-1} = s_1^{x_1} s_2^{x_2} \cdots s_n^{x_n} t_m^{-y_m} \cdots t_2^{-y_2} t_1^{-y_1}$, which is also a product of powers of elements of S. Thus, by Theorem 16, W is a subgroup. Clearly, W contains S, and therefore W contains the intersection of all subgroups containing S; that is, $W \supseteq \langle S \rangle$. On the other hand, if H is any subgroup containing S, H must contain all powers of elements in S and all products of such powers. Thus, $H \supseteq W$, and therefore W is contained in the intersection of all subgroups containing S; that is $\langle S \rangle \supseteq W$. Therefore, $W = \langle S \rangle$. $\square$

Examples

(i) For every positive integer n, since every element of S_n can be expressed as a product of transpositions, the set of transpositions is a set of generators for S_n.

(ii) A_4 is generated by the set of cycles of length 3.

(iii) In Z_8, $\langle \bar{2} \rangle = \{\bar{0}, \bar{2}, \bar{4}, \bar{6}\}$.

(iv) In Z_{24}, $\langle \bar{4}, \overline{10} \rangle = \{\bar{0}, \bar{2}, \bar{4}, \bar{6}, \bar{8}, \overline{10}, \overline{12}, \overline{14}, \overline{16}, \overline{18}, \overline{20}, \overline{22}\}$.

(v) $Z_7 = \langle \bar{1} \rangle = \langle \bar{2} \rangle = \langle \bar{3} \rangle = \langle \bar{4} \rangle = \langle \bar{5} \rangle = \langle \bar{6} \rangle$.

(vi) If the group is Z, the integers, then $Z = \langle 1 \rangle = \langle -1 \rangle = \langle \{2, 3\} \rangle$.

(vii) The additive group Q, the rational numbers, is generated by the set $F = \{1/n : n \in N\}$.

(viii) $S_3 = \langle \{(1 \quad 2), (1 \quad 2 \quad 3)\} \rangle$.

(ix) The additive group of complex numbers is generated by the set of all real numbers and the set of all pure imaginary numbers.

(x) The group of integers is generated by the set of all positive integers.

(xi) The set of all vectors in three-dimensional space is generated by $\{(x, y, z) : x \geq 0, y \geq 0, z \geq 0\}$.

(xii) In the multiplicative group of nonzero real numbers, the subgroup generated by the positive integers is the set of all positive rational numbers.

We strongly advise the reader to verify most of the above examples for himself by applying the definition or Propositions 19 or 20.

If H is a subgroup of the group G, the *order* of H is defined to be the number of elements in H. If H has an infinite number of elements, H is said to be of infinite order. We denote the order of H by $|H|$. If a is an element of the group G, the *order of a* is defined to be the order of the subgroup generated by a; that is, the order of a is $|\langle a \rangle|$. The element a is said to have *infinite order* if $\langle a \rangle$ is infinite and *finite order* if $\langle a \rangle$ is finite.

Proposition 21

If G is a group and a is an element of G with finite order, then the order of a is the smallest positive integer n such that $a^n = e$.

Proof: We know $\langle a \rangle = \{a^z : z \in \mathbf{Z}\}$. Consider the set $B = \{a^n : n \in \mathbf{N}\}$. Clearly, $B \subseteq \langle a \rangle$. If $\langle a \rangle$ is finite, then not all the powers of a are distinct. Therefore, there are positive integers i and j such that $i < j$ and $a^i = a^j$. Then $a^i a^{-i} = a^j a^{-i}$, and hence $e = a^{j-i}$. Thus, $a^t = e$ for some positive t, because $j - i > 0$. Let k be the smallest positive integer such that $a^k = e$. Let $a^z \in \langle a \rangle$. Then, for some integers q and r,

$$z = kq + r \quad \text{and} \quad 0 \le r < k$$

But $a^z = a^{kq+r} = a^{kq} a^r = (a^k)^q a^r = e^q a^r = a^r$, and therefore

$$a^z \in \{a^0, a^1, \ldots, a^{k-1}\}$$

Thus, we have shown that $\langle a \rangle \subseteq \{a^0, a^1, \ldots, a^{k-1}\}$. Since $\{a^0, a^1, \ldots, a^{k-1}\} \subseteq \langle a \rangle$, $\{a^0, a^1, \ldots, a^{k-1}\} = \langle a \rangle$. We need yet to show that the elements of $\{a^0, a^1, \ldots, a^{k-1}\}$ are distinct. Suppose $0 < p \le q < k$ and and $a^p = a^q$. Then $a^p a^{-p} = a^q a^{-p}$ and $e = a^{q-p}$. But $0 \le q - p < k$, because $0 < p \le q < k$. Since k is the *smallest positive integer* such that $a^k = e$, $q - p$ cannot be positive and thus must be 0. Thus, $q = p$, the elements $\{e, a^1, \ldots, a^{k-1}\}$ are distinct, and $|\langle a \rangle| = k$. $\square$

The proof of the following corollary is actually contained in the proof of the previous proposition. We leave finding it as an exercise.

Corollary 22

If a is an element in the group G $|\langle a \rangle| = k$, and $a^h = e$, then $k \mid h$.

Proof: Exercise.

EXERCISE SET E

1. (a) In S_3, what is the subgroup generated by $(1 \quad 2)$?
    (b) In S_4, what is the subgroup generated by $(1 \quad 2 \quad 3)$? By $(1 \quad 2 \quad 3 \quad 4)$? By $(1 \quad 2)$ and $(3 \quad 4)$?
    (c) In S_3, what is the subgroup generated by $(1 \quad 2 \quad 3)$ and $(1 \quad 3)$?
    (d) In the group of symmetries of a square, what is the subgroup generated by the elements of order 2? By those of order 4?
2. What is the subgroup of $\mathbf{Z}_{18}$ generated by $\bar{1}$? By $\bar{3}$? By $\{\bar{4}, \bar{6}\}$? By $\bar{5}$? By $\{\bar{3}, \bar{4}\}$?
3. What is the subgroup of a group G generated by the empty set?
4. Prove: A cyclic group is abelian.
5. Prove Corollary 22.
6. Prove that if H is any subgroup of $\mathbf{Z}$, then $H = m\mathbf{Z}$ for some integer m. (Hint: If $H = \{0\}$, $m = 0$. Otherwise H contains positive integers. Let m be the smallest positive integer in H and show that $H = m\mathbf{Z}$.)
7. Prove: A subgroup of a cyclic group is cyclic.
8. Prove: If G is a group such that every element of G is of order 2, then G is commutative.

9. Let G be a group, and let a be an element of G. The centralizer of a in G, denoted by $C(a)$, is the set of all elements of G that commute with a; that is, $C(a) = \{x \in G : xa = ax\}$. Prove that $C(a)$ is a subgroup of G.

**10. Let G be an abelian group, and let H and K be subgroups of G. Define $HK = \{hk : h \in H \text{ and } k \in K\}$. Prove that HK is a subgroup of G. Prove that $HK = \langle H \cup K \rangle$.

**11. (Especially challenging problem)
Prove: If G is a group, and A and B are subsets of G such that $A \cup B = G$, then either $\langle A \rangle = G$ or $\langle B \rangle = G$.

§6. Congruence Relations and Lagrange's Theorem

In Chapter 2 we discussed the equivalence relation of congruence modulo m in the integers. This concept can be generalized to arbitrary groups by the following definition.

Let H be a subgroup of the group G. If x and y are elements of G, we say that x *is congruent to y modulo H if* $xy^{-1} \in H$. That x is congruent to y modulo H is denoted by $x \equiv y \pmod{H}$. If the group is denoted additively, $a \equiv b \pmod{H}$ iff $a - b \in H$.

Theorem 23

If H is a subgroup of the group G, then congruence modulo H is an equivalence relation in G.

Proof: If a is an element of G, then since $aa^{-1} = e$ and $e \in H$, $a \equiv a \pmod{H}$ and the relation is reflexive. Suppose a and b are elements of G. If $a \equiv b \pmod{H}$, then $ab^{-1} \in H$. But H is a subgroup, so $(ab^{-1})^{-1} \in H$; that is, $ba^{-1} \in H$. Since $ba^{-1} \in H$, $b \equiv a \pmod{H}$, and thus the relation is symmetric. If a, b, and c are elements of G such that $a \equiv b \pmod{H}$ and $b \equiv c \pmod{H}$, then $ab^{-1} \in H$ and $bc^{-1} \in H$. Since H is a subgroup, $(ab^{-1})(bc^{-1}) \in H$. But $(ab^{-1})(bc^{-1}) = a[(b^{-1}b)c^{-1}] = a(ec^{-1}) = ac^{-1}$. Thus, $ac^{-1} \in H$, $a \equiv c \pmod{H}$, and the relation is transitive. $\square$

The reader should compare the above proof to the proof of Proposition 27 in Chapter 2.

If we examine again the definition of congruence modulo m in $\mathbf{Z}$, the statement $m \mid (a - b)$ means that $a - b$ is a multiple of m. The set of all multiples of m is the subgroup of $\mathbf{Z}$ generated by m, and therefore $a \equiv b \pmod{m}$ means (according to the new definition) $a \equiv b \pmod{\langle m \rangle}$.

When we examined the congruence classes modulo m in $\mathbf{Z}$ we found that each class could be obtained by adding one of its elements to all multiples of m. For example, in congruence modulo 7,

$$\bar{5} = \{\ldots, -16, -9, -2, 5, 12, 19, \ldots\}$$
$$= \{\ldots, 7(-3) + 5, 7(-2) + 5, 7(-1) + 5,$$
$$7 \cdot 1 + 5, 7 \cdot 2 + 5, \ldots\}$$

In congruence modulo 9,

$$\overline{4} = \{\ldots, -14, -5, 4, 13, 22, \ldots\}$$
$$= \{\ldots, 9(-2) + 4, 9(-1) + 4, 9 \cdot 0 + 4, 9 \cdot 1 + 4, 9 \cdot 2 + 4, \ldots\}$$

Let $\mathbf{R}^*$ be the multiplicative group of nonzero real numbers and let $\mathbf{Q}^+$ be the subgroup of all positive rational numbers. $\mathbf{Q}^+$ is a subgroup of $\mathbf{R}^*$ and $\pi \equiv 3\pi \pmod{\mathbf{Q}^+}$, because

$$\pi(3\pi)^{-1} = \tfrac{1}{3} \quad \text{and} \quad \tfrac{1}{3} \in \mathbf{Q}^+$$

If r is any nonzero real number and q is any positive rational number, then $r \equiv qr \pmod{\mathbf{Q}^+}$, because

$$r(qr)^{-1} = q^{-1} \quad \text{and} \quad q^{-1} \in \mathbf{Q}^+$$

Thus, the congruence class for r contains all the numbers that we can obtain by multiplying r by positive rationals.

On the other hand, if

$$t \equiv r \pmod{\mathbf{Q}^+}$$

then

$$tr^{-1} \in \mathbf{Q}^+$$

Thus

$$tr^{-1} = q$$

for some q in $\mathbf{Q}^+$, and therefore

$$t = qr$$

for some q in $\mathbf{Q}^+$.

Thus, the class for r is just the set of numbers we get by multiplying r by positive rationals; that is, if we denote the class for r by $\overline{r}$,

$$\overline{r} = \{qr : q \in \mathbf{Q}^+\}$$

Now that we have considered some specific examples, let us consider the most general case. Before going further, however, we need some notation and some definitions.

If H is a subgroup of the group G and a is an element of G, the *right coset of H in G determined by a* is denoted by Ha and defined by

$$Ha = \{ha : h \in H\}$$

The *left coset of H in G determined by a* is denoted by aH and defined by

$$aH = \{ah : h \in H\}$$

If the groups are denoted additively we have

$$H + a = \{h + a : h \in H\} \quad \text{and} \quad a + H = \{a + h : h \in H\}$$

Examples

(a) In the case above, $(\mathbf{Q}^+)\pi$ is the set of positive rational multiples of π.

(b) If C denotes the set of all constant functions, $x^2 + C$ is the set of all functions obtained by adding constants to x^2.

(c) See also Exercise 13 in Set D.

If H is a subgroup of the group G and a is an element of G, the class for a with respect to congruence modulo H will be denoted by $\bar{a}$.

Proposition 24

If H is a subgroup of the group G and a is an element of G, then $\bar{a} = Ha$; that is, the class for a with respect to congruence modulo H is precisely the right coset of H in G determined by a.

Proof: Let $h \in H$. Since $h = (ha)a^{-1}$, $(ha)a^{-1} \in H$, and therefore $ha \equiv a \pmod{H}$ for every element h in H. Thus, all elements of the group obtained by multiplying an element of H on the right by a are in the congruence class for a; that is, $\{ha: h \in H\} \subseteq \bar{a}$.

Now suppose $b \in \bar{a}$; that is, suppose $b \equiv a \pmod{H}$. Then $ba^{-1} \in H$, and therefore $ba^{-1} = h$ for some h in H. But, then, $b = ha$, which implies that $b \in Ha$. Hence, $\bar{a} \subseteq Ha$.

Since $Ha \subseteq \bar{a}$ and $\bar{a} \subseteq Ha$, $\bar{a} = Ha$. □

In many discussions concerning congruence classes, we prefer to use the coset notation, that is, Ha instead of $\bar{a}$, because no confusion can then arise as to the subgroup that determines the congruence relation under consideration.

Example. Let $\mathbf{Z}[x]$ denote the additive group of all polynomials $f(x) = a_n x^n + \cdots + a_1 x + a_0$, where each a_i is an integer, and let H be the set of all polynomials with a zero constant term, that is, $H = \{f \in \mathbf{Z}[x]: f(0) = 0\}$.

Since $x \in H$, $H \neq \phi$. If p and q are polynomials in H, then $(p - q)(0) = p(0) - q(0) = 0 - 0 = 0$, and therefore H is a subgroup of $\mathbf{Z}[x]$. Now let f and g be any polynomials in $\mathbf{Z}[x]$. Then $f \equiv g \pmod{H}$ iff $f - g \in H$ iff $(f - g)(0) = 0$ iff $f(0) - g(0) = 0$ iff $f(0) = g(0)$. Thus, $f \equiv g \pmod{H}$ iff $f(0) = g(0)$.

Now, if we view integers as constant polynomials, we see that if p is any polynomial in $\mathbf{Z}[x]$, $p \equiv p(0) \pmod{H}$, because $p(0) = p(0)$. If f is any polynomial in $\mathbf{Z}[x]$, the coset $H + f$ is the same as the coset $H + f(0)$, because $H + f = \bar{f} = \overline{f(0)} = H + f(0)$. Thus, every coset can be expressed in the form $H + z$ for some integer z.

Proposition 25

If H is a subgroup of G and a and b are elements of G, then Ha, Hb, and aH all have the same number of elements; that is, there is a one-to-one correspondence between the elements of Ha and those of Hb and aH.

Proof: Define ϕ from Ha to aH by

$$x\phi = axa^{-1}$$

Let $h \in H$. Then $ha \in Ha$, and $(ha)\phi = a(ha)a^{-1} = ah \in aH$, and therefore $(Ha)\phi \subseteq aH$.

That ϕ is one-to-one follows from Exercise 12 in Set D.

If $y \in aH$, $y = ah$ for some $h \in H$. Then $ha \in Ha$, and since

$$(ha)\phi = a(ha)a^{-1} = ah = y,$$

ϕ is onto. Therefore, ϕ is a one-to-one function from Ha onto aH.

We define λ from Ha to Hb by

$$x\lambda = xa^{-1}b$$

Let $h \in H$; then $(ha)\lambda = (ha)(a^{-1}b) = hb \in Hb$, and therefore $(Ha)\lambda \subseteq Hb$. That λ is one-to-one follows from Exercise 12 in set D. If $y \in Hb$, then $y = hb$ for some h in H and $ha \in Ha$. Thus, $(ha)\lambda = hb = y$ and λ is onto. Therefore, λ is a one-to-one function from Ha onto Hb. □

Another way of stating Proposition 25 is to say that if H is a subgroup of G, then all right cosets and all left cosets of H have the same number of elements. In the case where $a = e$, then $Ha = He = H$ and we see that every coset of H in G has the same number of elements that H has. If H is finite with order k, then each coset has k elements. Since the congruence classes modulo H are just the right cosets of H in G, *all the congruence classes modulo H have the same number of elements.* Furthermore, since distinct equivalence classes are always disjoint, *distinct right cosets are disjoint.*

$$Ha \neq Hb \quad \text{implies} \quad Ha \cap Hb = \emptyset$$

because

$$\bar{a} \neq \bar{b} \quad \text{implies} \quad \bar{a} \cap \bar{b} = \emptyset$$

Theorem 26 (Lagrange's Theorem)

If G is a finite group and H is a subgroup of G, then the order of H divides the order of G.

Proof: Let $|H| = k$ and $|G| = m$.

The decomposition of the group G into congruence classes modulo H is a partition of G and, since each class is a right coset of H in G, each class has k elements. Since G is finite, G has finitely many subsets, so the number of classes is finite. Let n be the number of classes. Then there are n distinct, disjoint classes, each with k elements, and therefore there are $n \cdot k$ elements in G. Then $n \cdot k = m$, and therefore $k \mid m$.

The following is a more arithmetic way of setting up the above proof. The n classes are all cosets, each with k elements. $G = Ha_1 \cup Ha_2 \cup Ha_3 \cup \cdots \cup Ha_n$, where $i \neq j$ implies $Ha_i \cap Ha_j = \emptyset$.

Then $|G| = k + k + \cdots + k = k \cdot n$, or $m = k \cdot n$, and therefore $k \mid m$. $\square$

The following is a special case of Lagrange's theorem where the subgroup is generated by one element.

Corollary 27

If G is a finite group and a is an element of G, then the order of a divides the order of G.

According to Corollary 27, a group with 15 elements could have no elements of order 2, 4, 6, 7, 8, 9, 10, 11, 12, 13, 14. The only possible orders for nontrivial elements in a group of order 15 are 3, 5, and 15 itself. We also see that a group of order 19 could have only $\{e\}$ and the group itself as subgroups.

If H is a subgroup of the group G, the number of right cosets of H in G (or the number of left cosets of H in G) is called the *index of H in G* and is denoted by $[G:H]$. Another way of stating Lagrange's theorem is by $[G:H] \, |H| = |G|$ or by $[G:H] = |G| / |H|$.

EXERCISE SET F

1. List all the right cosets of S_3 in S_4. List all the left cosets of S_3 in S_4. Is each left coset a right coset?
2. Since the right cosets of S_3 in S_4 are just the equivalence classes for congruence modulo S_3, if π and ρ are permutations in S_4, $\pi \equiv \rho \pmod{S_3}$ iff π and ρ are in the same right coset. Use your decomposition of S_4 into right cosets to determine which of the following statements are true.
   (a) $(1 \quad 2 \quad 3) \equiv (1 \quad 3 \quad 2) \quad (\bmod\, S_3)$
   (b) $(1 \quad 4 \quad 3 \quad 2) \equiv (1 \quad 2 \quad 3 \quad 4) \quad (\bmod\, S_3)$
   (c) $(1 \quad 2) \equiv (3 \quad 4) \quad (\bmod\, S_3)$
   (d) $(1 \quad 3 \quad 4) \equiv (2 \quad 3 \quad 4) \quad (\bmod\, S_3)$
   (e) $(1 \quad 3 \quad 2 \quad 4) \equiv (1 \quad 4) \quad (\bmod\, S_3)$
   (f) $(1 \quad 2 \quad 3) \equiv (2 \quad 3) \quad (\bmod\, S_3)$
3. The group of rotations of a square is a subgroup of the group of symmetries of a square. Determine all right cosets of this subgroup.
4. Let G be a group, let M and N be subgroups of G, and let b be an element of G. Prove that $(Mb) \cap (Nb) = (M \cap N)b$.
5. Let $\mathcal{L}$ denote the set of all polynomials of degree 1 or less.
   (a) In $\mathbf{R}[x]$, is $3x^2 \equiv 2x^2 \pmod{\mathcal{L}}$?
   (b) In $\mathbf{R}[x]$, is $4x^3 + 2x^2 - 5x + 3 \equiv 4x^3 + 2x^2 + 8x - 19 \pmod{\mathcal{L}}$?
   (c) In $\mathbf{R}[x]$, is $4x^7 + x^6 + 3x \equiv 4x^7 + x^6 + 17x - 19 \pmod{\mathcal{L}}$?
   (d) Can you give another description of congruence modulo $\mathcal{L}$?
6. Let $\mathcal{F}$ be the set of all polynomials that have 5 as a root. $\mathcal{F}$ is a subgroup of $\mathbf{R}[x]$. In $\mathbf{R}[x]$, $f \equiv g \pmod{\mathcal{F}}$ iff $f - g \in \mathcal{F}$, that is, if 5 is a root of $f - g$.
   (a) Is $x^3 \equiv 5x^2 \pmod{\mathcal{F}}$?
   (b) Is $3x^3 - 7x^2 \equiv 7x^3 - 3x^2 \pmod{\mathcal{F}}$?

(c) Is $x^2 + 2x \equiv 7x \pmod{\mathfrak{F}}$?

(d) Is $x^2 + 2x \equiv 35 \pmod{\mathfrak{F}}$?

(e) Is $x^3 - 2x^2 - 3x \equiv 0 \pmod{\mathfrak{F}}$?

(f) Is $x^3 - 2x^2 - 3x \equiv 60 \pmod{\mathfrak{F}}$?

(g) Prove that $f \equiv g \pmod{\mathfrak{F}}$ iff $f(5) = g(5)$.

(h) Prove that $f(x) \equiv f(5) \pmod{\mathfrak{F}}$.

(i) Describe the cosets of $\mathfrak{F}$ in $\mathbf{R}[x]$. Prove your assertion.

7. Let $\mathcal{C}$ denote the group of continuous functions from the closed interval $[0, 2]$ into the real numbers; the operation is the usual addition of functions. Let

$$W = \left\{ f \in \mathcal{C} \colon \int_0^2 f(x)\, dx = 0 \right\}.$$

(a) Determine three elements of W.

(b) Prove that W is a subgroup of $\mathcal{C}$.

(c) Prove: If $f \in \mathcal{C}$ and $g \in \mathcal{C}$, then $f(x) \equiv g(x) \pmod{W}$ iff

$$\int_0^2 f(x)\, dx = \int_0^2 g(x)\, dx$$

(d) Each real number r can be viewed as a constant function, for example, for all x, $g(x) = 4$. Prove: If $h(x) \in \mathcal{C}$, then $h(x) \equiv \frac{1}{2} \int_0^2 h(x)\, dx$ $\pmod{W}$.

(e) Describe the cosets of W in $\mathcal{C}$. Prove your assertion.

8. We defined congruence modulo H in a "right-handed" sense by $a \equiv b \pmod{H}$ iff $ab^{-1} \in H$. If we let H be a subgroup of G, we can define a relation $\sim$ in G by $a \sim b$ iff $a^{-1}b \in H$.

(a) Prove that $\sim$ is an equivalence relation in G.

(b) Prove that the equivalence classes for $\sim$ are the left cosets of H in G.

Some authors use the above relation rather than the congruence that we defined. The choice of right-handed or left-handed definitions is arbitrary. For abelian groups and certain important subgroups, the two relations turn out to be the same.

9. A subgroup K of a group G is said to be *normal* if, for each $a \in G$, $Ka = aK$.

(a) Show that $A_3 = \{i, (1 \ 2 \ 3), (1 \ 3 \ 2)\}$ is a normal subgroup of S_3.

(b) Show that S_3 is not a normal subgroup of S_4.

(c) Show that, although $A_3(1 \ 2) = (1 \ 2)A_3$, it is not true that $\rho(1 \ 2) = (1 \ 2)\rho$ for each ρ in A_3.

10. Show that if H is a normal subgroup of G, then congruence modulo H and the "left-handed" relation defined in Exercise 8 are the same; that is, prove that $a \equiv b \pmod{H}$ iff $a \sim b$.

11. Let H be a subgroup of G and let a be an element of G. Let $a^{-1}Ha = \{a^{-1}ha \colon h \in H\}$. Prove that $a^{-1}Ha$ is a subgroup of G.

12. Let H be a subgroup of G. The *normalizer* of H, denoted by $N(H)$, is defined to be the set of all elements a of G such that $a^{-1}Ha = H$. Thus, $N(H) = \{a \in G \colon a^{-1}Ha = H\}$.

Prove: (i) $N(H)$ is a subgroup of G.

(ii) H is a subgroup of $N(H)$.

§7. Factor Groups

In Chapter 2 we constructed the groups Z_m by defining an addition operation on the equivalence classes with respect to congruence modulo m. For example, in Z_{12},

$$\overline{8} + \overline{6} = \overline{14} \quad \text{and} \quad \overline{5} + \overline{10} = \overline{3}$$

Thus, starting with the group Z and the subgroup $\langle m \rangle$, we were able to construct the new group Z_m. Since the relation of congruence modulo m generalized to congruence modulo a subgroup, we might expect the process of constructing new groups from Z and $\langle m \rangle$ also to generalize to any group and one of its subgroups. This is, however, not the case.

Consider, for example, the group S_3 and the subgroup $B = \{i, (1 \quad 2)\}$. The classes for congruence modulo B in S_3 are just the right cosets of B in S_3.

$$B = \{i, (1 \quad 2)\}$$
$$B(1 \quad 3) = \{(1 \quad 3), (1 \quad 2 \quad 3)\}$$
$$B(2 \quad 3) = \{(2 \quad 3), (1 \quad 3 \quad 2)\}$$

Thus, $\overline{i} = (\overline{1 \quad 2})$, $(\overline{1 \quad 3}) = (\overline{1 \quad 2 \quad 3})$, and $(\overline{2 \quad 3}) = (\overline{1 \quad 3 \quad 2})$. Suppose, now, that we tried to define an operation in this set of classes as we did for Z_m. In the case of Z_m we defined $\overline{a} + \overline{b}$ by $\overline{a} + \overline{b} = \overline{a + b}$ and then showed that this rule did, in fact, give us an operation. If we tried such a rule in the above case it would not give us an operation, because different representatives of classes would, in some cases, yield different classes.

For example, if we tried to define $\overline{a} \circ \overline{b}$ by $\overline{a} \circ \overline{b} = \overline{a \circ b}$, look what happens in the case where $a = (1 \quad 3)$ and $b = (2 \quad 3)$. Then $(\overline{1 \quad 3}) = (\overline{1 \quad 2 \quad 3})$ and $(\overline{2 \quad 3}) = (\overline{1 \quad 3 \quad 2})$. If the rule is to give an operation, then $(\overline{1 \quad 3}) \circ (\overline{2 \quad 3})$ should be the same as $(\overline{1 \quad 3}) \circ (\overline{1 \quad 3 \quad 2})$; that is, $(\overline{1 \quad 3}) \circ (\overline{2 \quad 3})$ should be the same as $(\overline{1 \quad 3}) \circ (\overline{1 \quad 3 \quad 2})$, which implies that $(\overline{1 \quad 2 \quad 3})$ is the same as $(\overline{1 \quad 2})$ and this would require that $(1 \quad 2 \quad 3) \equiv (1 \quad 2)$ (modulo B), which does not hold.

Thus, this rule does not determine an operation on the set of congruence classes modulo B.

Consider now another subgroup of S_3, namely $A_3 = \{i, (1 \quad 2 \quad 3), (1 \quad 3 \quad 2)\}$. The classes for congruence modulo A_3 in S_3 are just the right cosets

$$A_3 \quad \text{and} \quad A_3(1 \quad 2) = \{(1 \quad 2), (2 \quad 3), (1 \quad 3)\}$$

In this set of classes, if we attempt to define an operation by $\overline{a} \circ \overline{b} = \overline{a \circ b}$, the rule does, in fact, give us an operation. Look again at the two classes,

$$A_3 = \{i, (1 \quad 2 \quad 3), (1 \quad 3 \quad 2)\}$$

and

$$A_3(1\ \ 2) = \{(1\ \ 2),(2\ \ 3),(1\ \ 3)\}$$

The product of any element from A_3 with any element from $A_3(1\ \ 2)$ is in $A_3(1\ \ 2)$, because the elements of A_3 are even permutations and those in $A_3(1\ \ 2)$ are odd. Thus, no matter which elements of the classes $(\overline{1\ \ 2\ \ 3})$ and $(\overline{1\ \ 2})$ we pick, we still get a product in $A_3(1\ \ 2) = (\overline{1\ \ 2})$. Similarly, the product of any two elements in A_3 is in A_3, and the product of any two elements in $A_3(1\ \ 2)$ is in A_3, because the product of two odd permutations is an even permutation.

Why does the rule work in one case and not another? There is a significant difference in the nature of the subgroups B and A_3. If we write down the right and left cosets of each in S_3, the difference begins to show itself:

$$A_3(1\ \ 2) = \{(1\ \ 2),(2\ \ 3),(1\ \ 3)\} = A_3(2\ \ 3) = A_3(1\ \ 3)$$
$$(1\ \ 2)A_3 = \{(1\ \ 2),(1\ \ 3),(2\ \ 3)\} = (2\ \ 3)A_3 = (1\ \ 3)A_3$$

Since

$$\{(1\ \ 2),(2\ \ 3),(1\ \ 3)\} = \{(1\ \ 2),(1\ \ 3),(2\ \ 3)\}$$
$$A_3(1\ \ 2) = (1\ \ 2)A_3, \quad A_3(1\ \ 3) = (1\ \ 3)A_3,$$
$$\text{and} \quad A_3(2\ \ 3) = (2\ \ 3)A_3$$

Also,

$$A_3 = A_3 i = A_3(1\ \ 2\ \ 3) = A_3(1\ \ 3\ \ 2) = (1\ \ 3\ \ 2)A_3$$
$$= (1\ \ 2\ \ 3)A_3$$

Thus, for each permutation p in S_3, $pA_3 = A_3 p$.

In the case of the subgroup B, however, since

$$B = \{i, (1\ \ 2)\}$$
$$B(1\ \ 3) = \{(1\ \ 3),(1\ \ 2\ \ 3)\}$$

and

$$(1\ \ 3)B = \{(1\ \ 3),(1\ \ 3\ \ 2)\}$$

Thus, $B(1\ \ 3) \neq (1\ \ 3)B$, and B is not a normal subgroup of S_3. Normal subgroups were defined in Exercise Set F and we repeat the definition here:

If G is a group and H is a subgroup of G, H is said to be *normal* if $aH = Ha$ for every element a in G.

That H is a normal subgroup of G is frequently denoted by $H \lhd G$ or $G \rhd H$. Clearly, every subgroup of an abelian group is normal. We just saw that A_3 is normal in S_3 and it is not too hard to show that, for each positive integer n, A_n is normal in S_n.

Notice that the definition of normality requires that $aH = Ha$ for *every element a in G*. It is always true that $aH = Ha$ for each a in H if H is a subgroup; normality requires that $aH = Ha$ for each *a in the entire group G*.

For convenience, at this point we extend the notation we used for cosets. If A and B are nonempty subsets of a group G, by AB we mean the set

$$AB = \{xy : x \in A \text{ and } y \in B\}$$

If A or B is a singleton, say $A = \{a\}$, then

$$aB = \{ay : y \in B\} \quad \text{and} \quad Ba = \{ya : y \in B\}$$

Subsets of groups are sometimes called *complexes*, and the above operation is called complex multiplication. This operation has several nice properties that we establish in the next three assertions.

Proposition 28

If G is a group and A, B, and C are nonempty subsets of G, then $(AB)C = A(BC)$.

Proof: Let $y \in (AB)C$.
Then $y = pq$ for some p in AB and q in C.
But $p = mn$ for some m in A and n in B.
Thus,

$$y = (mn)q \quad \text{and} \quad y = m(nq)$$

Since $n \in B$ and $q \in C$, $nq \in BC$, and therefore

$$m(nq) \in A(BC)$$

that is,

$$y \in A(BC)$$

Thus we have shown that $(AB)C \subseteq A(BC)$.
We leave it to the reader to complete the proof by showing that $(AB)C \supseteq A(BC)$. □

Proposition 29

If G is a group and A, B, and C are nonempty subsets of G such that $A \subseteq B$, then $AC \subseteq BC$ and $CA \subseteq CB$.

Proof: Exercise.

Proposition 30

If G is a group and H is a subgroup of G, then $HH = H$.

Proof: Exercise.

The following proposition gives us some other useful criteria for deciding on the normality of a subgroup.

Proposition 31

Let G be a group, and let H be a subgroup of G. The following statements are equivalent:
(i) H is normal in G.
(ii) $yHy^{-1} = H$ for every y in G.
(iii) $x^{-1}Hx = H$ for every x in G.
(iv) $z^{-1}Hz \subseteq H$ for every z in G.

Proof: We shall make use of Propositions 28, 29, and 30.

To show that (i) implies (ii):
If H is normal in G, then $yH = Hy$ for each y in G. Then

$$(yH)y^{-1} = (Hy)y^{-1}$$
$$yHy^{-1} = H(yy^{-1})$$
$$yHy^{-1} = He$$
$$yHy^{-1} = H$$

To show that (ii) implies (iii):
If $yHy^{-1} = H$ for each y in G, then
$(x^{-1})H(x^{-1})^{-1} = H$, because $x^{-1} \in G$.
Therefore, $x^{-1}Hx = H$ for each x in G.

To show that (iii) implies (iv):
If $z^{-1}Hz = H$ for each z in G, then $z^{-1}Hz \subseteq H$, because $H \subseteq H$.

To show that (iv) implies (i):
Let $a \in G$. Then $a^{-1} \in G$ also and, by (iv)
$a^{-1}Ha \subseteq H$ and $(a^{-1})^{-1}Ha^{-1} \subseteq H$,
that is
$a^{-1}Ha \subseteq H$ and $aHa^{-1} \subseteq H$.
Now, applying Proposition 28 we have
$a(a^{-1}Ha) \subseteq aH$ and $(aHa^{-1})a \subseteq Ha$
which implies
$(aa^{-1})Ha \subseteq aH$ and $(aH)(a^{-1}a) \subseteq Ha$.
Thus
$eHa \subseteq aH$ and $aHe \subseteq Ha$;
that is,
$Ha \subseteq aH$ and $aH \subseteq Ha$.
Therefore, $Ha = aH$ and H is normal in G. $\square$

Using condition (iv) in Proposition 31 we can show that a subgroup H of a group G is normal by showing that if h is an arbitrary element of H and g an arbitrary element of G, then $g^{-1}hg$ is in H. This is a very commonly used technique.

In Chapter 2, before showing that our rule for addition in $\mathbf{Z}_m$ did, in fact, define an operation, we showed that if $a \equiv b \pmod{m}$ and $c \equiv d \pmod{m}$, then $a + c \equiv b + d \pmod{m}$; that is, we showed that congruence modulo m was compatible with addition. This property of congruence modulo m generalizes to congruence modulo a *normal* subgroup, and we establish it in the following proposition.

Proposition 32

If a, b, c, and d are elements of the group G, and H is a normal subgroup of G such that $a \equiv b\,(mod\ H)$ and $c \equiv d\,(mod\ H)$, then $ac \equiv bd\,(mod\ H)$.

Proof: If $a \equiv b \pmod{H}$, then $ab^{-1} \in H$.

$$ab^{-1} = aeb^{-1} = a(cc^{-1})b^{-1} = (ac)(c^{-1}b^{-1}) = (ac)(bc)^{-1}$$

Therefore, $(ac)(bc)^{-1} \in H$ and $ac \equiv bc \pmod{H}$.
Since $c \equiv d \pmod{H}$, $cd^{-1} \in H$.
Then $b(cd^{-1})b^{-1} \in H$, because H is normal.
Hence, $(bc)(d^{-1}b^{-1}) \in H$; that is, $(bc)(bd)^{-1} \in H$, and therefore

$$bc \equiv bd \pmod{H}$$

Thus, we have $ac \equiv bc$ and $bc \equiv bd \pmod{H}$, and therefore

$$ac \equiv bd \pmod{H} \quad \square$$

The reader should restate Proposition 32 using additive notation and then reexamine the proof of Proposition 28 in Chapter 2.

Since the relation of congruence modulo H is comptabile with the group operation if H is a normal subgroup and in abelian groups, all subgroups are normal, the relation of congruence modulo a subgroup is always compatible with the operation in an abelian group.

If G is a group and H is a normal subgroup of G, then G/H denotes the set of all equivalence classes of G with respect to congruence modulo H, that is, the set of all congruence classes. Thus,

$$G/H = \{\bar{a}\!:\!a \in G\}$$

With what we know about congruence modulo a subgroup and normal subgroups, we know that $G/H = \{Ha\!:\!a \in G\} = \{aH\!:\!a \in G\}$.

Now we are ready to extend the technique by which we constructed Z_m.

If H is a normal subgroup of G, we define an operation in G/H by $\bar{a}\bar{b} = \overline{ab}$.

Before considering any of the properties of this operation, *we must show that we have, in fact, defined an operation.* Our rule says to take an element in the class for a and an element in the class for b, multiply them together, and let the product of the classes be the class to which this product belongs. How do we know that we will get the same resultant class no matter which elements we choose?

If H is a normal subgroup of G and a, b, c, and d are elements of G such that $\bar{a} = \bar{c}$ and $\bar{b} = \bar{d}$, then, since $a \equiv c \pmod{H}$ and $b \equiv d \pmod{H}$, $ab \equiv cd \pmod{H}$, and $\overline{ab} = \overline{cd}$. Thus, no matter which elements we use from $\bar{a}$ and $\bar{b}$, we get the same *class* as the product and our rule does, in fact, define an operation.

Theorem 33

If H is a normal subgroup of G, then G/H is a group with respect to the operation defined by $\bar{a}\bar{b} = \overline{ab}$ for all elements a and b in G.

Proof: Let $\bar{a}$, $\bar{b}$, and $\bar{c}$ be elements of G/H. (Remember that the elements of G/H are classes.) Then

$$\bar{a}(\bar{b}\bar{c}) = \bar{a}(\overline{bc})$$
$$= \overline{a(bc)}$$
$$= \overline{(ab)c}$$
$$= \overline{(ab)}\bar{c}$$
$$= (\bar{a}\bar{b})\bar{c}$$

Therefore, the operation is associative.

Let $\bar{a}$ be an element of G/H and consider the class $\bar{e}$, where e is the identity of G. Then $\bar{a}\bar{e} = \overline{ae} = \bar{a}$, and therefore $\bar{e}$ is a right identity element in G/H.

If $\bar{a}$ is an element of G/H, then a is an element of G, and thus there is an element a^{-1} in G and a class $\overline{a^{-1}}$ in G/H. Then $\bar{a}\overline{a^{-1}} = \overline{aa^{-1}} = \bar{e}$, and therefore $\overline{a^{-1}}$ is a right inverse for $\bar{a}$.

Since Axioms (i), (ii), and (iii) are satisfied, G/H is a group. (The reader should reexamine Theorem 29 in Chapter 2.) □

The group G/H is called the *factor group* of G with respect to H or just G modulo H.

Remember that, in a factor group G/H, each element is a congruence class that is, in fact, a coset. When we define $\bar{a}\bar{b} = \overline{ab}$ we are requiring that $Ha \cdot Hb = Hab$. Notice that if we use complex multiplication, the normality of H, and Propositions 28, 29, and 30, we see that

$$(Ha)(Hb) = H(aH)b = H(Ha)b = (HH)ab = Hab$$

Thus, we could define multiplication in G/H just by using complex multiplication, because the product of two of these cosets is again one of these cosets. It is, however, the normality of H that enables us to do so. The reader can easily verify the group axioms using complex multiplication, and we leave this verification as an exercise.

In examples of factor groups, the coset notation is frequently more convenient than the class notation, and most common factor groups are denoted using cosets.

Examine the following examples of groups, normal subgroups, and factor groups.

(i) Let the group G be the usual two-dimensional space and let $H = \{(x, y):x = y\}$. Geometrically, H is the line $y = x$ in the plane.

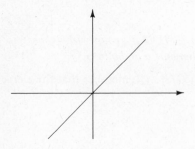

The cosets of H in G are just the translations of H. For example, $H + (0, 2)$ is the line $y = x + 2$.

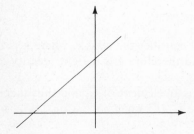

Notice that $(\overline{0, 2}) = H + (0, 2) = H + (1, 3) = (\overline{1, 3})$, because $(0, 2) \equiv (1, 3)$ (mod H); that is, $(1, 3) - (0, 2) = (1, 1)$, which is in H. In this case we can add classes just by adding points on the lines.

(ii) Let G be the group of symmetries of a square, and let H be the group of rotations of the square. G/H is a group with two elements.

(iii) Let $\mathcal{D}$ be the group of differentiable functions from $\mathbf{R}$ into $\mathbf{R}$ and C be the subgroup of constant functions. Then $f \equiv g$ (mod C) iff $f' = g'$. If h is a differentiable function, from calculus we know that $\bar{h} = h + C$ is the set of all functions with the same derivative as h. When we say that $\int (f + g) = \int f + \int g$ we mean that if $F' = f$, $G' = g$, then $\int f = F + C$, $\int g = G + C$, and

$$\int (f + g) = (F + C) + (G + C) = (F + G) + C$$

When we add indefinite integrals we are adding cosets that are, in fact, congruence classes modulo C. By $\int x^2 = (x^3/3) + C$ we mean that $\int x^2$ is the coset of C in $\mathcal{D}$ determined by $x^3/3$.

(iv) Consider the additive group $\mathbf{R}$ of real numbers and the subgroup generated by the integer 360. Then $\mathbf{R}/\langle 360 \rangle$ is a group and it is common to express complex numbers in the form $r(\cos \theta + i \sin \theta)$, where r is a real number, the absolute value of the complex number, and θ is an element of $\mathbf{R}/\langle 360 \rangle$. We do not use the coset or class notation but $\cos 15 + i \sin 15 = \cos 375 + i \sin 375$, because $15 \equiv 375 \pmod{\langle 360 \rangle}$.

De Moivre's theorem states that

$$[r(\cos a + i \sin a)][s(\cos b + i \sin b)] = (rs)[\cos(a + b) + i \sin(a + b)]$$

but addition of a and b is taken to be the addition modulo $\langle 360 \rangle$.

(v) For any positive integer n, consider the groups S_n and A_n. We leave as an exercise the proof that A_n is normal in S_n. The group S_n/A_n is the set $\{A_n, A_n(1 \quad 2)\}$ and the table is below.

	A_n	$A_n(1 \quad 2)$
A_n	A_n	$A_n(1 \quad 2)$
$A_n(1 \quad 2)$	$A_n(1 \quad 2)$	A_n

(vi) Let S be the usual three-dimensional space, and let M be any plane through the origin. M is a subgroup of S with respect to vector addition. Then S/M is just the set of all planes parallel to M, and the addition of any two of these planes is accomplished by adding a vector from one to a vector from the other and then determining the plane parallel to M that contains this sum. Thus,

$$[M + (1, 3, 5)] + [M + (2, 4, -1)] = M + (3, 7, 4)$$

EXERCISE SET G

1. Which of the following are normal subgroups?
   (a) the even integers in the integers
   (b) S_4 in S_5
   (c) $\{i,(1 \quad 2),(3 \quad 4),(1 \quad 2)(3 \quad 4)\}$ in S_4
   (d) A_3 in A_4
   (e) $\langle \{(1 \quad 2 \quad 3),(1 \quad 2 \quad 4),(1 \quad 3 \quad 4),(2 \quad 3 \quad 4)\} \rangle$ in S_4
2. Prove that if H is a subgroup of the finite group G and $[G:H] = 2$, then $H \lhd G$.
3. Prove that, for any positive integer n, $A_n \lhd S_n$.
4. Complete the proof of Proposition 28.
5. Prove Proposition 29.
6. Prove Proposition 30.
7. Consider the group $\mathbf{Z}_{12}$ and the subgroup generated by $\bar{3}$. List $\langle \bar{3} \rangle$ and the cosets of $\langle \bar{3} \rangle$ in $\mathbf{Z}_{12}$. Using coset notation rather than class notation, construct the table for $\mathbf{Z}_{12}/\langle \bar{3} \rangle$.

8. Show that $T = \{i, (1\ \ 2)(3\ \ 4), (1\ \ 3)(2\ \ 4), (1\ \ 4)(2\ \ 3)\}$ is a normal subgroup of A_4. Construct a table for A_4/T.

9. Prove that the set of rotations of a triangle is a normal subgroup of the group of symmetries of a triangle. (See Exercise Set B.) Form the factor group and make a table.

10. Prove Theorem 33 using complex multiplication of cosets; that is, prove that G/H is a group with respect to complex multiplication.

11. (a) Prove: If A and B are normal subgroups of the group G, then $A \cap B$ is a normal subgroup of G.
    (b) If $\mathcal{C}$ is a nonempty collection of normal subgroups of a group G, prove that $\cap \mathcal{C}$ is a normal subgroup of G.

**12. Let G be a group and let $\sim$ be an equivalence relation on G that is compatible with the operation; that is, if $a \sim b$ and $c \sim d$, then $ac \sim bd$. Prove that $\sim$ is congruence modulo some normal subgroup. [Hint: Prove that $\bar{e}$ is a normal subgroup and that $a \sim b$ iff $a \equiv b\,(\text{mod}\,\bar{e})$.]

13. If S is a subgroup of the group G, how would you define the normal subgroup of G generated by S? This subgroup is called the *normal closure* of S.

14. We say that a subset S of a group G is a *normal subset* of G if $x^{-1}Sx = S$ for each x in G. Prove that if S is a normal subset of G, then $\langle S \rangle$ is a normal subgroup of G.

15. Let G be a group, and let H be a normal subgroup of G. Define ϕ from G to G/H by $x\phi = \bar{x}$ for each x in G.
    (a) Show that, for each u and v in G, $(uv)\phi = (u\phi)(v\phi)$.
    (b) Show that ϕ is onto.

16. Prove that if a is an element of the group G of order 2 and y is an element of G, then $y^{-1}ay$ is of order 2. Now show that the subgroup of G generated by the elements of order 2 is a normal subgroup.

17. Generalize Exercise 16.

18. Let A and B be subgroups of G. Prove that, if A is normal or B is normal, then AB is a subgroup of G and $AB = \langle A \cup B \rangle$.

**19. Let G be a finite group with $m \cdot n$ elements, and suppose A is a subgroup of order m and B is a subgroup of order n and $A \cap B = \{e\}$. Prove that $G = AB$.

**20. (An especially challenging problem.)
Prove: If $n \geq 5$, then A_n has two normal subgroups. (Hints: First show that, if w and p are two cycles of length 3, then for some even permutation a, $a^{-1}wa = p$. Then show that any nontrivial normal subgroup of A_n contains a cycle of length 3. Hence, any nontrivial normal subgroup contains all cycles of length 3, and therefore contains A_n.)

§8. Other Ways of Making "New Groups from Old"

When we coordinatize two- or three-dimensional space, we usually use ordered pairs or ordered triples of numbers. Vector addition is just component addition in that, for example,

$$(2,\ 3) + (7,\ -2) = (9,\ 1)\quad \text{and}\quad (1,\ -2,\ 4) + (3,\ 5,\ -2) = (4,\ 3,\ 2)$$

This idea can be generalized for any groups or, for that matter, for any systems in which we have operations. For now we restrict ourselves to groups.

If A and B are groups, then the set $A \times B$ was defined by

$$A \times B = \{(a, b):a \in A \quad \text{and} \quad b \in B\}$$

We now define an operation in $A \times B$ by

$$(x, y)(u, v) = (xu, yv)$$

Theorem 34

If A and B are groups, then $A \times B$ with the operation as above is a group.

Proof: Let $(a, b),(s, t)$ and (p, q) be elements of $A \times B$. Then

$$\begin{aligned}
[(a, b)(s, t)](p, q) &= (as, bt)(p, q) \\
&= [(as)p, (bt)q] \\
&= [a(sp), b(tq)] \\
&= (a, b)(sp, tq) \\
&= (a, b)[(s, t)(p, q)]
\end{aligned}$$

Therefore, the operation is associative.

If we let e denote the identity element of A and θ the identity element of B, then (e, θ) is an identity element for $A \times B$, because

$$(a, b)(e, \theta) = (ae, b\theta) = (a, b)$$

for every pair (a, b) in $A \times B$.

If $(a, b) \in A \times B$, then there are elements a^{-1} in A and b^{-1} in B such that

$$aa^{-1} = e \quad \text{and} \quad bb^{-1} = \theta$$

Hence, $(a, b)(a^{-1}, b^{-1}) = (aa^{-1}, bb^{-1}) = (e, \theta)$, and therefore (a^{-1}, b^{-1}) is a right inverse for (a, b). $\square$

The set $A \times B$ with the operation defined above is called the *direct product* or *direct sum* of A and B and is denoted by $A \oplus B$. (Some writers use the symbol $\otimes$ or just the symbol $\times$ for direct product or direct sum; however, $\otimes$ is frequently used for tensor products and the symbol $\times$ has a set-theoretic meaning, and we shall not use it in any other way.) We can, in a similar way, form the group $A \oplus B \oplus C$, where the elements are ordered triples from the groups A, B, and C. Also, we can construct groups such as $A_1 \oplus A_2 \oplus A_3 \oplus \cdots \oplus A_n$, where $A_1, \ldots, A_n$ are groups and the elements are ordered n-tuples. The definitions and verifications are natural extensions of those for two groups. The student who has studied linear algebra is familiar with two- and three-dimensional space, where the vectors are ordered pairs and ordered triples of real numbers. In these cases he is studying the groups $\mathbf{R} \oplus \mathbf{R}$ and $\mathbf{R} \oplus \mathbf{R} \oplus \mathbf{R}$.

Another, but very similar, way to "make new groups from old" is to consider all the functions from some set into a group. Let G be a group, let S be a nonempty set, and let F be the set of all functions from S into G. In F define a "pointwise" multiplication by defining $(x)(fg) = (xf)(xg)$ for each element x in the set S. The proof that F, with this operation, is a group we leave as an exercise, but we shall examine some familiar cases.

Examples

(a) Let S be the set of all real numbers between 0 and 1 inclusive, that is, $[0, 1]$, and let G be the multiplicative group of positive real numbers. The operation is then the usual "pointwise" multiplication of functions.

(b) Let S be the set $\mathbf{N}$ of positive integers and let G be the additive group $\mathbf{R}$ of real numbers. Then the set of functions from $\mathbf{N}$ into $\mathbf{R}$ is just the set of all infinite sequences of real numbers, and our operation is the usual pointwise addition of sequences.

(c) Let $S = \mathbf{R}$ and $G = \mathbf{R}$. Then we have all functions of a real variable, and the operation is the usual addition of functions.

In this chapter we have begun a study of the theory of groups, and we shall return to this study in portions of Chapter 6. Before going further into group theory, however, we shall study two other types of algebraic structures. Many of the results obtained in this chapter have analogues that hold in other areas, and some of our assertions can be extended for other structures.

EXERCISE SET H

1. Construct a table for $\mathbf{Z}_2 \oplus \mathbf{Z}_3$. Show that $\mathbf{Z}_2 \oplus \mathbf{Z}_3$ is cyclic.
2. Show that $\mathbf{Z}_2 \oplus \mathbf{Z}_9$ is cyclic.
3. Show that $\mathbf{Z}_2 \oplus \mathbf{Z}_6$ is not cyclic.
4. Is $\mathbf{Z}_4 \oplus \mathbf{Z}_5$ cyclic? Is $\mathbf{Z}_4 \oplus \mathbf{Z}_6$ cyclic? Is $\mathbf{Z}_4 \oplus \mathbf{Z}_9$ cyclic?
5. Can you determine necessary and sufficient conditions on p and q for $\mathbf{Z}_p \oplus \mathbf{Z}_q$ to be cyclic? Prove your assertion.
6. Consider the group $\mathbf{Z}_4 \oplus \mathbf{Z}_9$ and the subgroup generated by $(\bar{2}, \bar{3})$. List $\langle(\bar{2}, \bar{3})\rangle$ and the cosets of $\langle(\bar{2}, \bar{3})\rangle$ in $\mathbf{Z}_4 \oplus \mathbf{Z}_9$. Construct a table for $\mathbf{Z}_4 \oplus \mathbf{Z}_9 / \langle(\bar{2}, \bar{3})\rangle$.
7. Consider the group $\mathbf{Z}_3 \oplus \mathbf{Z}_4$ and the subgroup generated by $(\bar{1}, \bar{0})$. Construct a table for $\mathbf{Z}_3 \oplus \mathbf{Z}_4 / \langle(\bar{1}, \bar{0})\rangle$.
8. Construct a table for $S_3 \oplus \mathbf{Z}_2$.
9. Prove: If A and B are groups, then $A \times \{e\}$ and $\{e\} \times B$ are normal subgroups of $A \oplus B$. Can you describe the cosets of $A \times \{e\}$ in $A \oplus B$?
10. Let A and B be groups, and define ϕ from $A \oplus B$ to A by $(a, b)\phi = a$.

Prove that $[(x, y)(u, v)]\phi = [(x, y)\phi][(u, v)\phi]$ for each x and u in A and each y and v in B. What is $e\phi^{-1}$?

11. Let G be a group, let S be a nonempty set, and let F be the set of all functions from S into G. Prove that F is a group with respect to pointwise multiplication, that is, $(x)(fg) = (xf)(xg)$ for each x in S.

12. Let S be the group of all sequences of real numbers with the usual pointwise addition, and let W be the set of all sequences that are 0 from some point on, that is, $W = \{\{a_n\}: \text{for some } k, \text{if } m > k, a_m = 0\}$. Show that W is a subgroup of S.

**13. Let G be a group, let S be a nonempty set, and let F be the group of all functions from S into G with pointwise multiplication. Define H by $H = \{f \in F: xf = e \text{ for all but finitely many elements } x \text{ in } S\}$. Prove that H is a normal subgroup of F.

**14. Let $\{G_i: i \in I\}$ be a nonempty collection of groups, one group G_i associated with each element i of the set I. Let $S = \cup G_i$, and now define ΠG_i to be the set of all functions f from I into S such that $(i)f \in G_i$ for each element i in I. Show that ΠG_i is a group under pointwise multiplication. The group ΠG_i is generally called the *complete direct product* or *sum* of the family $\{G_i: i \in I\}$.

**15. In the group ΠG_i above, define $\bigoplus G_i$ by $\bigoplus G_i = \{f \in G_i: (i)f = e \text{ for all but finitely many elements } i \text{ in } I\}$. Prove that $\bigoplus G_i \triangleleft \Pi G_i$. The group $\bigoplus G_i$ is usually called the *restricted direct product* or *sum* of the family $\{G_i: i \in I\}$.

5

RINGS
AND FIELDS

In the mathematics you studied before taking this course, you were usually concerned with systems in which you had two operations, and these were generally some form of addition and multiplication. In elementary school and high school you studied positive integers, positive rational numbers, integers, rational numbers, real numbers, complex numbers, and polynomials. In calculus you extended your knowledge to include differentiable functions and continuous functions. If you studied linear algebra, you learned to add and multiply matrices. Finally, in Chapter 2 of this book you learned to add and multiply in Z_m for any integer m. In studying algebraic structures it is quite natural for us to consider systems in which we have two operations that are generally called addition and multiplication. The most important of such structures are those known as rings and fields. In this chapter we develop the elementary properties of these structures and see how some of the theory of groups can be extended to other classes of structures.

§1. The Axioms, Definitions, and Elementary Properties

Most of the operations with which we deal in mathematics are associative. This is not to say that most operations are, in fact, associative, but just that many of the more interesting ones are. They may, however, be more interesting because the associativity makes them easier to work with and yields many results. In any case, we shall, for now, limit ourselves to systems with two associative operations. We shall call these two operations addition and multiplication and denote them by $+$ and $\cdot$. Every operation that we have denoted by "$+$" has been commutative and so, in the class of structures we wish to study, we shall assume that addition is commutative. Since Z, R, $R[x]$, and many other examples of two-operation systems have a zerolike element, we shall require our systems to have an additive identity. We saw in Chapter 4 that the existence of negatives

110

leads to many important results, and we shall continue to study only those structures that have such elements.

In studying systems with two operations the relationships between the two operations are of primary importance. In all the two-operation systems previously studied the addition and multiplication were connected by the two distributive laws:

$$a \cdot (b + c) = a \cdot b + a \cdot c \quad \text{and} \quad (x + y) \cdot z = x \cdot z + y \cdot z$$

These laws, then, seem reasonable as axioms.

In discussing two-operation systems we shall employ the notational conveniences commonly used in elementary algebra. We are now ready to condense our axioms and define the class of structures known as rings.

Axioms for a Ring

A set R with two operations, which we shall call addition and multiplication and denote by $+$ and $\cdot$, respectively, is said to be a *ring* if the following conditions are satisfied.

(i) If a, b, and c are elements of R, then $a + (b + c) = (a + b) + c$.

(ii) If a and b are elements of R, then $a + b = b + a$.

(iii) There is an element 0 in R such that $a + 0 = a$ for each a in R.

(iv) For each element a in R, there is an element $-a$ in R (called the negative of a) such that $a + (-a) = 0$.

(v) For any elements a, b, and c in R, $a \cdot (b \cdot c) = (a \cdot b) \cdot c$.

(vi) For any elements a, b, and c in R, $a \cdot (b + c) = a \cdot b + a \cdot c$ and $(a + b) \cdot c = a \cdot c + b \cdot c$.

If we examine the first four axioms we see that a ring is an abelian group with respect to the operation of addition. Thus, *every assertion that we have established for groups will hold for the additive structure of a ring.* As with groups, the symbol "0" is read "zero" but need not refer to the number zero.

A ring is said to be *commutative* if the multiplication operation is commutative. Most of the rings you have studied previously are commutative, with the notable exception of matrix rings.

A ring R is said to have an *identity or a "one"* if there is an element 1 in R such that $a \cdot 1 = 1 \cdot a = a$ for each element a in R. (The symbol 1 need not refer to the "number" one.)

We saw in Chapter 1 that if a system has a two-sided identity with respect to a particular operation, then this identity is unique. Therefore, *if a ring has a "1", the "1" is unique.*

If x and y are elements of a ring R such that $x \neq 0$ and $y \neq 0$ but $xy = 0$, then x is called a *left proper zero divisor* and y is called a *right proper zero divisor*. A commutative ring that has a 1 such that $1 \neq 0$ and that has no proper zero divisors is called an *integral domain*.

Examples

(a) The continuous functions from R into R with the usual functional addition and multiplication is a commutative ring with 1.

(b) The even integers provide an example of a commutative ring without a 1 that has no proper zero divisors.

(c) Z_6 is a commutative ring with 1 that is not an integral domain because, for example, $\bar{2} \cdot \bar{3} = \bar{0}$.

(d) The integers and the polynomials over the real numbers provide examples of integral domains.

(e) The 3×3 matrices over the real numbers provide an example of a ring that is not commutative, but that does contain a 1 and has proper zero divisors.

(f) The 2×2 matrices with entries from the set of even integers provide an example of a noncommutative ring without 1.

The real numbers, rational numbers, and complex numbers all are examples of commutative rings with 1. For each nonzero real number r, there is a real number r^{-1} such that $rr^{-1} = 1$. Similarly, nonzero rational numbers and nonzero complex numbers also have multiplicative inverses. Rings with this property are very important.

A *field* is a commutative ring with 1 in which $1 \neq 0$ and for each nonzero element a in the ring, there is an element b in the ring such that $ab = 1$.

Examples of fields are Z_2, Z_3, Z_5, and Z_7. Also, the set $R(x)$ of all rational functions, that is,

$$R(x) = \{p(x)/q(x) : p(x) \text{ and } q(x) \text{ are polynomials in } R[x]$$
$$\text{and } q(x) \neq 0\}$$

is a field under the usual addition and multiplication. Other examples are given in the exercises.

Since every ring satisfies essentially the first six axioms for the integers, any assertion that we proved about Z using only these first six axioms will be a valid theorem for every ring.

In Chapter 2 the proofs of the first seven assertions depended only on Axioms (i)–(vi) and can be adapted to provide proofs of the following seven assertions. One needs only to replace the word "integer" by the phrase "element of R."

Proposition 1

If R is a ring and a, b, and c are elements of R such that $a + c = b + c$, then $a = b$.

Corollary 2

If R is a ring and x is an element of R such that $x + x = x$, then $x = 0$.

Corollary 3

If R is a ring and u and v are elements of R such that $u + v = 0$, then $u = -v$.

Corollary 4

If R is a ring with elements h and k, then $-(h + k) = (-h) + (-k)$.

Actually, we have two sets of proofs for the above assertions available to us. We can translate the proofs from $\mathbf{Z}$ or just notice that since a ring is an abelian group with respect to addition, the above assertions must be true because they are true in any abelian group.

Proposition 5

If R is a ring and a is an element of R, then $a \cdot 0 = 0$ and $0 \cdot a = 0$.

Theorem 6 (The Law of Signs)

If R is a ring with elements a and b, then $a \cdot (-b) = -(a \cdot b) = (-a) \cdot b$ and $(-a) \cdot (-b) = a \cdot b$.

In a ring we define subtraction just as we did for $\mathbf{Z}$. If R is a ring with elements a and b, then

$$a - b = a + (-b)$$

Proposition 7

If R is a ring with elements a, b, and c, then $a \cdot (b - c) = a \cdot b - a \cdot c$ and $(a - b) \cdot c = a \cdot c - b \cdot c$.

EXERCISE SET A

1. Which of the following are rings under the usual addition and multiplication? If not, why not?
   (a) the positive integers
   (b) the odd integers
   (c) the differentiable functions from $\mathbf{R}$ into $\mathbf{R}$
   (d) all differentiable functions f such that $f'(5) = 0$
   (e) $17\mathbf{Z}$
   (f) the set of all rational numbers that can be expressed as finite decimals
   (g) $\{p/2^n : p \in \mathbf{Z}$ and $n \in \mathbf{N}\}$
   *(h) the 3×2 matrices over the real numbers
   *(i) the set of all bounded functions from $\mathbf{R}$ into $\mathbf{R}$.
   *(j) the set of all functions from $\mathbf{R}$ into $\mathbf{R}$ that are differentiable at least twice.
*2. Is Euclidean three-dimensional space a ring with respect to vector addition and vector cross product?
*3. Is the set of continuous functions from $\mathbf{R}$ into $\mathbf{R}$ a ring with respect to the usual addition and composition for the multiplication?

4. By direct computation show that Z_8 is not a field.
5. By direct computation show that Z_5 is a field.
6. Show that Z is not a field.
7. Prove Proposition 5.
8. Prove Theorem 6.
9. Prove Proposition 7.
10. Show that a ring R is a field if the nonzero elements of R form an abelian group with respect to multiplication.
11. Let G be an abelian group denoted additively. In G define multiplication by $ab = 0$ for every pair of elements a and b in G. Prove that G is a ring. Rings in which the product of any two elements is zero are called *zero rings*.
12. Let $Z[\sqrt{2}] = \{p + q\sqrt{2}:p$ and q are integers$\}$. Show that $Z[\sqrt{2}]$ is a ring under the usual addition and multiplication.
13. Let $Q(\sqrt{2}) = \{a + b\sqrt{2}:a$ and b are rational numbers$\}$. Show that $Q(\sqrt{2})$ is a field under the usual addition and multiplication.
**14. Let $Q(\sqrt[3]{2}) = \{a + b\sqrt[3]{2} + c\sqrt[3]{4}:a, b,$ and c are rational numbers$\}$. Show that $Q(\sqrt[3]{2})$ is a field with respect to the usual addition and multiplication.
15. Show that the set of differentiable functions from R into R is not an integral domain.
**16. Let R denote the real numbers. In $R \times R$ define addition by $(a, b) + (c, d) = (a + c, b + d)$ and multiplication by $(a, b)(c, d) = (ac - bd, ad + bc)$. Prove that $R \times R$ is a field with respect to these operations.
**17. Let S be a set and let $\mathcal{S}$ be the set of all subsets of S. In $\mathcal{S}$ define ∇ by $A \nabla B = (A \cup B) - (A \cap B)$. Prove that $\mathcal{S}$ is a ring with ∇ as the addition and $\cap$ as the multiplication.
**18. A ring R is said to be *Boolean* if $aa = a$ for each element a in R.
 (i) Prove that $b + b = 0$ for each element b in a Boolean ring.
    (Hint: Show that $b + b = b + b + b + b$.)
 (ii) Prove that a Boolean ring is commutative.

§2. Zero Divisors and Units

Since not every ring is an integral domain, the generalization of Theorem 13, Chapter 2, cannot be expected to hold in arbitrary rings. The following assertion is, however, very important.

Proposition 8

Let R be a ring. The following conditions are logically equivalent.
(i) R has no proper zero divisors.
(ii) For any elements a, b, and c of R: If $a \neq 0$ and $ab = ac$, then $b = c$, and if $a \neq 0$ and $ba = ca$, then $b = c$.

Proof: We shall show that (i) implies (ii) and (ii) implies (i).
   First assume (i), that R has no proper zero divisors.
If $a \neq 0$ and $ab = ac$, then

$$ab - ac = 0$$

and, by Proposition 7,

$a(b - c) = 0$

Since $a \neq 0$ and R has no proper zero divisors,

$b - c = 0$

and therefore $b = c$.
A similar argument holds if $ba = ca$.

Now assume that (ii) holds, and let x and y be elements of R such that $xy = 0$.
If $x \neq 0$, then $xy = x0$ and, by (ii), $y = 0$.
If $y \neq 0$, then $xy = 0y$ and, by (ii), $x = 0$.
Thus, if $xy = 0$, $x = 0$ or $y = 0$ and R has no proper zero divisors. $\square$

The reader should make sure that he realizes that *the conditions of Proposition 8 may not hold in a ring.* All the assertion says is that if one holds, the other must.

We now apply Proposition 8 to show that a field possesses no proper zero divisors.

Proposition 9

Every field is an integral domain.

Proof: Let F be a field. We shall show that condition (ii) of Proposition 8 holds in F.

Let a, b, and c be elements of F such that $c \neq 0$ and $ac = bc$. Then, since F is a field and $c \neq 0$, $cd = 1$ for some element d in F. Thus,

$(ac)d = (bc)d$

and therefore

$a(cd) = b(cd)$

which implies

$a \cdot 1 = b \cdot 1$

and hence

$a = b$

Since F is commutative, if $ca = cb$, $ac = bc$, and therefore $a = b$. $\square$

Since $\mathbf{Z}$ is not a field, *not all integral domains are fields.* In the case of finite rings, however, the following important theorem holds.

Theorem 10

A finite integral domain is a field.

Proof: Let R be a finite integral domain with n distinct nonzero elements and denote the set of nonzero elements by R^*. Thus,

$$R^* = \{a_1, a_2, \ldots, a_n\}$$

We need only show that every element of R^* has a right inverse.

Consider an arbitrary element a_i in R^*. If we multiply each element of R^* by a_i we get

$$a_i R^* = \{a_i a_1, a_i a_2, \ldots, a_i a_n\}$$

Since R is an integral domain and $a_i \neq 0$, $a_i R^* \subseteq R^*$.

By Proposition 8, the elements $a_i a_1, a_i a_2, \ldots, a_i a_n$ are all distinct. (For example, since $a_i \neq 0$ and $a_1 \neq a_2$, $a_i a_1 \neq a_i a_2$.)

Thus, $a_i R^*$ contains n elements. Since R^* also has exactly n elements and $a_i R^* \subseteq R^*$, $a_i R^* = R^*$. Since $1 \in R^*$, $1 \in a_i R^*$, and therefore, for some a_j, $1 = a_i a_j$.

Since for each a_i in R^* there is an element a_j in R such that $a_i a_j = 1$, by definition, R is a field. $\square$

If R is a ring with 1 and a is an element of R, we say that a is a *unit* if, for some element b in R, $ab = ba = 1$.

Examples

(a) In $\mathbf{Z}$, the only units are 1 and -1.

(b) In $\mathbf{R}$, all nonzero elements are units.

(c) In any field, all nonzero elements are units.

(d) In $\mathbf{Z}_{10}$, $\bar{1}, \bar{3}, \bar{7}$, and $\bar{9}$ are units because

$$\bar{3} \cdot \bar{7} = \bar{1} \quad \text{and} \quad \bar{9} \cdot \bar{9} = \bar{1}$$

(e) In the ring of 3×3 matrices over $\mathbf{R}$, all nonsingular matrices are units.

(f) In the ring of 2×2 matrices over $\mathbf{Z}$, the units are the matrices with determinant 1 or -1.

In the proof of Proposition 9, we actually used only the fact that non-zero field elements are units. The assertion and the proof can be generalized to yield the following proposition. We leave the proof as an exercise.

Proposition 11

If R is a ring with 1, a is a unit in R, and x and y are elements of R such that $ax = ay$ or $xa = ya$, then $x = y$. Furthermore, a is not a zero divisor.

Proof: Exercise (Hint: Adapt the proofs from Proposition 8 and Proposition 9.)

Proposition 12

If R is a ring with 1, then the set U of units in R is a group with respect to multiplication.

Proof: Since 1 is a unit, the set U is nonempty. If p and q are units in R, then there exist elements u and v in R such that $pu = up = 1$ and $qv = vq = 1$. Then vu is in R, and

$$(pq)(vu) = p(qv)u = p1u = pu = 1$$
$$(vu)(pq) = v(up)q = v1q = vq = 1$$

and pq is a unit.

Thus, U is closed under multiplication, and multiplication is an operation on the set of units. Associativity in U is inherited from R. Since 1 is a unit, there is a multiplicative identity in U. If $a \in U$, then $ab = ba = 1$ for some b in R and, by definition, b is a unit. Thus, every unit has an inverse in U. □

Corollary 13

The set of nonzero elements of a field forms an abelian group with respect to multiplication.

Proof: Every nonzero element of a field is a unit and multiplication in a field is commutative. □

By Corollary 13 and group theory the multiplicative inverse of an element of a field is unique. We shall, as before, denote the unique multiplicative inverse of a by a^{-1}.

EXERCISE SET B

1. Prove that $\mathbf{Z}_p$ is a field iff p is prime. (Hint: Prove that $\mathbf{Z}_p$ is an integral domain iff p is prime; then apply Theorem 10.)
2. Prove: If R is an integral domain and a is an element of R such that $aa = a$, then $a = 0$ or $a = 1$.
3. Determine the set of all units in $\mathbf{Z}_6$, in $\mathbf{Z}_8$, in $\mathbf{Z}_9$, and in $\mathbf{Z}_{12}$.
4. Prove Proposition 11.
5. Let R be a ring with 1, and let a, b, and c be elements of R such that $ab = 1$ and $bc = 1$. Prove that $a = c$.
6. (a) Prove: In $\mathbf{Z}_m$, $\bar{n}$ is a unit iff m and n are relatively prime.
   (b) Prove: In $\mathbf{Z}_m$, if $\bar{n}$ is not a unit, then $\bar{n}$ is a zero divisor.
   (c) By example, show that (b) does not hold, in general, for all rings, that is, in some ring exhibit a nonunit that is not a zero divisor.

§3. Subrings and Subfields

If R is a ring and S is a subset of R, we say that S is a *subring* of R if S is closed under the addition and multiplication of R and is, in fact, a ring with respect to the operations of R restricted to S.

The theory of subrings is similar to the theory of subgroups. What do we need to show to prove that a particular subset of a ring is a subring? The Associative and Commutative laws for addition, the Associative law for multiplication, and the Distributive laws are all hereditary; since they hold for all elements of the ring, they will hold for those in a subset. Thus, we need to establish only Axioms (iii) and (iv) and closure under the two operations. Axioms (iii) and (iv) are actually two of the group axioms for addition. Thus, if we show that a subset of a ring is a subgroup under addition and closed under multiplication we shall have shown that it is a subring. We state this as a theorem.

Theorem 14

If R is a ring and S is a subset of R that is a subgroup of R with respect to addition and is closed under multiplication, then S is a subring of R.

Corollary 15

If R is a ring and S is a nonempty subset of R that is closed under subtraction and multiplication, then S is a subring of R.

Proof: If S is closed under subtraction, then if a and b are in S, $a - b$ is in S.

Since $a + (-b) \in S$ for each a and b in S, S is an additive subgroup of R by Theorem 16, Chapter 4. Since S is a subgroup with respect to addition and closed under multiplication, S is a subring. □

Examples.   In the ring **C**, of complex numbers,

(a) **R**, the real numbers, is a subring of **C**,
(b) **Q**, the rational numbers, is a subring of **R**,
(c) D, the finite decimals, is a subring of **Q**,
(d) **Z**, the integers, is a subring of D, and
(e) E, the even integers, is a subring of **Z**.
(f) The ring of polynomials over **R** is a subring of $\mathfrak{D}$, the differentiable functions from **R** to **R**.
(g) $\mathfrak{D}$ is a subring of $\mathfrak{C}$, the continuous functions from **R** into **R**.
(h) The 2×2 matrices with integral entries is a subring of the 2×2 matrices with rational entries.
(i) The set of all polynomials that have 8 for a root is a subring of the ring of all polynomials over **R**.
(j) Note also that R and $\{0\}$ are always subrings of R for any ring R. Many of the structures considered in Exercise Set A are, in fact, subrings of some other ring.

Just as Theorem 16 in Chapter 4 gave us simpler criteria for determining when a subset of a group was a subgroup, Theorem 14 and Corollary 15 give us simpler criteria for showing that a subset of a ring is a subring. We know that the intersection of subgroups is a subgroup, and we now show that the intersection of subrings is a subring.

Proposition 16

If R is a ring and α is a nonempty collection of subrings of R, then $\cap \alpha$ is a subring of R.

Proof: Since α is a collection of subgroups of R with respect to addition, $\cap \alpha$ is a subgroup of R by Proposition 18 in Chapter 4. If a and b are elements of $\cap \alpha$, then $a \in S$ and $b \in S$ for every subring S in α. Hence, $ab \in S$ for each S in α, and therefore $ab \in \cap \alpha$. Since $\cap \alpha$ is an additive subgroup of R and is closed under multiplication, $\cap \alpha$ is a subring. $\square$

It now follows, from Proposition 16, that we can describe a subring of a ring by describing a collection of subrings and forming the intersection of this collection. If W is any subset of a ring R, the intersection of all subrings containing W will again be a subring.

If R is a ring and W is a subset of R, the intersection of all subrings containing W is called the *subring of R generated by W*.

We denote the subring generated by W by the symbol $\langle W \rangle$. A ring is *cyclic* if it is generated by one element. If $S = \langle W \rangle$, we call W a *set of generators* of S.

Examples

(a) The subring of $\mathbf{Q}$ generated by $\{1\}$ is just $\mathbf{Z}$ because, as a group, $\{1\}$ generates $\mathbf{Z}$, and therefore any subring containing 1 would have to contain $\mathbf{Z}$. Since $\mathbf{Z}$ is, in fact, a ring and is a subset of every ring containing 1, $\mathbf{Z}$ is the subring of $\mathbf{Q}$ generated by 1.

(b) The set D of finite decimals is a subring of $\mathbf{Q}$ generated by $\{0.1\}$. To see this, first note that D is, in fact, a subring. Any ring that contains 0.1 will have to contain $n(0.1)$ for every integer n, because it will have to contain the subgroup of $\mathbf{Q}$ generated by 0.1. Hence, $10(0.1)$ will be in any such ring; that is, 1 will be in every such ring. But if 1 is in a subring of $\mathbf{Q}$, every integer is in the subring. Thus, $\mathbf{Z}$ is a subset of every subring that contains 0.1. Furthermore, since subrings are closed under multiplication, every subring that contains 0.1 must contain $(0.1)^n$ for all positive integers n. Thus, every subring of $\mathbf{Q}$ that contains 0.1 must contain $\{p(0.1)^n : p$ is an integer and n is a positive integer$\}$. This set is, however, just the set D. Since D is a subring of every subring that contains 0.1, D is the intersection of all such subrings, and thus is the subring generated by 0.1. D is also the subring generated by $\{\frac{1}{5}, \frac{1}{2}\}$. This follows from our argument above and the fact that $0.1 \in \langle\{\frac{1}{5}, \frac{1}{2}\}\rangle$ and $\langle\{\frac{1}{5}, \frac{1}{2}\}\rangle \subseteq \langle 0.1 \rangle$.

(c) The ring $\mathbf{Z}[\sqrt{2}]$ is generated by $\{1, \sqrt{2}\}$. It is also generated by $\{1 + \sqrt{2}, \sqrt{2}\}$. It is not generated by $\sqrt{2}$ alone.

(d) The ring $\mathbf{Q}(\sqrt{2})$ is generated by $\mathbf{Q}^+ \cup \{\sqrt{2}\}$.

(e) The ring of 2×2 matrices over $\mathbf{R}$ is generated by the nonsingular 2×2 matrices over $\mathbf{R}$.

(f) The ring of polynomials over $\mathbf{Z}$ is generated by $\{1, x\}$.

Since a ring is a group with respect to addition, any set that, as a group, generates the additive structure of a ring will, of course, generate the ring. In general, however, the additive subgroup generated by a subset of a ring is not the same as the subring generated by the subset. In the example above the subgroup of Q generated by $\{0.1\}$ is the set of all rational numbers that can be expressed with one or less decimal places, whereas the subring generated by $\{0.1\}$ is D.

A theory of substructures for fields can be developed in much the same manner as for groups and rings.

If F is a field and K is a subset of F, we say that K is a *subfield* of F if K is closed under the addition and multiplication of F and K is a field with respect to these operations restricted to K.

Since a field F can contain only one nonzero element a such that $aa = a$, the multiplicative identity in any subfield of F must be the identity of F.

A ring R is a field if the nonzero elements of R form an abelian group, and therefore a subset K of a field F is a subfield if K is a subgroup with respect to addition and the nonzero elements of K form a subgroup with respect to multiplication. Applying Theorem 16 from Chapter 4, we obtain the following assertion.

Proposition 17

If F is a field and K is a subset of F such that K contains nonzero elements, is closed under subtraction, and for each pair x and y of nonzero elements of K, $xy^{-1} \in K$, then K is a subfield of F.

Proof: Since K is closed under subtraction, K is a subgroup of F with respect to addition.

Since $xy^{-1} \in K$ for every pair x and y of nonzero elements of K, the nonzero elements of K form a subgroup of the nonzero elements of F.   □

One can easily use this criteria to prove the following proposition, which is analogous to Proposition 16.

Proposition 18

If F is a field and $\mathfrak{F}$ is a nonempty collection of subfields of F, then $\cap \, \mathfrak{F}$ is a subfield of F.

Proof: Exercise.

By Proposition 18, if S is a subset of a field F, then the intersection of all subfields of F containing S is again a subfield of F. If S is a subset of a field F, the *subfield of F generated by S* is the intersection of all subfields of F containing S.

The subring of a field generated by a set and the subfield of a field generated by a set are not, in general, the same. For example, in the real numbers R, the subring generated by $\{1\}$ is Z, but the subfield generated by $\{1\}$ is Q. To see that the subfield of R generated by $\{1\}$ is, in fact Q, note that since $\{1\}$ generates Z as a subring, Z must be a subset of the subfield generated by $\{1\}$. But then, for each nonzero integer q, $1/q$ must be in the field generated by $\{1\}$. This implies, however, that $p(1/q)$ is in the subfield generated by $\{1\}$ for every integer p and nonzero integer q, and therefore every rational number is in the subfield generated by $\{1\}$. Since Q itself is a field, Q is the intersection of all subfields of R that contain 1; that is, Q is the subfield of R generated by 1.

Other examples include the following:

(a) The subfield of Z_7 generated by $\{\bar{2}\}$ is Z_7.
(b) The subfield of R generated by $\sqrt{2}$ is $Q(\sqrt{2})$.
(c) The subfield of the real numbers generated by $\sqrt[3]{2}$ is $\{a + b \sqrt[3]{2} + c\sqrt[3]{4}:a, b,$ and c are rational$\}$.
(d) The subfield of R generated by $\{\sqrt{2}, \sqrt{3}\}$ is the set $\{a + b\sqrt{2} + c\sqrt{3} + d\sqrt{6}:a, b, c,$ and d are rational$\}$.

These examples are not obvious but are not too difficult to demonstrate.

In the case of Z_7, the additive subgroup generated by 2 is, in fact, Z_7 since $\{n\bar{2}:n \in N\} = \{\bar{2}, \bar{4}, \bar{6}, \bar{1}, \bar{3}, \bar{5}, \bar{0}\}$.

In R, every subfield must contain 1. Any subfield that contains 1 must contain Q, and therefore Q is a subfield of every subfield contained in R, because Q itself is a field. Now, any subfield containing $\sqrt{2}$ must contain Q and $\sqrt{2}$ and be closed under addition and multiplication. Thus, any subfield containing $\sqrt{2}$ must contain $\{a + b\sqrt{2}:a$ and b are rational$\}$. In Exercise 13 of Set A, we called this set $Q(\sqrt{2})$ and showed that it was, in fact, a field. Since $Q(\sqrt{2})$ is contained in every subfield containing $\sqrt{2}$ and is itself a subfield, $Q(\sqrt{2})$ is the intersection of all subfields containing $\sqrt{2}$.

Any subfield of R must contain Q and, since subfields are closed under multiplication, any subfield of S containing $\sqrt{2}$ and $\sqrt{3}$ must contain $\sqrt{6}$. The set $\{a + b\sqrt{2} + c\sqrt{3} + d\sqrt{6}:a, b, c,$ and d are rational$\}$ must then be a subset of every subfield of R containing $\sqrt{2}$ and $\sqrt{3}$. This set is easily shown to be a subgroup under addition and closed under multiplication. We leave these verifications to the reader. To show that $1/(a + b\sqrt{2} + c\sqrt{3} + d\sqrt{6})$ is also a number of this form requires a significant amount of manipulation, which we shall also omit and leave to the reader. Once these arguments are completed, it will have been shown that the set is a field. Since it is contained in every subfield containing $\{\sqrt{2}, \sqrt{3}\}$, it must be the intersection of all such fields.

EXERCISE SET C

1. Let $\mathcal{F}$ be the ring of all functions from the real numbers to the real numbers with the usual addition and multiplication of functions as the operations.

Which of the following are subrings? If not, why not?

(a) $\{g \in \mathfrak{F} : 4g = 0\}$
(b) $\{f \in \mathfrak{F} : \mathbf{Z}f \subseteq \mathbf{Z}\}$
(c) the increasing functions
(d) the functions that are differentiable at least twice
(e) the differentiable functions g such that $g' = g$
*(f) the functions with at most finitely many discontinuities

2. Determine all subrings of $\mathbf{Z}_9$. Of $\mathbf{Z}_8$. Of $\mathbf{Z}_{12}$. Of $\mathbf{Z}_7$.
3. In $\mathbf{Z}_{12}$, what is the subring generated by $\bar{8}$? By $\bar{2}$? By $\{\bar{3}, \bar{4}\}$?
4. Show that if p is prime, $\mathbf{Z}_p$ contains no subrings other than $\{0\}$ and $\mathbf{Z}_p$.
5. Prove Proposition 18.
6. Viewing the real numbers $\mathbf{R}$ as an additive group, what is the subgroup of $\mathbf{R}$ generated by $\sqrt{3}$?
7. Viewing the real numbers $\mathbf{R}$ as a ring, what is the subring of $\mathbf{R}$ generated by $\sqrt{3}$? (Note that the subring generated by $\sqrt{3}$ would have to contain the subgroup generated by $\sqrt{3}$.)
8. Viewing the real numbers $\mathbf{R}$ as a field, what is the subfield of $\mathbf{R}$ generated by $\sqrt{3}$? (Note that the subfield generated by $\sqrt{3}$ would have to contain the subring generated by $\sqrt{3}$.)
9. Viewing $\mathbf{R}[x]$ as an additive group, what is the subgroup generated by $\{1, x^2\}$?
10. Viewing $\mathbf{R}[x]$ as a ring, what is the subring generated by $\{1, x^2\}$?
11. Viewing $\mathbf{R}(x)$ as a field, what is the subfield generated by $\{1, x^2\}$?
**12. What is the subfield of $\mathbf{R}$ generated by $\sqrt[3]{3}$? Prove your assertion.
**13. What is the subfield of $\mathbf{R}$ generated by $\sqrt{5}$? Prove your assertion.
**14. What is the subfield of $\mathbf{R}$ generated by $\{\sqrt{5}, \sqrt{7}\}$? Prove your assertion.
15. Consider the ring $\mathbf{R}$ of real numbers and the subring $\mathbf{Z}$ of integers and the relation of congruence modulo $\mathbf{Z}$. We know from the study of groups that this relation is compatible with addition. Is it compatible with multiplication in that if $a \equiv b$ and $c \equiv d$, must $ac \equiv bd \pmod{\mathbf{Z}}$?
16. A subring A of a ring R is called an *ideal* if, whenever $r \in R$ and $a \in A$, $ra \in A$ and $ar \in A$. Which of the following subrings are ideals?
(a) $\mathbf{Q}$ in $\mathbf{R}$
(b) the even integers in $\mathbf{Z}$
(c) the differentiable functions in the ring of continuous functions
(d) in $\mathbf{R}[x]$, the set of all polynomials that have 5 for a root
17. Which subrings of $\mathbf{Z}_8$ are ideals?

§4. Ideals

The relation of congruence modulo a subgroup extends simply to congruence modulo a subring.

If R is a ring, S is a subring of R, and a and b are elements of R, we say that *a is congruent to b modulo S* if $a - b \in S$. As before we write $a \equiv b \pmod{S}$.

Since S is a subgroup of R with respect to addition, and since the definition of congruence involves only the additive structures, congruence modulo a subring is an equivalence relation. Furthermore, the congruence classes are again the cosets. Remember, however, that the notation is additive. Thus,

$$\bar{a} = S + a$$

All the results that we obtained concerning cosets and congruence classes for groups will also hold for rings. Thus, for example, if S is a subring of the finite ring R, the order of S divides the order of R.

In Exercise Set C, Exercise 15 we saw that congruence modulo a subring need not be compatible with multiplication; that is, in general, if S is a subring of the ring R, $a \equiv b \pmod{S}$, and $c \equiv d \pmod{S}$, it need not happen that $ac \equiv bd \pmod{S}$.

As another example, consider the subring $\mathbf{Z}[\sqrt{2}]$ in $\mathbf{Q}[\sqrt{2}]$. Although

$$0.5 + \sqrt{2} \equiv 1.5 + \sqrt{2} \pmod{\mathbf{Z}[\sqrt{2}]}$$

and

$$0.7 \equiv 0.7 \pmod{\mathbf{Z}[\sqrt{2}]}$$
$$0.7(0.5 + \sqrt{2}) \neq 0.7(1.5 + \sqrt{2})$$

that is,

$$0.35 + 0.7\sqrt{2} \neq 1.05 + 0.7\sqrt{2}$$

because

$$(1.05 + 0.7\sqrt{2}) - (0.35 + 0.7\sqrt{2}) = 0.7$$

and

$$0.7 \notin \mathbf{Z}[\sqrt{2}]$$

In Chapter 4 we saw that congruence modulo an arbitrary subgroup need not be compatible with the group operation, and now we have seen that congruence modulo an arbitrary subring need not be compatible with ring multiplication. The subgroups that yielded such compatibility in group theory were normal subgroups. The analogous structure in ring theory is an *ideal*.

If R is a ring and A is a subring of R, we say that A is an *ideal* if for each element r in R and each a in A, $ra \in A$ and $ar \in A$.

Ideals are, then, subrings that are closed under multiplication by arbitrary elements of the ring. Notice that if R is a ring and A is a nonempty subset of R such that A is closed under subtraction and whenever $r \in R$ and $a \in A$, $ra \in A$ and $ar \in A$, then A is an ideal because A is a subgroup of R with respect to addition and is closed under multiplication. This is the criteria we shall usually employ to show that a subset of a ring is an ideal.

The following are examples of ideals.

(a) The even integers E is an ideal in Z, since E is a subring and if n is even and z is any integer, nz is also even.
(b) In the ring of polynomials over R, the subring of all polynomials having 5 for a root is an ideal, since if $f(5) = 0$ and g is any polynomial $(g \cdot f)(5) = g(5)f(5) = g(5) \cdot 0 = 0$, and therefore $g \cdot f$ is a polynomial with 5 for a root.
(c) If m is any integer, mZ is an ideal in Z.
(d) The set of all continuous functions that are 0 at every integer is an ideal in the ring of continuous functions.

The following subrings are not ideals:

(a) Z in Q.
(b) The differentiable functions in the ring of continuous functions.
(c) The even integers in R.
(d) Q in R.

The importance of ideals in ring theory arises from the fact that congruence modulo an ideal is compatible with multiplication. We establish this in the following assertion.

Proposition 19

If A is an ideal in the ring R, and if a, b, c, and d are elements of R such that $a \equiv b \pmod{A}$ and $c \equiv d \pmod{A}$, then $ac \equiv bd \pmod{A}$.

Proof: If $a \equiv b \pmod{A}$ and $c \equiv d \pmod{A}$, then $a - b \in A$ and $c - d \in A$.

Since A is an ideal,

$$(a - b)c \in A \quad \text{and} \quad b(c - d) \in A$$

Then

$$(ac - bc) \in A \quad \text{and} \quad (bc - bd) \in A$$

Thus,

$$[(ac - bc) + (bc - bd)] \in A$$

which implies

$$ac - bd \in A$$

and therefore

$$ac \equiv bd \pmod{A} \quad \square$$

The reader should compare the above proof with the proof of Proposition 28 in Chapter 2.

EXERCISE SET D

1. List all the ideals of Z_{12}.
2. Justify the last two examples of ideals given in the text; that is,
   (a) show that $m Z$ is an ideal in Z for every integer m,
   (b) show that the set of all continuous functions that are 0 at every integer is an ideal in the ring of continuous functions.
3. Prove that the even integers is not an ideal in the ring of real numbers.
4. Show that the ring of differentiable functions is not an ideal in the ring of continuous functions.
5. Let R be a ring, and let $\mathcal{C}$ be a nonempty collection of ideals in R. Prove that $\cap \, \mathcal{C}$ is an ideal in R.
6. Prove: If R is a *commutative ring* and a is an element of R, then $\{ax : x \in R\}$ is an ideal in R.
7. Prove that every subring of Z is an ideal. (Hint: What form must a subgroup of Z take?)

§5. New Rings from Old

In Chapter 4 we constructed factor groups G/H, where H was a normal subgroup of G. In ring theory we can construct *factor rings* in the same manner, where an ideal plays the role previously played by a normal subgroup. If A is an ideal in the ring R, we know that R/A is a group with respect to the addition of congruence classes:

$$\bar{a} + \bar{b} = \overline{a + b}$$

In R/A, we now define multiplication by

$$\bar{a} \cdot \bar{b} = \overline{ab}$$

That this rule does, in fact, define an operation follows from the fact that congruence modulo A is compatible with multiplication. For suppose

$$\bar{a} = \bar{c} \quad \text{and} \quad \bar{b} = \bar{d}$$

Then $a \equiv c \pmod{A}$ and $b \equiv d \pmod{A}$ and, by Proposition 19,

$$ab \equiv cd \pmod{A}$$

and therefore

$$\overline{ab} = \overline{cd}$$

Thus, no matter which elements in a and b we choose, the resulting class will still be the same and the rule does define an operation on R/A.

Theorem 20

If A is an ideal in the ring R, then R/A is a ring with operations defined by

$$\bar{a} + \bar{b} = \overline{a + b} \quad \text{and} \quad \bar{a} \cdot \bar{b} = \overline{ab}$$

Proof: We know that R/A is a group with respect to addition, because R is a group with respect to addition and A is a normal subgroup of R. We need only show that multiplication is associative and that the Distributive laws are satisfied.

Let a, b, and c be elements of R/A. Then

$$\bar{a}(\bar{b}\bar{c}) = \bar{a}(\overline{bc}) = \overline{a(bc)} = \overline{(ab)c} = (\overline{ab})\bar{c} = (\bar{a}\bar{b})\bar{c}$$

and therefore the multiplication is associative. Also,

$$\bar{a}(\bar{b} + \bar{c}) = \bar{a}(\overline{b + c}) = \overline{a(b + c)} = \overline{ab + ac} = \overline{ab} + \overline{ac} = \bar{a}\bar{b} + \bar{a}\bar{c}$$

The other distributive law is demonstrated similarly. □

The following are examples of factor rings:

(a) In Chapter 2 we saw that $\mathbf{Z}_m = \mathbf{Z}/\langle m \rangle$ is a ring for each integer m.

(b) If B is the set of all polynomial multiples of $x^2 + 1$ in $\mathbf{R}[x]$, then B is an ideal in $\mathbf{R}[x]$ and $\mathbf{R}[x]/B$ is a factor ring. Since $x^2 - (-1) = x^2 + 1$ and $x^2 + 1 \in B$, $x^2 \equiv -1 \pmod{B}$, and therefore $\overline{x^2} = -\bar{1}$ in $\mathbf{R}[x]/B$. Since $x^2 \equiv -1$, every polynomial is congruent to a polynomial of the form $ax + b$. The elements of $\mathbf{R}[x]/B$ are elements of the form $\bar{a} \cdot \bar{x} + \bar{b}$. Notice that in $\mathbf{R}[x]/B$, the equation $y^2 = -\bar{1}$ has a solution, namely $\bar{x}$.

(c) In $\mathbf{Z}_4$, $\{\bar{0}, \bar{2}\}$ is an ideal and the tables for $\mathbf{Z}_4/\{\bar{0}, \bar{2}\}$ are given below. $\mathbf{Z}_4/\{0, 2\} = \{\{\bar{0}, \bar{2}\}, \{\bar{1}, \bar{3}\}\}$.

+	$\{\bar{0}, \bar{2}\}$	$\{\bar{1}, \bar{3}\}$
$\{\bar{0}, \bar{2}\}$	$\{\bar{0}, \bar{2}\}$	$\{\bar{1}, \bar{3}\}$
$\{\bar{1}, \bar{3}\}$	$\{\bar{1}, \bar{3}\}$	$\{\bar{0}, \bar{2}\}$

$\cdot$	$\{\bar{0}, \bar{2}\}$	$\{\bar{1}, \bar{3}\}$
$\{\bar{0}, \bar{2}\}$	$\{\bar{0}, \bar{2}\}$	$\{\bar{0}, \bar{2}\}$
$\{\bar{1}, \bar{3}\}$	$\{\bar{0}, \bar{2}\}$	$\{\bar{1}, \bar{3}\}$

A ring R is always an ideal in itself and $\{0\}$ is also always an ideal in the ring R. An ideal is said to be *proper* if it is neither R nor $\{0\}$. Notice that if R is a ring with 1 and A is a proper ideal in R, then $1 \notin A$ for if $1 \in A$, then $1r \in A$ for every r in R. But this means that $r \in A$ for each r in R, that is $R \subseteq A$, and therefore $R = A$, which is impossible if A is a proper ideal. Using this idea, one can establish the following assertion rather easily.

Proposition 21

A field contains no proper ideals.

Proof: Exercise. (Hint: Show that if an ideal contains a nonzero element, then it must contain 1.)

The proof of Proposition 21 can be generalized to establish the following assertion.

Proposition 22

If R is a ring with 1 and A is a proper ideal in R, then A contains no units.

Proof: Exercise.

The concept of the direct product and direct sum of groups also extends easily to rings. If R and S are rings, we know that $R \oplus S$ is an abelian group. Define a multiplication in $R \times S$ by

$$(x, y)(u, v) = (xu, yv)$$

Theorem 23

If R and S are rings, then $R \times S$ with addition, defined by

$$(x, y) + (u, v) = (x + u, y + v)$$

and multiplication, defined by

$$(x, y)(u, v) = (xu, yv)$$

is a ring.

Proof: We leave the verification of the Associative law of multiplication and the Distributive laws as exercise.

If R and S are rings, by $R \oplus S$ we mean $R \times S$ with the two "component-wise" operations defined above. We call $R \oplus S$ the *direct sum* of R and S.

As we can for groups, we can extend this definition to three rings $R, S,$ and T and form $R \oplus S \oplus T$, where the elements are ordered triples and for any finite collection of rings $R_1, R_2, \ldots, R_n$, where the elements of the new ring are ordered n-tuples. The proof for Theorem 23 extends easily to these cases.

If R is a ring and W is any nonempty set, the set of all functions from W into R is a ring with respect to the usual pointwise addition and multiplication of functions; that is,

$$(x)(f + g) = xf + xg$$
$$(x)(f \cdot g) = (xf)(xg) \qquad \text{for each } x \text{ in } R$$

We leave as an exercise the verification that this structure is a ring and point out in passing that if $\mathbf{R}$ is the real numbers and $\mathbf{N}$ is the positive integers, this construction gives us all infinite sequences of real numbers with the usual addition and multiplication of sequences.

EXERCISE SET E

1. Complete the proof of Theorem 20.
2. Construct addition and multiplication tables for all rings that can be formed from Z_{12}, that is, all rings Z_{12}/A, where A is a proper ideal in Z_{12}.
3. Construct the multiplication tables for $Z_4 \oplus Z_3$ and for $Z_2 \oplus Z_6$.
4. Prove that if R and S are rings, $R \oplus \{0\}$ is an ideal in $R \oplus S$.
5. List all the ideals in $Z_4 \oplus Z_3$ and $Z_2 \oplus Z_6$.
6. Construct addition and multiplication tables for all the factor rings you can construct from $Z_4 \oplus Z_3$.
7. Construct the addition and multiplication tables for all the factor rings you can construct from $Z_2 \oplus Z_6$.
8. Prove Proposition 21.
9. Prove Proposition 22.
10. Prove: If $R \oplus S$ is an integral domain, then either $R = \{0\}$ or $S = \{0\}$.
11. Prove: If R is a commutative ring and A is an ideal in R, then R/A is commutative.
12. Prove: If R and S are commutative rings with 1, then $R \oplus S$ is a commutative ring with 1.
13. Prove: If R is a ring and W is a nonempty set, then the set of all functions from W into R is a ring with respect to pointwise addition and multiplication. Show also that if R has a "1," then this ring will have a "1."
14. Use Exercise 6 from Exercise Set D to prove the following *partial* converse to Proposition 21.

    If R is a *commutative* ring with $1 \neq 0$, and R contains no proper ideals, then R is a field.

**15. Let R be a ring, and let $\sim$ be an equivalence relation on R that is compatible with addition and multiplication; that is, if $a \sim b$ and $c \sim d$, then $a + c \sim b + d$ and $ac \sim bd$. Prove that $\sim$ is congruence modulo some ideal. (Hint: Prove that $\bar{0}$, the class for 0 with respect to $\sim$, is an ideal and that $x \sim y$ iff $x \equiv y \pmod{\bar{0}}$.)

16. Let R be a ring and let A be an ideal in R. Define ϕ from R to R/A by $x\phi = \bar{x}$. Prove that if u and v are in R, $(u + v)\phi = u\phi + v\phi$ and $(uv)\phi = (u\phi)(v\phi)$. What is $\bar{0}\phi^{-1}$?

17. Let R and S be rings, and define ϕ from $R \oplus S$ to R by $(x, y)\phi = x$. Prove that if (a, b) and (c, d) are in $R \oplus S$, then

$$[(a, b) + (c, d)]\phi = (a, b)\phi + (c, d)\phi \quad \text{and}$$
$$[(a, b)(c, d)]\phi = [(a, b)\phi][(c, d)\phi]$$

18. Let R and S be rings, and define λ from R into $R \oplus S$ by $a\lambda = (a, 0)$ for each a in R. Prove that if x and y are in R,

$$(x + y)\lambda = x\lambda + y\lambda \quad \text{and} \quad (x, y)\lambda = (x\lambda)(y\lambda)$$

Show also that λ is one-to-one.

19. Show that if a, b, and c are elements of the field F and $a \neq 0$, then the equation $ax + b = c$ has a unique solution in F.

20. Show that if a, b, c, α, β, and γ are elements of the field F the system of equations

$$ax + by = c$$
$$\alpha x + \beta y = \gamma$$

either has no solution, has a unique solution, or has as many solutions as there are elements in F.

6

OPERATION-PRESERVING FUNCTIONS

In the study of groups, rings, and fields we are, as pointed out in Chapter 4, working at what could be called the second level of abstraction. A natural question that arises, then, is under what circumstances might we really be working only at the first level or even be working with specific structures rather than abstract ideas? If we have some class of structures that all *behave* exactly alike, then we can study all of them by just studying one of them. In such a case, then, we can work at a lower abstraction level. The idea of *behaving alike* is made precise by the concept of an isomorphism. In this chapter we shall study isomorphisms and homomorphisms and see how seemingly different structures may behave completely or partially alike.

§1. Isomorphism

Below are the tables for the groups $\mathbf{Z}_2$, $\{1, -1\}$, and $\{i, (1 \quad 2)\}$.

$+$	$\bar{0}$	$\bar{1}$
$\bar{0}$	$\bar{0}$	$\bar{1}$
$\bar{1}$	$\bar{1}$	$\bar{0}$

$\cdot$	1	-1
1	1	-1
-1	-1	1

$\circ$	i	$(1 \quad 2)$
i	i	$(1 \quad 2)$
$(1 \quad 2)$	$(1 \quad 2)$	i

If we examine these three tables we see that in each case the identity element occupies the upper left and lower right corners and the other element occupies the upper right and lower left corners. If G is a group

130

and b is an element of order 2 in G, the table for $\langle b \rangle$ must have the form below, because $b^2 = e$.

	e	b
e	e	b
b	b	e

This table looks much like those above, except that the symbols are different.

In a finite group the operation is completely described by the table. Since all four tables are very much alike, the four groups must be very much alike. In what sense are they similar? The elements certainly are different; it is the operations that are alike. Consider now the tables for Z_3 and A_3.

$+$	$\bar{0}$	$\bar{1}$	$\bar{2}$
$\bar{0}$	$\bar{0}$	$\bar{1}$	$\bar{2}$
$\bar{1}$	$\bar{1}$	$\bar{2}$	$\bar{0}$
$\bar{2}$	$\bar{2}$	$\bar{0}$	$\bar{1}$

$\circ$	i	$(1\ 2\ 3)$	$(1\ 3\ 2)$
i	i	$(1\ 2\ 3)$	$(1\ 3\ 2)$
$(1\ 2\ 3)$	$(1\ 2\ 3)$	$(1\ 3\ 2)$	i
$(1\ 3\ 2)$	$(1\ 3\ 2)$	i	$(1\ 2\ 3)$

If we take the first table and in it we replace $+$ by $\circ$, $\bar{0}$ by i, $\bar{1}$ by $(1\ 2\ 3)$, and $\bar{2}$ by $(1\ 3\ 2)$, we shall get the table for A_3. Thus, any equation or other group-theoretic statement that holds for Z_3 can be translated into a statement that will hold for A_3. For example,

In Z_3:

$\bar{2} + \bar{2} = \bar{1}$

$\bar{1} + \bar{2} = \bar{0}$

Z_3 has no elements of order 2

Z_3 is commutative

In A_3:

$(1\ 3\ 2) \circ (1\ 3\ 2) = (1\ 2\ 3)$

$(1\ 2\ 3) \circ (1\ 3\ 2) = i$

A_3 has no elements of order 2

A_3 is commutative

All true statements concerning Z_3 and the group operation $+$ translate to true statements concerning A_3 and the operation $\circ$, because the table for Z_3 translates to the table for A_3 and the table completely describes the operation.

Consider now the two rings Z_6 and $Z_2 \oplus Z_3$. Following are the tables for Z_6.

+	$\bar{0}$	$\bar{1}$	$\bar{2}$	$\bar{3}$	$\bar{4}$	$\bar{5}$
$\bar{0}$	$\bar{0}$	$\bar{1}$	$\bar{2}$	$\bar{3}$	$\bar{4}$	$\bar{5}$
$\bar{1}$	$\bar{1}$	$\bar{2}$	$\bar{3}$	$\bar{4}$	$\bar{5}$	$\bar{0}$
$\bar{2}$	$\bar{2}$	$\bar{3}$	$\bar{4}$	$\bar{5}$	$\bar{0}$	$\bar{1}$
$\bar{3}$	$\bar{3}$	$\bar{4}$	$\bar{5}$	$\bar{0}$	$\bar{1}$	$\bar{2}$
$\bar{4}$	$\bar{4}$	$\bar{5}$	$\bar{0}$	$\bar{1}$	$\bar{2}$	$\bar{3}$
$\bar{5}$	$\bar{5}$	$\bar{0}$	$\bar{1}$	$\bar{2}$	$\bar{3}$	$\bar{4}$

$\cdot$	$\bar{0}$	$\bar{1}$	$\bar{2}$	$\bar{3}$	$\bar{4}$	$\bar{5}$
$\bar{0}$	$\bar{0}$	$\bar{0}$	$\bar{0}$	$\bar{0}$	$\bar{0}$	$\bar{0}$
$\bar{1}$	$\bar{0}$	$\bar{1}$	$\bar{2}$	$\bar{3}$	$\bar{4}$	$\bar{5}$
$\bar{2}$	$\bar{0}$	$\bar{2}$	$\bar{4}$	$\bar{0}$	$\bar{2}$	$\bar{4}$
$\bar{3}$	$\bar{0}$	$\bar{3}$	$\bar{0}$	$\bar{3}$	$\bar{0}$	$\bar{3}$
$\bar{4}$	$\bar{0}$	$\bar{4}$	$\bar{2}$	$\bar{0}$	$\bar{4}$	$\bar{2}$
$\bar{5}$	$\bar{0}$	$\bar{5}$	$\bar{4}$	$\bar{3}$	$\bar{2}$	$\bar{1}$

If, in both of the above tables we replace $\bar{0}$ by $(\bar{0}, \bar{0})$, $\bar{1}$ by $(\bar{1}, \bar{1})$, $\bar{2}$ by $(\bar{0}, \bar{2})$, $\bar{3}$ by $(\bar{1}, \bar{0})$, $\bar{4}$ by $(\bar{0}, \bar{1})$ and $\bar{5}$ by $(\bar{1}, \bar{2})$ we obtain the following tables:

+	$(\bar{0}, \bar{0})$	$(\bar{1}, \bar{1})$	$(\bar{0}, \bar{2})$	$(\bar{1}, \bar{0})$	$(\bar{0}, \bar{1})$	$(\bar{1}, \bar{2})$
$(\bar{0}, \bar{0})$	$(\bar{0}, \bar{0})$	$(\bar{1}, \bar{1})$	$(\bar{0}, \bar{2})$	$(\bar{1}, \bar{0})$	$(\bar{0}, \bar{1})$	$(\bar{1}, \bar{2})$
$(\bar{1}, \bar{1})$	$(\bar{1}, \bar{1})$	$(\bar{0}, \bar{2})$	$(\bar{1}, \bar{0})$	$(\bar{0}, \bar{1})$	$(\bar{1}, \bar{2})$	$(\bar{0}, \bar{0})$
$(\bar{0}, \bar{2})$	$(\bar{0}, \bar{2})$	$(\bar{1}, \bar{0})$	$(\bar{0}, \bar{1})$	$(\bar{1}, \bar{2})$	$(\bar{0}, \bar{0})$	$(\bar{1}, \bar{1})$
$(\bar{1}, \bar{0})$	$(\bar{1}, \bar{0})$	$(\bar{0}, \bar{1})$	$(\bar{1}, \bar{2})$	$(\bar{0}, \bar{0})$	$(\bar{1}, \bar{1})$	$(\bar{0}, \bar{2})$
$(\bar{0}, \bar{1})$	$(\bar{0}, \bar{1})$	$(\bar{1}, \bar{2})$	$(\bar{0}, \bar{0})$	$(\bar{1}, \bar{1})$	$(\bar{0}, \bar{2})$	$(\bar{1}, \bar{0})$
$(\bar{1}, \bar{2})$	$(\bar{1}, \bar{2})$	$(\bar{0}, \bar{0})$	$(\bar{1}, \bar{1})$	$(\bar{0}, \bar{2})$	$(\bar{1}, \bar{0})$	$(\bar{0}, \bar{1})$

$\cdot$	$(\bar{0}, \bar{0})$	$(\bar{1}, \bar{1})$	$(\bar{0}, \bar{2})$	$(\bar{1}, \bar{0})$	$(\bar{0}, \bar{1})$	$(\bar{1}, \bar{2})$
$(\bar{0}, \bar{0})$	$(\bar{0}, \bar{0})$	$(\bar{0}, \bar{0})$	$(\bar{0}, \bar{0})$	$(\bar{0}, \bar{0})$	$(\bar{0}, \bar{0})$	$(\bar{0}, \bar{0})$
$(\bar{1}, \bar{1})$	$(\bar{0}, \bar{0})$	$(\bar{1}, \bar{1})$	$(\bar{0}, \bar{2})$	$(\bar{1}, \bar{0})$	$(\bar{0}, \bar{1})$	$(\bar{1}, \bar{2})$
$(\bar{0}, \bar{2})$	$(\bar{0}, \bar{0})$	$(\bar{0}, \bar{2})$	$(\bar{0}, \bar{1})$	$(\bar{0}, \bar{0})$	$(\bar{0}, \bar{2})$	$(\bar{0}, \bar{1})$
$(\bar{1}, \bar{0})$	$(\bar{0}, \bar{0})$	$(\bar{1}, \bar{0})$	$(\bar{0}, \bar{0})$	$(\bar{1}, \bar{0})$	$(\bar{0}, \bar{0})$	$(\bar{1}, \bar{0})$
$(\bar{0}, \bar{1})$	$(\bar{0}, \bar{0})$	$(\bar{0}, \bar{1})$	$(\bar{0}, \bar{2})$	$(\bar{0}, \bar{0})$	$(\bar{0}, \bar{1})$	$(\bar{0}, \bar{2})$
$(\bar{1}, \bar{2})$	$(\bar{0}, \bar{0})$	$(\bar{1}, \bar{2})$	$(\bar{0}, \bar{1})$	$(\bar{1}, \bar{0})$	$(\bar{0}, \bar{2})$	$(\bar{1}, \bar{1})$

If we examine these tables we see that they are precisely the tables for addition and multiplication in $\mathbf{Z}_2 \oplus \mathbf{Z}_3$. Thus, any equation or other ring-theoretic statement that holds in $\mathbf{Z}_6$ can be translated into a statement that holds in $\mathbf{Z}_2 \oplus \mathbf{Z}_3$. For example,

In $\mathbf{Z}_6$:
$\bar{2} + \bar{3} = \bar{5}$
$\mathbf{Z}_6 = \langle \bar{1} \rangle$
The equation $x^2 = x$ has four solutions in $\mathbf{Z}_6$
$\mathbf{Z}_6$ has two proper ideals

In $\mathbf{Z}_2 \oplus \mathbf{Z}_3$:
$(\bar{0}, \bar{2}) + (\bar{1}, \bar{0}) = (\bar{1}, \bar{2})$
$\mathbf{Z}_2 \oplus \mathbf{Z}_3 = \langle (\bar{1}, \bar{1}) \rangle$
The equation $x^2 = x$ has four solutions in $\mathbf{Z}_2 \oplus \mathbf{Z}_3$
$\mathbf{Z}_2 \oplus \mathbf{Z}_3$ has two proper ideals

Since their operations are essentially the same, Z_6 and $Z_2 \oplus Z_3$ have all the same ring-theoretic properties even though their elements are entirely different.

If we look again at the axioms for a group we see that each axiom is a statement involving the group operation. Thus, purely group-theoretic statements are statements involving only the operation. Since a group is, in fact, a set we frequently make assertions, such as Lagrange's theorem, that are both set theoretic and group theoretic in nature. The axioms for a ring are also all statements involving the operations. All purely ring-theoretic statements are also, then, statements involving the ring operations. As with groups, we also frequently make assertions that are both set theoretic and ring theoretic in nature.

If we have two groups, such as Z_3 and A_3 above, in which we know that group-theoretic statements about one have logically equivalent translations in the other, then everything we know, in a group-theoretic sense, about the first must hold for the second and vice versa. As groups, then, they have all the same properties. Similarly, if we have two rings R and S and we know that every ring-theoretic statement we can make about R has a logically equivalent translation into a statement about S, then we know that R and S have all the same ring-theoretic properties.

To make such translations we must first be able to replace every element of the first structure with an element of the second. Thus, for such translations to exist, there must be a one-to-one correspondence between the two sets. Furthermore, this correspondence must preserve statements about the operations. Rather than discuss correspondences, it is more convenient to discuss one-to-one functions and their inverses.

If G and H are groups and ϕ is a one-to-one function from G onto H, then ϕ is an *isomorphism* (or *group isomorphism*) from G onto H if, for each pair of elements x and y in G,

$$(xy)\phi = (x\phi)(y\phi)$$

If G and H are denoted additively, the equation above becomes

$$(x + y)\phi = (x\phi) + (y\phi)$$

In general, if the operation of G is denoted by $*$ and that of H by $\circ$, the equation becomes

$$(x * y)\phi = (x\phi)\circ(y\phi)$$

If R and S are rings and ϕ is a one-to-one function from R onto S, then ϕ is an *isomorphism* (or *ring isomorphism*) from R onto S if, for each pair of elements x and y in R,

$$(x + y)\phi = x\phi + y\phi \quad \text{and} \quad (xy)\phi = (x\phi)(y\phi)$$

An isomorphism from a group or ring onto itself is called an *automorphism*.

The two definitions above require that a translation process is possible, that is, that each element of the first structure corresponds to exactly one element of the second and that all elements of the second structure are used. They also require the function to preserve operations. Once operations are preserved, all group-theoretic and ring-theoretic statements must be preserved, because all such statements are formulated using only the operations of the structures.

If the structures in question are finite and not too large to handle, one can, as in the examples, use a function to translate the tables for the first structure into new tables and then compare to see whether or not these are the same as the tables for the second structure. If they are, then the operations are preserved and the function is an isomorphism. This is not, however, the general technique for showing that a particular function is an isomorphism. As with most functions, an isomorphism is usually defined by means of some rule, and then we use this definition to show that we do, in fact, have an isomorphism; that is, we prove that applying this rule yields a one-to-one, onto, operation-perserving function.

Examples

(a) The identity function is always an isomorphism from a group or ring onto itself.

(b) Consider the multiplicative group of positive real numbers and the additive group of all real numbers. From calculus we know that

$$\ln xy = \ln x + \ln y$$

and that ln is a one-to-one function from the set of positive real numbers onto the set of all real numbers. Therefore, ln is a group isomorphism. In this case we use the function to translate multiplicative problems into additive ones.

(c) Since $e^{\ln x} = x$ for every positive real number x, the exponential function is the inverse of ln. We know that the exponential function is one-to-one from the set of all real numbers onto the set of positive real numbers. Since $e^{x+y} = e^x e^y$, the exponential function also preserves the group operation and therefore is a group isomorphism.

(d) Define ϕ from S_4 to S_4 by

$$p\phi = (1 \ \ 2 \ \ 3)\, p(1 \ \ 3 \ \ 2) \qquad \text{for each permutation } p \text{ in } S_3$$

From previous exercises we know that ϕ is one-to-one, onto, and operation preserving. We observe that ϕ is operation preserving by noticing that

$$
\begin{aligned}
(pq)\,\phi &= (1 \ \ 2 \ \ 3)\, pq(1 \ \ 3 \ \ 2) \\
&= (1 \ \ 2 \ \ 3)\, piq(1 \ \ 3 \ \ 2) \\
&= (1 \ \ 2 \ \ 3)\, p(1 \ \ 3 \ \ 2)(1 \ \ 2 \ \ 3)\, q(1 \ \ 3 \ \ 2) \\
&= (p\phi)(q\phi)
\end{aligned}
$$

(e) Define λ from $\mathbf{R} \oplus \mathbf{Z}$ to $\mathbf{Z} \oplus \mathbf{R}$ by $(a, b)\lambda = (b, a)$ for each (a, b) in $\mathbf{R} \oplus \mathbf{Z}$. One can easily verify that λ is a ring isomorphism.

(f) Let $\mathcal{C}$ be the group of continuous functions, and let $\mathcal{D}_0$ be the group of all continuously differentiable functions that are 0 at 0. Define I from $\mathcal{C}$ to $\mathcal{D}_0$ by $fI = \int_0^x f(t)$. The function I is a group isomorphism.

(g) Let $\mathbf{R}[x]$ be the ring of all polynomials in x, and let $\mathbf{R}[x^2]$ be the subring of all polynomials in x^2. Define T by

$$(a_0 + a_1 x + \cdots + a_n x^n)\, T = a_0 + a_1 x^2 + a_2 x^4 + \cdots + a_n x^{2n}$$

The function T is a ring isomorphism from $\mathbf{R}[x]$ to $\mathbf{R}[x^2]$.

(h) Define ϕ from $\mathbf{Z}$ to $\mathbf{Z}$ by $x\phi = -x$. Then ϕ is a group isomorphism that is not a ring isomorphism, because $1\phi = -1$ but $(1\phi)(1\phi) = 1$, and therefore $(1 \cdot 1)\,\phi \neq (1\phi)(1\phi)$.

*(i) Let L be the ring of all linear transformations from $\mathbf{R}^2$ into $\mathbf{R}^2$, that is, from two dimensional space to two dimensional space, and let M be the ring of all 2×2 matrices with entries from $\mathbf{R}$. Define ϕ from M to L by letting the image of

$$\begin{bmatrix} a & b \\ c & d \end{bmatrix}$$

under ϕ be the function defined by

$$(x, y)\left(\begin{bmatrix} a & b \\ c & d \end{bmatrix} \phi \right) = (ax + cy,\ bx + dy)$$

Then ϕ is a ring isomorphism.

Since a ring is always a group with respect to addition, a ring isomorphism is always a group isomorphism with respect to the additive structure. Once we have established an assertion for group isomorphisms, we can frequently prove an analogous assertion for ring isomorphisms by extending our result to the multiplicative structure. The next pair of assertions provides an example of such an extension.

Proposition 1

If ϕ is an isomorphism from the group G onto the group H, then ϕ^{-1} is an isomorphism from H onto G.

Proof: We know that ϕ^{-1} is a one-to-one function from H onto G. We need to show that it preserves the group operation.

Let x and y be elements of H. Then $p\phi = x$ and $q\phi = y$ for some elements p and q in G; that is,

$$p = x\phi^{-1} \quad \text{and} \quad q = y\phi^{-1}$$

Then

$$
\begin{aligned}
(xy)\phi^{-1} &= [(p\phi)(q\phi)]\phi^{-1} \\
&= [(pq)\phi]\phi^{-1} \qquad \text{because } \phi \text{ preserves the group operation} \\
&= (pq)(\phi \circ \phi^{-1}) \\
&= (pq)I \\
&= pq \\
&= (x\phi^{-1})(y\phi^{-1}) \qquad \square
\end{aligned}
$$

Corollary 2

If ϕ is an isomorphism from the ring R onto the ring S, then ϕ^{-1} is an isomorphism from S onto R.

Proof: We know, from Proposition 1, that ϕ^{-1} is an isomorphism from the additive structure of S onto the additive structure of R. We need only show that ϕ^{-1} preserves multiplication. The proof is exactly the same as the latter part of the proof of Proposition 1.

Let x and y be elements of S. Then $p\phi = x$ and $q\phi = y$ for some elements p and q in R; that is,

$$
p = x\phi^{-1} \quad \text{and} \quad q = y\phi^{-1}
$$

Then

$$
\begin{aligned}
(xy)\phi^{-1} &= [(p\phi)(q\phi)]\phi^{-1} \\
&= [(pq)\phi]\phi^{-1} \\
&= (pq)(\phi \circ \phi^{-1}) \\
&= (pq)I \\
&= pq \\
&= (x\phi^{-1})(y\phi^{-1}) \qquad \square
\end{aligned}
$$

If G and H are groups (or rings) and there exists an isomorphism from G onto H, we write $G \simeq H$, and say that G is *isomorphic* to H. By Proposition 1 and Corollary 2, if $G \simeq H$, then $H \simeq G$.

If G is a group (or ring) the identity function I_G from G onto G is an isomorphism, and therefore $G \simeq G$. Since $H \simeq G$ whenever $G \simeq H$, it seems reasonable to ask if $G \simeq H$ and $H \simeq K$, must $G \simeq K$? The answer is affirmative and is established by the following assertions.

Proposition 3

If G, H, and K are groups such that $G \simeq H$ and $H \simeq K$, then $G \simeq K$.

Proof: Let ϕ be an isomorphism from G onto H and let λ be an isomorphism from H onto K.
We know that $\phi \circ \lambda$ is a one-to-one function from G onto K.
We need to show that $\phi \circ \lambda$ preserves the operation.
Let x and y be elements of G.

Then

$$(xy)(\phi \circ \lambda) = [(xy)\phi]\lambda \qquad \text{by definition of composition}$$
$$= [(x\phi)(y\phi)]\lambda \qquad \text{because } \phi \text{ is an isomorphism,}$$
$$\phi \text{ preserves the operation}$$
$$= [(x\phi)\lambda][(y\phi)\lambda] \qquad \text{because } \lambda \text{ is an isomorphism}$$
$$= [(x)(\phi \circ \lambda)][(y)(\phi \circ \lambda)] \qquad \text{by definition of composition}$$

Therefore, $\phi \circ \lambda$ preserves the group operation and $\phi \circ \lambda$ is an isomorphism. □

Corollary 4

If R, S, and T are rings such that $R \cong S$ and $S \cong T$, then $R \cong T$.

Proof: Exercise.

From Proposition 1, Proposition 3, and the fact that every group is isomorphic to itself, it follows that if $\mathcal{G}$ is any collection of groups, then isomorphism or the relation of "isomorphic to" is an equivalence relation on $\mathcal{G}$. Similarly, if $\mathcal{R}$ is any collection of rings, then isomorphism is an equivalence relation on $\mathcal{R}$. We can use these facts to prove that two structures are isomorphic by showing that the structures in question are isomorphic to some third structure. An example of this technique is used in proving the following assertion.

Proposition 5

Any two infinite cyclic groups are isomorphic.

Proof: Let G be an infinite cyclic group and let $G = \langle d \rangle$.
Define ϕ from $\mathbf{Z}$ to G by $x\phi = d^x$.
We know that ϕ is onto, since $\langle d \rangle = \{d^n : n \in \mathbf{Z}\}$.
If $x\phi = y\phi$, then $d^x = d^y$, which implies $d^{x-y} = e$ and $d^{y-x} = e$. If $x - y \neq 0$, d is of finite order and $\langle d \rangle$ is finite, which is impossible. Therefore, if $x\phi = y\phi$, $x = y$ and hence ϕ is one-to-one.
Since $(x + y)\phi = d^{x+y} = d^x d^y = (x\phi)(y\phi)$, ϕ is operation preserving.
Thus, $\mathbf{Z} \cong G$.
Since every infinite cyclic group is isomorphic to $\mathbf{Z}$, any two infinite cyclic groups are isomorphic. □

EXERCISE SET A

1. Let $\mathbf{Z}_5^*$ be the multiplicative group of nonzero elements of $\mathbf{Z}_5$, and define ϕ from $\mathbf{Z}_5^*$ to $\mathbf{Z}_4$ by

   $$\bar{1}\phi = \bar{0}, \quad \bar{2}\phi = \bar{1}, \quad \bar{3}\phi = \bar{3}, \quad \text{and} \quad \bar{4}\phi = \bar{2}$$

   Prove that ϕ is an isomorphism by constructing the multiplication table for $\mathbf{Z}_5^*$, translating it, and comparing it to the addition table for $\mathbf{Z}_4$.

2. Let Z_7^* be the group of nonzero elements of Z_7. Prove that $Z_7^* \cong Z_6$.

3. Let $G = \{i, (1\ 2), (3\ 4), (1\ 2)(3\ 4)\}$. Prove that $G \cong Z_2 \oplus Z_2$.

4. Prove that the group of symmetries of a triangle is isomorphic to S_3.

5. Prove that $\langle (1\ 2\ 3\ 4) \rangle \cong Z_4$.

6. Define ϕ from Z_{12} to $Z_3 \oplus Z_4$ by $\bar{0}\phi = (\bar{0}, \bar{0})$, $\bar{1}\phi = (\bar{1}, \bar{1})$, $\bar{2}\phi = (\bar{2}, \bar{2})$, $\bar{3}\phi = (\bar{0}, \bar{3})$, $\bar{4}\phi = (\bar{1}, \bar{0})$, $\bar{5}\phi = (\bar{2}, \bar{1})$, $\bar{6}\phi = (\bar{0}, \bar{2})$, $\bar{7}\phi = (\bar{1}, \bar{3})$, $\bar{8}\phi = (\bar{2}, \bar{0})$, $\bar{9}\phi = (\bar{0}, \bar{1})$, $\overline{10}\phi = (\bar{1}, \bar{2})$, $\overline{11}\phi = (\bar{2}, \bar{3})$. Prove that ϕ is an isomorphism.

7. Prove that $Z_{10} \cong Z_2 \oplus Z_5$ (as rings).

8. Let R^* denote the multiplicative group of all nonzero real numbers, let R^+ denote the multiplicative group of all positive real numbers, and let $\{1, -1\}$ be a group under the usual multiplication. Define ϕ from $R^+ \oplus \{1, -1\}$ to R^* by $(a, b)\phi = ab$. Prove that ϕ is an isomorphism.

9. Let R^*, R^+, and $\{1, -1\}$ be as in Exercise 8. The group $R^*/\{1, -1\}$ consists of right cosets of the form $\{1, -1\}y$ or $\{y, -y\}$, where y is a nonzero real number. Define λ from $R^*/\{1, -1\}$ to R^+ by $\{a, -a\}\lambda = |a|$. Prove first that this rule does define a function from $R^*/\{1, -1\}$ to R^+, and then prove that λ is an isomorphism.

10. (a) Let A and B be groups. Prove that $A \oplus B \cong B \oplus A$.
    (b) State and prove an analogous assertion for rings.

11. Let $T = \{(x, x):x \in Q\}$.
    (a) Prove that T is a subring of $Q \oplus Q$.
    (b) Prove that $T \cong Q$.

*12. Let M be the set of all 2×2 matrices that are of the form

$$\begin{bmatrix} x & 0 \\ 0 & x \end{bmatrix}$$

where x is a rational number. Prove that M is a ring and $M \cong Q$.

13. Prove Corollary 4.

14. Using only the definition of isomorphism prove: If ϕ is an isomorphism from the group (ring) G onto the group (ring) H and G is commutative, then H is commutative.

15. Prove: If ϕ is an isomorphism from the group G onto the group H and e_G denotes the identity of G and e_H denotes the identity of H, then $e_G\phi = e_H$. (Hint: Show that $(e_G\phi)(e_G\phi) = e_G\phi$ and apply Proposition 1 from Chapter 4.)
    (b) State and prove an analogous assertation for rings.

16. Let G be a group (or ring). By $A(G)$ we mean the set of all automorphisms of G. Prove that $A(G)$ is a group with respect to the operation of composition of functions. (Hint: $A(G) \subseteq P(G)$.)

17. Prove: If ϕ is an isomorphism from the group G onto the group H and b is an element of G, then $(b\phi)^{-1} = (b^{-1})\phi$.

18. Prove: If ϕ is an isomorphism from the group G onto the group H and b is an element of G of order m, then $b\phi$ is of order m in H.

19. As groups, is $Z_2 \oplus Z_2 \cong Z_4$? Is $Z_2 \oplus Z_4 \cong Z_8$? Justify your answers.

20. Prove that any two groups of order 2 are isomorphic. (Hint: It suffices to prove that any group of order 2 is isomorphic to $\mathbf{Z}_2$.)
21. Prove: If ϕ is an isomorphism from the ring R onto the ring S and b is a proper zero divisor in R, then $b\phi$ is a proper zero divisor in S.
22. Prove: If R is a ring with 1 and ϕ is an isomorphism from R onto the ring S, then 1ϕ is a multiplicative identity in S.
23. Prove: If F is a field, R is a ring, and ϕ is an isomorphism from F onto R, then R is a field.
24. Prove that if F is a field with exactly two elements, then $F \simeq \mathbf{Z}_2$.
**25. Prove that if F is a field with exactly three elements, then $F \simeq \mathbf{Z}_3$.

§2. Some Special Isomorphisms

In Exercise Set A, Exercise 20, the student was asked to prove that any two groups of order 2 are isomorphic. The simplest way to do so is to show that any such group is isomorphic to $\mathbf{Z}_2$. We shall now show that any two groups of order 3 are isomorphic.

Since a group of order 3 cannot contain an element of order 2 and can contain only one element of order 1 (the element e), a group of order 3 must contain an element of order 3. Thus, if G is a group of order 3, G is cyclic and $G = \langle b \rangle$ for some element b in G. Hence, $G = \{e, b, b^2\}$. Define ϕ from G to $\mathbf{Z}_3$ by $e\phi = \overline{0}$, $b\phi = \overline{1}$, and $b^2\phi = \overline{2}$.

The table for G is

	e	b	b^2
e	e	b	b^2
b	b	b^2	e
b^2	b^2	e	b

The table resulting from ϕ is

	$\overline{0}$	$\overline{1}$	$\overline{2}$
$\overline{0}$	$\overline{0}$	$\overline{1}$	$\overline{2}$
$\overline{1}$	$\overline{1}$	$\overline{2}$	$\overline{0}$
$\overline{2}$	$\overline{2}$	$\overline{0}$	$\overline{1}$

We know that the table on the right is the table for $\mathbf{Z}_3$, and therefore ϕ preserves the operation. Thus, ϕ is an isomorphism.

In the exercises we saw that $\mathbf{Z}_2 \oplus \mathbf{Z}_2$ was not isomorphic to $\mathbf{Z}_4$. The tables for these groups follow.

	$\overline{0}$	$\overline{1}$	$\overline{2}$	$\overline{3}$
$\overline{0}$	$\overline{0}$	$\overline{1}$	$\overline{2}$	$\overline{3}$
$\overline{1}$	$\overline{1}$	$\overline{2}$	$\overline{3}$	$\overline{0}$
$\overline{2}$	$\overline{2}$	$\overline{3}$	$\overline{0}$	$\overline{1}$
$\overline{3}$	$\overline{3}$	$\overline{0}$	$\overline{1}$	$\overline{2}$

	$(\bar{0}, \bar{0})$	$(\bar{1}, \bar{0})$	$(\bar{0}, \bar{1})$	$(\bar{1}, \bar{1})$
$(\bar{0}, \bar{0})$	$(\bar{0}, \bar{0})$	$(\bar{1}, \bar{0})$	$(\bar{0}, \bar{1})$	$(\bar{1}, \bar{1})$
$(\bar{1}, \bar{0})$	$(\bar{1}, \bar{0})$	$(\bar{0}, \bar{0})$	$(\bar{1}, \bar{1})$	$(\bar{0}, \bar{1})$
$(\bar{0}, \bar{1})$	$(\bar{0}, \bar{1})$	$(\bar{1}, \bar{1})$	$(\bar{0}, \bar{0})$	$(\bar{1}, \bar{0})$
$(\bar{1}, \bar{1})$	$(\bar{1}, \bar{1})$	$(\bar{0}, \bar{1})$	$(\bar{1}, \bar{0})$	$(\bar{0}, \bar{0})$

The group Z_4 contains elements of order 4, whereas every element of $Z_2 \oplus Z_2$ is of order 2. What are the possibilities for a group with four elements? If G is a group with four elements, then each nontrivial element is of order 2 or order 4 (by the corollary to Lagrange's theorem). If G has an element of order 4, then $G = \{e, b, b^2, b^3\}$, and it is easy to verify that $G \cong Z_4$ under the function $e\phi = \bar{0}$, $b\phi = \bar{1}$, $b^2\phi = \bar{2}$, and $b^3\phi = \bar{3}$. We leave the verification to the reader. Suppose every element of G is of order 2, and let G be denoted by $G = \{e, a, b, c\}$. If we start to construct the table, then the fact that every element is of order 2 requires that we begin with the following:

	e	a	b	c
e	e	a	b	c
a	a	e		
b	b		e	
c	c			e

Now, what are the possibilities for ab? Since a and e already appear in the row following a and the element b already appears in the column under b, the only possibility for ab is c and then the only possibility for ac is b. By similar reasoning we must have $ba = c$ and $ca = b$.

	e	a	b	c
e	e	a	b	c
a	a	e	c	b
b	b	c	e	
c	c	b		e

Now, the only possibility for bc is a and the only possibility for cb is also a. Thus, we must have the table

	e	a	b	c
e	e	a	b	c
a	a	e	c	b
b	b	c	e	a
c	c	b	a	e

If we define ϕ from $\mathbf{Z}_2 \oplus \mathbf{Z}_2$ to G by

$$(\bar{0}, \bar{0})\phi = e, \quad (\bar{1}, \bar{0})\phi = a, \quad (\bar{0}, \bar{1})\phi = b, \quad \text{and} \quad (\bar{1}, \bar{1})\phi = c$$

it is not hard to show that ϕ is an isomorphism. We leave the verification to the reader.

Realize that, in this case, after deciding that there was no element of order 4, we had no choices left. The entire table was determined once we knew that every element was of order 2.

Thus far in the exercises and discussions we have shown that

(a) Any two groups of order 2 are isomorphic.

(b) Any two groups of order 3 are isomorphic.

(c) A group of order 4 is either isomorphic to $\mathbf{Z}_4$ or $\mathbf{Z}_2 \oplus \mathbf{Z}_2$. (A group that is isomorphic to $\mathbf{Z}_2 \oplus \mathbf{Z}_2$ is called a Klein 4-group.)

If G is a group of order 5, then the order of each nontrivial element of G must be 5 and if b is any nontrivial element of G, then $G = \{e, b, b^2, b^3, b^4\}$. It is not hard to show that $G \simeq \mathbf{Z}_5$. We leave this verification as an exercise.

(d) Any two groups of order 5 are isomorphic.

We are familiar with two groups of order 6, S_3 and $\mathbf{Z}_6$. We know that they are not isomorphic, because $\mathbf{Z}_6$ is abelian and S_3 is not. Now let G be a group of order 6. If G has an element q of order 6, then $G = \{e, q, q^2, q^3, q^4, q^5\}$ and it is easy to show that $G \simeq \mathbf{Z}_6$. Now let us assume that G has no element of order 6 and see what the table must look like. Since the order of an element must divide the order of the group, every nontrivial element must be of order 2 or order 3. Is it possible for every element to be of order 2? Let a and b be distinct elements of G. If every element is of order 2, part of the table must be of the form

	e	a	b
e	e	a	b
a	a	e	
b	b		e

Since $ab \neq a$ and $ab \neq b$ and $ab \neq e$, we must have another element of G for the product of a and b. Let us just denote this element by ab. Our table now takes the form

	e	a	b	ab		
e	e	a	b	ab		
a	a	e	ab	b		
b	b	ab	e	a		
ab	ab	b	a	e		
					e	
						e

We know that G is commutative, because every element is of order 2 (see Exercise 8, Exercise Set E, Chapter 4). But now $\{e, a, b, ab\}$ is a subgroup of order 4 in a group of order 6, and this is impossible by Lagrange's theorem. Thus, in a group of order 6 not every element can be of order 2, and therefore, if G is not isomorphic to $\mathbf{Z}_6$, it must have an element of order 3. Thus, part of our table must have the form below, where a is an element of order 3.

	e	a	a^2
e	e	a	a^2
a	a	a^2	e
a^2	a^2	e	a

Now let b be another element of G. Since $ab \neq a$, $ab \neq a^2$, $ab \neq e$, and $ab \neq b$, ab must be another element of G. Furthermore, by similar reasoning a^2b is still another element of G. Thus, part of the table must take the form below.

	e	a	a^2	b	ab	a^2b
e	e	a	a^2	b	ab	a^2b
a	a	a^2	e	ab	a^2b	b
a^2	a^2	e	a	a^2b	b	ab
b	b					
ab	ab					
a^2b	a^2b					

Now, what are the possibilities for b^2? If $b^2 = a$, then b is of order 6, because $b^2 = a \neq e$ and $b^3 = ab \neq e$. Our assumption was that G had no elements of order 6, and therefore $b^2 \neq a$. Similarly, if $b^2 = a^2$, b will again be of order 6. Since $b^2 \neq ab$, $b^2 \neq b$, and $b^2 \neq a^2b$, b^2 must be e. Now, what are the possibilities for ba? If $ba = ab$, then G is commutative. But if $ba = ab$, then $(ab)^2 = a^2b^2 = a^2e = a^2$ and $(ab)^3 = (ab)^2ab = a^2ab = b$. Thus, if $ab = ba$, ab is not of order 2 or 3 and hence must be of order 6. This contradicts our assumption about G, and therefore $ab \neq ba$. What are the possibilities for ba? We know $ba \neq e$, $ba \neq a$, $ba \neq a^2$, $ba \neq b$, and $ba \neq ab$, and therefore the only possibility for ba is a^2b. If we let $ba = a^2b$, the table is completely determined. Since $ba = a^2b$,

$$ba^2 = (ba)a = (a^2b)a = a^2(ba) = a^2(a^2b) = ab$$
$$b(ab) = (ba)b = (a^2b)b = a^2b^2 = a^2$$
$$b(a^2b) = (ba^2)b = (ab)b = ab^2 = a$$

Thus, we complete the fourth row of the table,

	e	a	a^2	b	ab	a^2b
e	e	a	a^2	b	ab	a^2b
a	a	a^2	e	ab	a^2b	b
a^2	a^2	e	a	a^2b	b	ab
b	b	a^2b	ab	e	a^2	a
ab	ab					
a^2b	a^2b					

By associativity, the fifth row is obtained by multiplying each entry in the fourth row on the left by a. Similarly, the sixth row is obtained by multiplying each entry in the fifth row on the left by a. Thus, we have the following table:

	e	a	a^2	b	ab	a^2b
e	e	a	a^2	b	ab	a^2b
a	a	a^2	e	ab	a^2b	b
a^2	a^2	e	a	a^2b	b	ab
b	b	a^2b	ab	e	a^2	a
ab	ab	b	a^2b	a	e	a^2
a^2b	a^2b	ab	b	a^2	a	e

The table was completely determined once we decided that there were no elements of order 6 in the group. If we examine the table for S_3 we see that there are no elements of order 6 in S_3 and thus expect that $G \simeq S_3$. We leave as an exercise the finding of such an isomorphism.

A group of order 7 can easily be shown to be isomorphic to Z_7. There are five types of groups of order 8, and these are Z_8, $Z_2 \oplus Z_4$, $Z_2 \oplus Z_2 \oplus Z_2$, the group of symmetries of a square, and another group known as the quaternian group. The proof of this last assertion is long and drawn out and not of sufficient importance to discuss here; the interested reader should consult one of the reference texts listed for group theory. It can also be shown that a group of order 9 is isomorphic either to Z_9 or $Z_3 \oplus Z_3$.

Determining all of the nonisomorphic groups of some particular order is a common type of problem in finite group theory. Many of the results are concerned with the prime factorization of the order of the group. For example, if p is a prime and G is a group of order p^2, it can be shown that either $G \cong Z_{p^2}$ or $G \cong Z_p \oplus Z_p$. The treatment of such topics is beyond the scope of this text and is, in fact, usually studied only at the graduate level in mathematics.

In a sense, finite rings are not as easy to describe as finite groups. The fact that we have two operations leads us to expect that more variance is possible, and we see this if we consider the case of a ring with two elements: Z_2 is a ring with two elements and the zero ring $\{0, a\}$ with $a + a = 0$ and $a \cdot a = 0$ is a ring with two elements that is not isomorphic to Z_2. If R is any ring with two elements, however, R is isomorphic to one of the above, because if $R = \{0, b\}$, as groups, R is isomorphic to Z_2 and the only possibilities for a multiplication table for R are below.

	0	b
0	0	0
b	0	b

	0	b
0	0	0
b	0	0

In the first case $R \cong Z_2$, and in the second case R is a zero ring with two elements.

If we consider the set $\{0, a, b\}$ with the addition below

+	0	a	b
0	0	a	b
a	a	b	0
b	b	0	a

and zero multiplication, we have a zero ring with three elements, and it is not hard to show that every zero ring with three elements is isomorphic to this ring. We know that Z_3 is not a zero ring and we now have two nonisomorphic rings of order 3. Are there any other possibilities? Let R

be a ring with elements 0, x, and y. As an additive group $R \simeq Z_3$, and therefore the addition table for R must be

	0	x	y
0	0	x	y
x	x	y	0
y	y	0	x

What are the possibilities for the multiplication?

The table must start out as shown below.

	0	x	y
0	0	0	0
x	0		
y	0		

What are the possibilities for x^2? If $x^2 = 0$, then $(y + y)x = xx = 0$, and therefore

$$yx + yx = 0$$

By the addition table we see that yx must be 0. Similarly, $x(y + y) = xx = 0$, which implies $xy + xy = 0$, and therefore $xy = 0$. Since $y = x + x$, $yy = (x + x)y = xy + xy = 0 + 0$, and therefore $y^2 = 0$. Thus, if $x^2 = 0$ (and similarly, if $y^2 = 0$), the ring is a zero ring. Similarly, if $xy = 0$, $xy + xy = 0$, $x(y + y) = 0$ and $xx = 0$. Thus, if R has any proper zero divisors, R is a zero ring. If R is not a zero ring, then R has no zero divisors, and therefore either $xy = y$ or $xy = x$. If $xy = y$, then $xx = x$, because otherwise

$$xx = y \quad \text{and} \quad x(y - x) = 0$$

which is impossible. If $xy = y$, then $yx = y$, because otherwise $yx = x$ and $yx - xx = 0$, which implies $(y - x)x = 0$, which is impossible. Furthermore, if $xy = y$, $yy = x$, because otherwise $(x - y)y = xy - yy = y - y = 0$, which is impossible. Thus, if $xy = y$, it follows that the table must take the form shown below.

	0	x	y
0	0	0	0
x	0	x	y
y	0	y	x

This is easily seen to be the form of the multiplication table for Z_3. In this case x is the multiplicative identity. If, on the other hand, in R we have $xy = x$, y will be the identity and $xx = y$. In either case $R \cong Z_3$.

Thus, we have seen that a ring with three elements is either isomorphic to Z_3 or to a zero ring with three elements.

A ring with five elements is either isomorphic to Z_5 or to a zero ring with five elements having the same additive structure as Z_5.

Rings with four or six elements are more complicated than those with a prime number of elements.

To prove most theorems concerning isomorphisms we need more tools than just the definitions. Some of these tools will be developed in the next section and will be applied in the latter part of the chapter.

EXERCISE SET B

1. Show that any group of order 5 is isomorphic to Z_5.
2. Show that any group of order 7 is isomorphic to Z_7.
3. Prove that S_3 is isomorphic to the nonabelian group G of order 6 that was constructed in the text.
4. Let H be a group of order 8 containing elements x and y such that $x^4 = e$, $y^2 = x^2$, $yx = x^3 y$. Construct the table for H.

**5. Show that any group of order 9 is isomorphic to Z_9 or to $Z_3 \oplus Z_3$.

**6. Show that any abelian group of order 8 is isomorphic to Z_8, $Z_2 \oplus Z_4$, or $Z_2 \oplus Z_2 \oplus Z_2$.

7. Prove that a ring with five elements is either a zero ring or is isomorphic to Z_5.
8. State and prove an analogous assertion for a ring with seven elements.

**9. Prove that if G is a group of order p and p is a prime, then $G \cong Z_p$.

**10. Prove that if R is a ring with p elements and p is prime, then either R is a zero ring or $R \cong Z_p$.

§3. Homomorphisms

In studying isomorphisms we were concerned with one-to-one operation-preserving functions from one group or ring onto another. The one-to-one and onto conditions guarantee that the translation process goes both ways. Since the inverse of an isomorphism is an isomorphism, if G and H are isomorphic groups (or rings) any group-theoretic (or ring-theoretic) statement that holds in one of the structures must hold in the other. Suppose now that G and H are groups and that ϕ is a function from G into H such that $(xy)\phi = (x\phi)(y\phi)$ for every pair of elements in G. In this case we no longer are assured of a complete translation process and, in fact, all we can be sure of is that every equation in G will give rise to an equation in H, because the operation is preserved. Still, however, the fact that the operation is preserved is very useful. For example, consider the group $\mathfrak{D}$ of all differentiable functions from $\mathbf{R}$ into $\mathbf{R}$ and the group $\mathfrak{F}$ of all functions from $\mathbf{R}$ into $\mathbf{R}$. The derivative operator "D" preserves the addition operation, because $(f + g)' = f' + g'$ for any two differentiable functions f and g. This property of the derivative operator enables us to differentiate sums by differentiating "term by term." Clearly this operator (or function) is not one-to-one because, for example,

$$(x^2 + 3)' = (x^2)'$$

Neither is it onto because there are functions from $\mathbf{R}$ into $\mathbf{R}$ that are not the derivative of any function.

Functions from one algebraic structure to another that preserve operations are generally called *homomorphisms*. We make the following two formal definitions.

If G and H are groups and ϕ is a function from G into H, then ϕ is a *homomorphism* (or *group homomorphism*) from G into H if for each pair of elements x and y in G,

$$(xy)\phi = (x\phi)(y\phi).$$

If the groups G and H were denoted additively, the equation would have become

$$(x + y)\phi = x\phi + y\phi$$

In general, if the operation of G were denoted by $*$ and that of H by $\circ$, the equation would become

$$(x * y)\phi = (x\phi) \circ (y\phi)$$

If R and S are rings and ϕ is a function from R into S, then ϕ is a *homomorphism* (or *ring homomorphism*) from R into S if, for each pair of elements x and y in R,

$$(x + y)\phi = x\phi + y\phi \quad \text{and} \quad (xy)\phi = (x\phi)(y\phi)$$

A homomorphism from a group or ring into itself is called an *endomorphism*.

The above definitions are similar to the definitions of isomorphism; only the one-to-one and onto restrictions have been removed. A homomorphism that is one-to-one is sometimes called an *injection*, an *embedding*, a *monomorphism*, or an "*isomorphism into*." A homomorphism that is onto is sometimes called a *surjection* or an *epimorphism*. Notice that to prove that a homomorphism is an isomorphism we need to show that it is one-to-one and onto.

The following are examples of homomorphisms:

(a) Let $\mathbf{R}^+$ denote the multiplicative group of positive real numbers and let $\mathbf{R}^*$ denote the multiplicative group of all nonzero real numbers.

Define A from $\mathbf{R}^*$ to $\mathbf{R}^+$ by $xA = |x|$, the absolute value of x. Since the absolute value of a product is the product of the absolute values, A is a homomorphism.

(b) Define a function T from $\mathbf{Z}$ into $\mathbf{Z}$ by

$$xT = 3x$$
$$(x + y)T = 3(x + y) = 3x + 3y = xT + yT$$

Thus, T is a group homomorphism, is one-to-one, but is not onto. Notice that T is not a ring homomorphism because, for example, $(2 \cdot 4)3 \neq (2 \cdot 3)(4 \cdot 3)$.

(c) Consider the group $\mathbf{R}[x]$ of all polynomials over the real numbers with the operation of addition. The derivative operator D is a homomorphism from $\mathbf{R}[x]$ into $\mathbf{R}[x]$, since the derivative of a sum is the sum of the derivatives. D is not one-to-one but it is onto.

(d) In the same group $\mathbf{R}[x]$, D^2, the second derivative operator, is also a homomorphism.

(e) With the same group as above, consider the operator I_0 defined by

$$fI_0 = \int_0^x f.$$ Thus, fI_0 is the function whose derivative is f and that

takes the value of 0 at 0. For example, $x^3 I_0 = x^4/4$.

$$(f + g)I_0 = \int_0^x (f + g) = \int_0^x f + \int_0^x g = fI_0 + gI_0$$

Thus, I_0 is a group homomorphism. I_0 is one-to-one but not onto.

(f) Let $\mathcal{C}$ be all the continuous functions from $[0, 1]$ into $\mathbf{R}$. Define I from

$\mathcal{C}$ into $\mathbf{R}$ by $fI = \int_0^1 f.$ From calculus we know that I is a homo-

morphism (the integral of the sum is the sum of the integrals). Although $\mathbf{R}[x]$ and $\mathcal{C}$ are also rings, the examples (c), (d), (e), and (f) of homomorphisms are *not* ring homomorphisms.

(g) If we define α from $\mathbf{Z}$ to $\mathbf{Z}_7$ by $x\alpha = \bar{x}$, then α is a ring homomorphism.

(h) If we define β from $\mathbf{R}[x]$ to $\mathbf{R}$ by $f\beta = f(2)$, then β is a ring homomorphism.

(i) Define γ from $\mathbf{Q}[x]$ to $\mathbf{R}$ by $f\gamma = f(\pi)$. ($\mathbf{Q}[x]$ denotes all polynomials over the rational numbers.) Then γ is a ring homomorphism, is one-to-one, but is not onto.

(j) Let $\mathcal{C}$ be the ring of all continuous functions from $\mathbf{R}$ into $\mathbf{R}$. Define δ from $\mathcal{C}$ into $\mathbf{R}$ by $f\delta = f(\pi)$. Then δ is an onto ring homomorphism but δ is not one-to-one, because $(x - \pi)\delta = 0$ and $[(x - 2)(x - \pi)]\delta = 0$.

(k) Let G be a group, and let H be a normal subgroup. Define η from G to G/H by $x\eta = \bar{x}$. The function η is called the *natural function* from G to G/H and is a group homomorphism.

(l) If R is a ring and A is an ideal, the natural function η from R to R/A is a ring homomorphism.

The reader should verify for himself that each of the above functions is, in fact, a homomorphism.

Many of the assertions we proved for isomorphisms have analogues that hold for homomorphisms. We list two of these below and leave it to the reader to adapt the proofs.

Proposition 6

If G, H, and K are groups, ϕ is a homomorphism from G into H, and λ is a homomorphism from H into K, then $\phi \circ \lambda$ is a homomorphism from G into K.

Proof: Exercise.

Corollary 7

If R, S, and T are rings, ϕ is a homomorphism from R into S, and λ is a homomorphism from S into T, then $\phi \circ \lambda$ is a homomorphism from R into T.

Proof: Exercise.

In Exercise Set B the student was asked to show that, under an isomorphism, the image of the identity element in a group is the identity element in the image group; we now prove a similar result for homomorphisms.

Proposition 8

If ϕ is a homomorphism from the group G into the group H, then $e_G\phi = e_H$, where e_G denotes the identity of G and e_H denotes the identity of H.

Proof

$$e_G\phi = (e_G e_G)\phi = (e_G\phi)(e_G\phi)$$

Therefore, $e_G\phi = e_H$ (by Proposition 1 in Chapter 4; in any group, if $tt = t$, then $t = e$.) □

From now on we shall not bother to indicate which identity element we are discussing, since the context will indicate which group is under discussion. The above assertion could be restated as

$$e\phi = e$$

where it is understood that the e on the left refers to the identity of G, while the e on the right refers to the identity of H.

Corollary 9

If ϕ is a homomorphism from the ring R into the ring S, then

$$0\phi = 0$$

Proof: The function ϕ is a group homomorphism with respect to the additive structure of R and therefore, by Proposition 8, $0\phi = 0$. □

Since homomorphisms preserve identity elements and operations in general, we might expect that inverses also are preserved. This is established by the next two assertions.

Proposition 10

If ϕ is a homomorphism from the group G into the group H and a is an element of G, then $(a^{-1})\phi = (a\phi)^{-1}$.

Proof

$$(a\phi)(a^{-1}\phi) = (aa^{-1})\phi \qquad \text{because } \phi \text{ is a homomorphism}$$
$$= e\phi$$
$$= e$$

Therefore, $(a\phi)^{-1} = a^{-1}\phi$. □

In additive notation the above assertion becomes

$$(-a)\phi = -(a\phi)$$

The next corollary then follows immediately.

Corollary 11

If ϕ is a homomorphism from the ring R into the ring S and a is an element of R, then $(-a)\phi = -(a\phi)$.

Since products are preserved by a group homomorphism, we expect powers of an element also to be preserved. This is, in fact the case, and we leave the proof as an exercise.

Proposition 12

If ϕ is a homomorphism from the group G into the group H, b is an element of G, and m is an integer, then

$$(b^m)\phi = (b\phi)^m$$

Proof: Exercise. (Hints: Use induction, Proposition 8, and Proposition 10.)

Since homomorphisms preserve operations, equations and many other group- and ring-theoretic statements are transformed into new statements in the image structure. Not all such statements are carried over, however. For example, the inequality $a \neq b$ would not carry over in the case where $a\phi = b\phi$. Many of the statements we can make about the domain of a homomorphism will hold in the range. We establish the following two assertions as illustrations of such statements and leave some others as exercises.

Proposition 13

If ϕ is a homomorphism from the group G into the group H, b is an element of G, and m is a positive integer such that $b^m = e$, then $(b\phi)^m = e$.

Proof: By Proposition 12

$$(b^m)\phi = (b\phi)^m \qquad \text{for any integer } m$$

Then

$$e\phi = (b\phi)^m$$

and therefore,

$$e = (b\phi)^m \quad \square$$

Proposition 14

If ϕ is a homomorphism of the group G onto the group H and G is commutative, then H is commutative.

Proof: Let u and v be elements of H.
Then, because ϕ is onto, there are elements a and b in G such that $a\phi = u$ and $b\phi = v$.
Then

$$(ab)\phi = (a\phi)(b\phi) = uv$$

and

$$(ba)\phi = (b\phi)(a\phi) = vu$$

Since $ab = ba$, $uv = vu$ and H is commutative. $\quad \square$

EXERCISE SET C

1. Let $\mathbf{Q}^*$ denote the multiplicative group of all nonzero rational numbers. Define f from $\mathbf{Q}^*$ to $\mathbf{Q}^*$ by $xf = x^2$. Prove that f is a homomorphism. Is f one-to-one? Is f onto?

2. Let $\mathbf{R}^*$ denote the multiplicative group of all nonzero real numbers. Define g from $\mathbf{R}^*$ into $\mathbf{R}^*$ by $xg = x^3$. Is g a homomorphism? Proof? Is g one-to-one? Is g onto?

3. Define the function q from the additive group of real numbers into itself by $xq = 2x + 3$. Is q a homomorphism? Is q one-to-one? Is q onto?

4. Define the function p from the additive group of rational numbers into the additive group of real numbers by $xp = x\pi$. Is p a homomorphism? Is p one-to-one? Is p onto? Justify your answers.

5. Let G and H be groups. Define f from G into H by $xf = e$ for every element x in G. Prove that f is a homomorphism. State and prove an analogous assertion for rings.

6. Define ϕ from S_5 to $\mathbf{Z}_2$ by $p\phi = \bar{0}$ if p is even and $p\phi = \bar{1}$ if p is odd.

Prove that ϕ is a homomorphism. What is $\bar{0}\phi^{-1}$? Is $\bar{0}\phi^{-1}$ a normal subgroup of S_5?

7. Let T be defined from $\mathbf{R} \oplus \mathbf{R}$ to $\mathbf{R}$ by $(x, y)T = 2x + 3y$. Prove that T is a group homomorphism. Can you describe $\{(x, y):(x, y)T = 0\}$? Is this set a subgroup?

8. Let $\mathfrak{D}$ be the group of differentiable functions from $\mathbf{R}$ into $\mathbf{R}$. Define γ from $\mathfrak{D}$ into $\mathbf{R}$ by $f\gamma = f'(5)$ for each function f in $\mathfrak{D}$. Prove that γ is a homomorphism. What is $0\gamma^{-1}$? Show that $0\gamma^{-1}$ is a subgroup of $\mathfrak{D}$.

9. Let $\mathfrak{C}$ be the ring of all continuous functions from $\mathbf{R}$ into $\mathbf{R}$. Define λ from $\mathfrak{C}$ to the ring $\mathbf{R}$ by $g\lambda = g(3)$ for each function g in $\mathfrak{C}$. Prove that λ is a ring homomorphism. Describe $\{g \in \mathfrak{C} : g\lambda = 0\}$. Is this set an ideal in $\mathfrak{C}$?

10. Prove Proposition 6.

11. Prove Corollary 7.

12. Prove Proposition 12.

13. Prove: If G and H are abelian groups denoted additively and ϕ is a homomorphism from G into H, then $(a - b)\phi = a\phi - b\phi$ for each pair of elements a and b in G.

14. If R is a ring with 1 and λ is a homomorphism from the ring R *onto* the ring S, then 1λ is a multiplicative identity in S.

15. Define ϕ from $\mathbf{Z}$ into $\mathbf{Z} \oplus \mathbf{Z}$ by $a\phi = (a, 0)$. Prove that ϕ is a homomorphism. Show that 1ϕ is not an identity in $\mathbf{Z} \oplus \mathbf{Z}$.

16. Prove: If λ is a homomorphism from the ring R *onto* the ring S and R is commutative, then S is commutative.

17. (a) Let ϕ be a homomorphism from the group G into the group H, and let K be a subgroup of H. Prove that $K\phi^{-1}$ is a subgroup of G.

(b) State and prove an analogous assertion for rings.

§4. Cayley's Theorem

In each of the examples given of a homomorphism, even though the homomorphism was not onto, the image of the domain was either a subgroup or a subring. That this is, in general, the case is established by the following two propositions.

Proposition 15

If ϕ is a homomorphism from the group G into the group H, then $G\phi$, the image of G, is a subgroup of H.

Proof: $G\phi \neq \emptyset$ because $G \neq \emptyset$. Let x and y be elements of $G\phi$, the image of G in H.

Then for some elements a and b in G, $a\phi = x$ and $b\phi = y$.

Since $ab^{-1} \in G, (ab^{-1})\phi \in G\phi$.

But $(ab^{-1})\phi = (a\phi)(b^{-1}\phi) = (a\phi)(b\phi)^{-1} = xy^{-1}$.

Since $xy^{-1} \in G\phi$, $G\phi$ is a subgroup of H by Theorem 16, Chapter 4. $\square$

Corollary 16

If ϕ is a homomorphism from the ring R into the ring S, then $R\phi$, the image of R, is a subring of S.

Proof: From Proposition 15 we know $R\phi$ is a subgroup of S with respect to addition. If $x, y \in R\phi$, then for some elements a and b in R, $a\phi = x$ and $b\phi = y$. Since $ab \in R$, $(ab)\phi \in R\phi$. But $(ab)\phi = (a\phi)(b\phi) = xy$, and therefore $xy \in R\phi$. Since $R\phi$ is a subgroup with respect to addition and closed under multiplication, $R\phi$ is a subring of S. $\square$

We shall make use of Proposition 15 in the proof of Cayley's theorem. Before attempting the proof, however, let us look at a special case. (Recall Exercise 16, Set D, Chapter 4.)

For each integer z, define the function f_z from $\mathbf{Z}$ to $\mathbf{Z}$ by

$$xf_z = x + z \qquad \text{for each integer } x$$

For example, $xf_2 = x + 2$ for each integer x and
$xf_{-4} = x + (-4)$ for each integer x.
Clearly, each function f_z is a permutation of $\mathbf{Z}$.
Also,

$$(x)(f_n \circ f_m) = (xf_n)f_m = (x + n)f_m = (x + n) + m = x + (n + m) = xf_{n+m}.$$

Furthermore, $(x)(f_n \circ f_{-n}) = xf_{n+(-n)} = xf_0 = x + 0 = x = xI$,
and therefore $f_n \circ f_{-n} = I$ and $f_n^{-1} = f_{-n}$. If we let $W = \{f_z : z \in \mathbf{Z}\}$ the above arguments show that W is a subgroub of $P(\mathbf{Z})$, the group of all permutations of $\mathbf{Z}$.
If we define the function ϕ from $\mathbf{Z}$ to W by

$$c\phi = f_c$$

for each integer c, then if a and b are integers

$$(a + b)\phi = f_{a+b} \quad \text{and} \quad xf_{a+b} = x + (a + b) \qquad \text{for each integer } x$$

Also, $(a\phi) \circ (b\phi) = f_a \circ f_b$ and

$$\begin{aligned}
(x)(f_a \circ f_b) &= (xf_a)f_b \\
&= (x + a)f_b \\
&= (x + a) + b \\
&= x + (a + b) \qquad \text{for each integer } x \\
&= (x)f_{a+b}
\end{aligned}$$

Since $f_{a+b} = f_a \circ f_b$, $(a + b)\phi = (a\phi) \circ (b\phi)$, and therefore ϕ is a homomorphism.
Clearly, ϕ is one-to-one for if $a\phi = b\phi$, then $f_a = f_b$ and for all integers

$x, xf_a = xf_b$. But then, in particular, $0f_a = 0f_b$; that is, $0 + a = 0 + b$ and $a = b$.

Thus, the function ϕ is an isomorphism of $\mathbf{Z}$ onto W. Since W is a subgroup of $P(\mathbf{Z})$, ϕ is an isomorphism from $\mathbf{Z}$ into $P(\mathbf{Z})$ and W is a subgroup of $P(\mathbf{Z})$ that is isomorphic to $\mathbf{Z}$.

Cayley's theorem generalizes the above result to any group and the proof follows the above pattern.

Theorem 17 (Cayley's Theorem)

If G is a group, then G is isomorphic to a subgroup of $P(G)$, the group of permutations of the set G.

Proof: For each a in G, define the function p_a by

$$xp_a = xa \qquad \text{for each } x \text{ in } G$$

Let a be an arbitrary but fixed element of G.

If x and y are elements of G such that $xp_a = yp_a$, then $xa = ya$, and therefore $x = y$. Thus, p_a is a one-to-one function.

If y is an element of G, then $ya^{-1} \in G$ and $(ya^{-1})p_a = (ya^{-1})a = y$. Hence, p_a is onto.

Therefore, for each a in G, p_a is a permutation of G. (You were asked to make the above argument in Exercise 12, Set D, Chapter 4.)

Now define ϕ from G into $P(G)$ by $a\phi = p_a$ for each a in G.

Let u and v be elements of G.

For each element x in G,

$$xp_{uv} = (x)(uv) = (xu)v = (xu)p_v = (xp_u)p_v = (x)(p_u \circ p_v)$$

Thus, $p_{uv} = p_u \circ p_v$ for each pair of elements u and v in G.

Since $(uv)\phi = p_{uv}$ and $(u\phi)\circ(v\phi) = p_u \circ p_v$, it follows that $(uv)\phi = (u\phi)\circ(v\phi)$ for each pair of elements u and v in G, and therefore ϕ is a homomorphism.

Suppose u and v are elements of G such that $u\phi = v\phi$.

Then $p_u = p_v$, which in turn implies that $xp_u = xp_v$ for every element x in G. In particular, then, $ep_u = ep_v$, so $eu = ev$. Therefore, $u = v$ and ϕ is one-to-one.

Thus, ϕ is a one-to-one homomorphism onto $G\phi$, which is a subgroup of $P(G)$; that is, ϕ is an isomorphism from G onto the subgroup $G\phi$ of $P(G)$. $\square$

The image of G under the injection ϕ above is known as the *right regular representation* of G. If we have the table for a finite group it is not difficult to write this representation using the permutation notation from Chapter 3. We just list the elements and under them the result of multiplying them on the right by the element being represented. For example, consider the multiplication table for $\mathbf{Z}_5{}^*$, the set of nonzero elements of $\mathbf{Z}_5$.

	$\bar{1}$	$\bar{2}$	$\bar{3}$	$\bar{4}$
$\bar{1}$	$\bar{1}$	$\bar{2}$	$\bar{3}$	$\bar{4}$
$\bar{2}$	$\bar{2}$	$\bar{4}$	$\bar{1}$	$\bar{3}$
$\bar{3}$	$\bar{3}$	$\bar{1}$	$\bar{4}$	$\bar{2}$
$\bar{4}$	$\bar{4}$	$\bar{3}$	$\bar{2}$	$\bar{1}$

The permutation representing $\bar{1}$ would be

$$\begin{pmatrix} \bar{1} & \bar{2} & \bar{3} & \bar{4} \\ \bar{1} & \bar{2} & \bar{3} & \bar{4} \end{pmatrix}$$

The permutation representing $\bar{2}$ would be

$$\begin{pmatrix} \bar{1} & \bar{2} & \bar{3} & \bar{4} \\ \bar{2} & \bar{4} & \bar{1} & \bar{3} \end{pmatrix}$$

The permutation representing $\bar{3}$ would be

$$\begin{pmatrix} \bar{1} & \bar{2} & \bar{3} & \bar{4} \\ \bar{3} & \bar{1} & \bar{4} & \bar{2} \end{pmatrix}$$

and the permutation representing $\bar{4}$ would be

$$\begin{pmatrix} \bar{1} & \bar{2} & \bar{3} & \bar{4} \\ \bar{4} & \bar{3} & \bar{2} & \bar{1} \end{pmatrix}$$

Using cyclic notation we would have $p_{\bar{1}} = (\bar{1})$, $p_{\bar{2}} = (\bar{1}\ \ \bar{2}\ \ \bar{4}\ \ \bar{3})$, $p_{\bar{3}} = (\bar{1}\ \ \bar{3}\ \ \bar{4}\ \ \bar{2})$, $p_{\bar{4}} = (\bar{1}\ \ \bar{4})(\bar{2}\ \ \bar{3})$.

If G is a Klein 4-group with elements e, a, b, and c we have the table below and the representations given using cyclic notation.

	e	a	b	c
e	e	a	b	c
a	a	e	c	b
b	b	c	e	a
c	c	b	a	e

$$p_e = (e)$$
$$p_a = (e\ \ a)(b\ \ c)$$
$$p_b = (e\ \ b)(a\ \ c)$$
$$p_c = (e\ \ c)(a\ \ b)$$

In the case of tables that might be the table for a finite group,

Cayley's theorem provides a check to see if the table is the table for some group. We can write down the associated permutations and see if they form a subgroup of the group of permutations of the set. If they do, then we have a group. If they do not, it is not a group.

Consider, for example, the table below. If we write down the representation we have:

	e	A	B	C	D
e	e	A	B	C	D
A	A	B	e	D	C
B	B	C	D	A	e
C	C	D	A	e	B
D	D	e	C	B	A

$$p_e = (e)$$
$$p_A = (e\ A\ B\ C\ D)$$
$$p_B = (e\ B\ D\ C\ A)$$
$$p_C = (e\ C)(A\ D\ B)$$
$$p_D = (e\ D\ A\ C\ B)$$

As permutations

$$p_A p_C = (e\ A\ B\ C\ D)(e\ C)(A\ D\ B)$$
$$= (e\ D\ C\ B)(A)$$
$$= (e\ D\ C\ B)$$

which is not in our set. Thus, although the set $\{e, A, B, C, D\}$ with the operation defined by the table satisfies Axioms (ii) and (iii), it is not a group.

EXERCISE SET D

1. Using the group of symmetries of the square, write down the permutations of this group associated with each element according to Cayley's theorem. Express them as products of cycles.

2. Now, in the same group write down the left cosets determined by the subgroups generated by each element. For example,

$$\langle R \rangle = \{I, R, S, T\} \qquad A\langle R \rangle = \{A, M, B, N\}$$
$$\langle A \rangle = \{I, A\} \qquad R\langle A \rangle = \{R, N\} \qquad S\langle A \rangle = \{S, B\} \qquad T\langle A \rangle = \{T, M\}$$

3. Let G be a finite group, and let a be an element of G. In G define a relation by $x \sim y$ iff $xa^k = y$ for some positive integer k. Prove that $\sim$ is an equivalence relation. Show that the classes are just the left cosets of $\langle a \rangle$ in G.

**4. Let G be a finite group and let a be an element of order n in G. Show that the permutation p_a as defined in the proof of Cayley's theorem can be expressed as the product of disjoint cycles of length n, where each cycle is of the form $(x \ xa, xa^2 \ xa^3 \ \ldots \ xa^{n-1})$.

**5. Construct a 7×7 table in which Axioms (ii) and (iii) for a group are satisfied but Axiom (i) is not. Use a right representation to show that the structure is not a group.

§5. Kernels of Homomorphisms

If ϕ is a homomorphism from the group G into the group H, the set of all elements in G that are mapped to e by ϕ is of special interest. This set is called the *kernel* of ϕ. Thus,

$$\text{kernel of } \phi = \{x \in G : x\phi = e\} \quad \text{or} \quad \text{kernel of } \phi = e\phi^{-1}$$

We abbreviate "kernel of ϕ" by "ker ϕ."

If ϕ is a homomorphism from the ring R into the ring S, then the kernel of ϕ is $\{x \in R : x\phi = 0\}$ or just $0\phi^{-1}$.

Examples. We refer to the examples of homomorphisms given previously in §3.

In (a), the kernel of A is $\{1, -1\}$.

In (b), ker $T = \{0\}$.

In (c), ker D is all constant polynomials.

In (d), ker D^2 is all polynomials of degree 1 or less.

In (e), ker I_0 is the zero polynomial.

In (f), ker $I = \left\{ f : \int_0^1 f = 0 \right\}$.

In (g), ker $\alpha = \langle 7 \rangle$.

In (h), ker $\beta = \{f \in \mathbf{Q}[x] : f(2) = 0\}$.

In (i), ker $\gamma = \{f \in \mathbf{Q}[x] : f(\pi) = 0\} = \{0\}$, because π is transcendental over $\mathbf{Q}$.

In (j), ker $\delta = \{f \in \mathcal{C} : f(\pi) = 0\} \neq \{0\}$, because $g(x) = x - \pi$ is in the kernel of δ.

In (k), ker $\eta = H$ because $x\eta = \bar{e}$ iff $\bar{x} = \bar{e}$ iff $x \equiv e \pmod{H}$ iff $x \in H$.

In (l), ker $\eta = A$ by the same reasoning.

The following assertion provides a tool to use in showing that a particular homomorphism is an isomorphism or an injection.

Proposition 18

If ϕ is a homomorphism from the group G into the group H, then ϕ is one-to-one if and only if ker $\phi = \{e\}$.

Proof: If ϕ is one-to-one, then $e\phi^{-1}$ contains only one element and, by Proposition 8, this element must be e. Thus, $e\phi^{-1} = \{e\}$; that is, ker $\phi = \{e\}$.

Suppose now that $\ker \phi = \{e\}$. Let x and y be elements of G such that $x\phi = y\phi$. Then

$$(x\phi)(y\phi)^{-1} = e$$

which implies that

$$(x\phi)(y^{-1}\phi) = e$$

and therefore

$$(xy^{-1})\phi = e$$

Thus,

$$xy^{-1} \in \ker \phi$$

Since $\ker \phi = \{e\}$, $xy^{-1} = e$, $x = y$, and therefore ϕ is one-to-one. $\square$

Since a ring homomorphism is also a group homomorphism, a ring homomorphism is one-to-one iff its kernel is $\{0\}$.

In Chapter 4 normal subgroups played an important role in constructing factor groups and ideals played an analogous role in the construction of factor rings. These concepts are also closely related to homomorphisms.

Proposition 19

If ϕ is a homomorphism from the group G into the group H, then the kernel of ϕ is a normal subgroup of G.

Proof: $\ker \phi \neq \emptyset$, because $e \in \ker \phi$.

Let x and y be elements of $\ker \phi$, that is $x\phi = e$ and $y\phi = e$. Then

$$(xy^{-1})\phi = (x\phi)(y^{-1}\phi) = (x\phi)(y\phi)^{-1} = ee^{-1} = ee = e$$

Therefore, $xy^{-1} \in \ker \phi$ and, by Theorem 16 from Chapter 4, $\ker \phi$ is a subgroup of G.

Now let $x \in \ker \phi$ and $a \in G$.

$$\begin{aligned}
(a^{-1}xa)\phi &= (a^{-1}\phi)(x\phi)(a\phi) \\
&= (a^{-1}\phi)e(a\phi) \\
&= (a^{-1}\phi)(a\phi) \\
&= (a^{-1}a)\phi \\
&= e\phi \\
&= e
\end{aligned}$$

Thus, if $x \in \ker \phi$, $a^{-1}xa \in \ker \phi$ and, by Proposition 31, Chapter 4, $\ker \phi$ is a normal subgroup. $\square$

Proposition 20

If ϕ is a homomorphism from the ring R into the ring S, then the kernel of ϕ is an ideal in R.

Proof: We know from Proposition 19 that ker ϕ is a subgroup of R with respect to addition. We need yet to show that ker ϕ is closed under multiplication by arbitrary elements of R; that is, we need to show that if $a \in$ ker ϕ and $r \in R$, then $ar \in$ ker ϕ and $ra \in$ ker ϕ.

If $a \in$ ker ϕ and $r \in R$, then

$$(ar)\phi = (a\phi)(r\phi) = 0 \cdot (r\phi) = 0$$

and therefore $ar \in$ ker ϕ.

Also

$$(ra)\phi = (r\phi)(a\phi) = (r\phi) \cdot 0 = 0$$

and therefore $ra \in$ ker ϕ. $\square$

The derivative operator D is a group homomorphism from the additive group of differentiable functions into the additive group of all functions from the real numbers into the real numbers. The inverse image of a function g under D, that is, gD^{-1} or $D^{-1}g$, is usually denoted by $\int g$ or $\int g(x)\,dx$ and is the set of all antiderivatives of g. How do we go about finding, say, $\int x^2\,dx$? We know that $(\frac{1}{3}x^3)' = x^2$ and therefore $\frac{1}{3}x^3$ is one of the antiderivatives of x^2. We obtain the rest of them by adding constant functions to $\frac{1}{3}x^3$. There is a theorem in calculus that states that two functions have the same derivative if and only if they differ by a constant. Thus, $f' = x^2$ iff $f = \frac{1}{3}x^3 + k$ for some constant k and

$$\int x^2\,dx = \tfrac{1}{3}x^3 + C$$

where C denotes the subgroup of all constant functions. The set $\frac{1}{3}x^3 + C$ is just a coset of C in the group of differentiable functions.

How does one prove this theorem? Its proof is based on two assertions about derivatives whose proofs require calculus.

(i) If f and g are differentiable functions, then $(f + g)' = f' + g'$.
(ii) If $f'(x) = 0$ for all x, then f is a constant function.

Translated into the language of this section, the first assertion says that the derivative operator is a homomorphism and the second says that its kernel is the set of all constant functions.

The rest of the proof of the theorem is purely algebraic using these assertions.

Theorem

If f and g are differentiable functions, then $f' = g'$ iff f and g differ by a constant.

Proof

$$f' = g' \qquad \text{iff } f' - g' = 0 \qquad \text{(here 0 denotes the constant function 0)}$$

$$\text{iff } (f - g)' = 0 \quad \text{because the derivative operator is a}$$
$$\text{homomorphism by (i)}$$
$$\text{iff } f - g \text{ is a constant function} \quad \text{by (ii).} \quad \square$$

If we let C denote the group of all constant functions, our last statement becomes

$$f - g \in C$$

But

$$f - g \in C \quad \text{iff } f \equiv g \,(\text{mod } C)$$
$$\text{iff } f \in g + C$$

Putting these ideas together we now have, for any functions h and f,

$$\text{if } h \in \int f, \quad \text{then} \quad h + C = \int f = D^{-1}f$$

The second derivative operator D^2 is also a homomorphism. The kernel of D^2 is the set of all polynomials of degree 1 or less. Call this set $\mathcal{L}$. If f and g are two twice-differentiable functions, then

$$f'' = g'' \quad \text{iff } f'' - g'' = 0$$
$$\text{iff } (f - g)'' = 0$$
$$\text{iff } f - g \text{ is in the kernel of } D^2$$
$$\text{iff } f - g \in \mathcal{L}$$
$$\text{iff } f \in g + \mathcal{L}$$

To find $(D^2)^{-1} \sin x$ or $\int \int \sin x$ we first recall that

$$(-\sin x)'' = \sin x$$

Therefore,

$$\int \int \sin x = -\sin x + \mathcal{L}$$

or all functions one can get by adding $-\sin x$ to a polynomial of degree 1 or less.

Consider the function T from three dimensional space or $\mathbf{R} \oplus \mathbf{R} \oplus \mathbf{R}$ to $\mathbf{R}$ defined by

$$(x, y, z)T = 2x + 3y + 4z$$

[or $(x, y, z) \cdot (2, 3, 4)$]. It is easy to show that T is a homomorphism onto $\mathbf{R}$. The kernel of T is the set $\{(x, y, z):2x + 3y + 4z = 0\}$, which is a plane through the origin. Now let us find $5T^{-1}$.

By observation, we might notice that $(1, 1, 0)T = 5$. Therefore, $(1, 1, 0) \in 5T^{-1}$. We might also notice that $(0, 3, -1) \in 5T^{-1}$ and that

$$(1, 1, 0) - (0, 3, -1) = (1, -2, 1) \in \ker T$$

If $(x, y, z) \in 5T^{-1}$, then

$(x, y, z)T = 5$

and since $(1, 1, 0)T = 5$,

$(x, y, z)T - (1, 1, 0)T = 5 - 5 = 0$

Then

$[(x, y, z) - (1, 1, 0)]T = 0$ because T is a homomorphism

and therefore

$(x, y, z) - (1, 1, 0) \in \ker T$

But if

$(x, y, z) - (1, 1, 0) \in \ker T$

then

$(x, y, z) \in \ker T + (1, 1, 0)$

Thus, $5T^{-1} \subseteq \ker T + (1, 1, 0)$.

Conversely, if $(x, y, z) \in \ker T + (1, 1, 0)$, for some (a, b, c) in $\ker T$,

$(x, y, z) = (a, b, c) + (1, 1, 0)$ and $(a, b, c)T = 0$

Then

$$\begin{aligned}
(x, y, z)T &= [(a, b, c) + (1, 1, 0)]T \\
&= (a, b, c)T + (1, 1, 0)T \\
&= 0 + 5 \\
&= 5
\end{aligned}$$

Therefore, $(x, y, z) \in 5T^{-1}$, and we have shown $\ker T + (1, 1, 0) \subseteq 5T^{-1}$.

Thus, we see that $5T^{-1}$ is the coset of $\ker T$ in $\mathbf{R} \oplus \mathbf{R} \oplus \mathbf{R}$ determined by $(1, 1, 0)$. Since $\ker T$ is a plane, $\ker T + (1, 1, 0)$ is also a plane. It is just the plane parallel to $\ker T$ through $(1, 1, 0)$. Before going further see if you can determine $6T^{-1}$.

In each of the above cases we saw that the inverse image of an element under a homomorphism can be obtained by finding one element in the inverse image and then forming the coset of the kernel determined by this element. We now show that this principle holds in general.

Proposition 21

If ϕ is a homomorphism from the group G into the group H, h is an element of H, K is the kernel of ϕ, and g is an element of G such that $g\phi = h$, then $h\phi^{-1} = Kg$.

Proof: If $k \in K$, then $(kg)\phi = (k\phi)(g\phi) = eh = h$.

Therefore,

$$Kg \subseteq h\phi^{-1}$$

Now suppose $w \in h\phi^{-1}$.
Since $w\phi = h$ and $g\phi = h$, $w\phi = g\phi$.
Then

$$(w\phi)(g\phi)^{-1} = e$$
$$(w\phi)(g^{-1}\phi) = e$$

that is,

$$(wg^{-1})\phi = e$$

Thus,

$$wg^{-1} \in K$$

and

$$w = wg^{-1}g \in Kg$$

Therefore,

$$h\phi^{-1} \subseteq Kg$$

Since $Kg \subseteq h\phi^{-1}$ and $h\phi^{-1} \subseteq Kg$,

$$h\phi^{-1} = Kg \quad \square$$

Corollary 22

If ϕ is a homomorphism from the ring R into the ring S, h is an element of S, K is the kernel of ϕ, and g is an element of R such that $g\phi = h$, then $h\phi^{-1} = K + g$.

The principle asserted in Proposition 21 has many applications in mathematics. In solving linear differential equations, one first solves the associated homogeneous equation, then finds one particular solution, and finally adds this particular solution to the set of all solutions of the homogeneous equation to obtain the set of all solutions.
For example:
Find all functions f such that

$$f''(x) - f'(x) - 2f(x) = -4$$

This is equivalent to finding all functions f such that

$$(D^2 - D - 2)f = -4$$

It is easy to show that the operator $D^2 - D - 2$ is a homomorphism.
When we say first find all functions f such that

$$f'' - f' - 2f = 0$$

we are saying first find the kernel of $D^2 - D - 2$.

By using techniques from differential equations one can show that the kernel of $D^2 - D - 2$ is the group of all functions of the form $f(x) = ae^{-x} + be^{2x}$, where a and b are real numbers, that is,

$$\ker D^2 - D - 2 = \{g(x) = ae^{-x} + be^{2x} : a \in \mathbf{R} \text{ and } b \in \mathbf{R}\}$$

Now, since $(D^2 - D - 2)(2) = -4$, the constant function 2 is a solution of the equation. Thus, the set of all functions f such that

$$f'' - f' - 2f = 4$$

is just

$$\{g(x) = ae^{-x} + be^{2x} : a \in \mathbf{R} \text{ and } b \in \mathbf{R}\} + 2$$

or

$$\{f : f(x) = ae^{-x} + be^{2x} + 2 \text{ and } a \in \mathbf{R} \text{ and } b \in \mathbf{R}\}$$

More complicated linear equations require more complicated techniques, but the general process is illustrated above:

(i) Find the kernel.
(ii) Find one solution.
(iii) Determine the coset to get all solutions.

An alternative proof to Proposition 21 is available using some ideas developed in Chapter 1. In Exercise Set C of Chapter 1, for any function f from the set S into the set T we defined a relation $\sim_f$ on S by $x \sim_f y$ iff $xf = yf$ and showed that this relation was an equivalence relation. We also showed that if $\bar{x}$ denotes the class for x with respect to $\sim_f$ and $xf = p$, then $pf^{-1} = \bar{x}$.

Now let ϕ be a homomorphism from the group G into the group H. Then in G,

$$
\begin{aligned}
x \sim_\phi y \quad & \text{iff } x\phi = y\phi \\
& \text{iff } (x\phi)(y\phi)^{-1} = e \\
& \text{iff } (x\phi)(y^{-1}\phi) = e \\
& \text{iff } (xy^{-1})\phi = e \\
& \text{iff } xy^{-1} \in \ker \phi \\
& \text{iff } x \equiv y \,(\mathrm{mod}\ \ker \phi)
\end{aligned}
$$

Thus, $x \sim_\phi y$ iff $x \equiv y \,(\mathrm{mod}\ \ker \phi)$; that is, the equivalence relation determined by a homomorphism is the same as congruence modulo its kernel.

Now if $h \in H$ and $g\phi = h$, by our result from Chapter 1, $h\phi^{-1} = \bar{g}$. Since $\bar{g}$ is the class for g with respect to $\sim_\phi$ and the relation $\sim_\phi$ is the same as congruence modulo ker ϕ, $\bar{g}$ is the class for g with respect to congruence modulo ker ϕ. Since $\bar{g} = (\ker \phi)g$, $h\phi^{-1} = (\ker \phi)g$.

EXERCISE SET E

1. Let $\mathcal{D}$ denote the additive group of functions that are differentiable infinitely many times, that is, those functions f such that $D^n f$ exists for every positive integer n. What is ker D^3? What is ker D^4? What is ker D^5? What is ker D^n for a positive integer n?

2. Define ϕ from $\mathbf{Z}$ to $\mathbf{Z}_{12}$ by $x\phi = \overline{2x}$. Prove that ϕ is a group homomorphism. What is ker ϕ?

3. If we define ϕ from $\mathbf{Q}[x]$ to $\mathbf{R}$ by $f\phi = f(\pi)$, ϕ is a homomorphism. Prove that ϕ is one-to-one by using Proposition 18.

4. If ϕ is a homomorphism from $\mathbf{Z}$ into a ring R, what are the possibilities for ker ϕ?

5. Prove: If ϕ is a homomorphism from the group G onto the group H and $|G| = 17$, then either ϕ is an isomorphism or $|H| = 1$.

6. Prove: If λ is a homomorphism from the field F onto the ring R, then either λ is an isomorphism or $R = \{0\}$.

7. Define ϕ from $\mathbf{R} \oplus \mathbf{R}$ into $\mathbf{R} \oplus \mathbf{R}$ by

   $(x, y)\phi = (x + y, x - y)$

   Use Proposition 18 to prove that ϕ is a group isomorphism.

8. Define ϕ from $\mathbf{Q}$ into $\mathbf{Q}$ by $x\phi = x^2$. Then $0\phi^{-1} = \{0\}$ but ϕ is not one-to-one. Does this contradict Proposition 18? Explain.

*9. Let M be the set of all 2×2 matrices

   $$\begin{bmatrix} a & b \\ c & d \end{bmatrix}$$

   such that $ad - bc \neq 0$. M is known to be a group under multiplication. Define ϕ from M to $\mathbf{R}$ by

   $$\begin{bmatrix} a & b \\ c & d \end{bmatrix}\phi = ad - bc$$

   (a) Prove that ϕ is a homomorphism. (Can you give a common name for ϕ?)
   (b) What is the kernel of ϕ?
   (c) Prove that $A^{-1}B^{-1}AB \in \ker \phi$ for every pair of matrices A and B in M.

10. Define ϕ from $\mathbf{R} \oplus \mathbf{R}$ to $\mathbf{R}$ by $(x, y)\phi = x - y$. Prove that ϕ is a group homomorphism. What is ker ϕ? What is $3\phi^{-1}$? What is $(-2)\phi^{-1}$?

11. Define ϕ from the ring $\mathbf{R} \oplus \mathbf{R}$ into $\mathbf{R}$ by $(x, y)\phi = x$. What is the kernel of ϕ? What is $3\phi^{-1}$?

12. Define ϕ from S_4 to S_3 by $p\phi = i$ if p is even and $p\phi = (1\ 2)$ if p is odd. Prove that ϕ is a homomorphism. What is ker ϕ? What is $(2\ 3)\phi^{-1}$? What is $[(1\ 2\ 3\ 4)\phi]\phi^{-1}$?

13. Define λ from $\mathbf{Z}$ to $\mathbf{Z}_{10}$ by $x\lambda = \overline{3x}$. Prove that λ is a group homomorphism. What is ker λ? What is $(4\lambda)\lambda^{-1}$? What is $(7\lambda)\lambda^{-1}$?

14. Use Proposition 21 and your results from Exercise 1 to determine:
    (a) $\int\int\int x^4$
    (b) $\int\int\int \sin x$
    (c) $\int\int\int\int e^{2x}$
    (d) $\int\int\int\int\int \cos 3x$

15. (a) Let ϕ be a homomorphism from the group G into the group H. Prove: If N is a normal subgroup of H, then $N\phi^{-1}$ is a normal subgroup of G containing ker ϕ.
    (b) State and prove an analogous assertion for rings.

16. Let ϕ be a homomorphism from the ring R into the ring S with $K = $ ker ϕ. If p is an element of R, what is $(p\phi)\phi^{-1}$? Prove your assertion.

**17. Let ϕ be a homomorphism from the group G onto the group H with $K = $ ker ϕ and let M be a subgroup of G. Show that $(M\phi)\phi^{-1} = MK$.

18. Let G, H, and M be groups (rings) with ϕ as homomorphism from G into H and λ a homomorphism from G into M. Define μ from G into $H \oplus M$ by $x\mu = (x\phi,\ x\lambda)$ for each element x in G.
    (a) Prove that μ is a homomorphism.
    (b) Prove that ker $\mu = $ (ker ϕ) $\cap$ (ker λ).

**19. Let ϕ be a homomorphism from the group G onto the group H. In G define the *derived group* G' by $G' = \langle\{a^{-1}b^{-1}ab : a \in G \text{ and } b \in G\}\rangle$. Elements of the form $a^{-1}b^{-1}ab$ are called *commutators*. Prove: H is abelian iff $G' \subseteq$ ker ϕ.

**20. Let ϕ be a homomorphism from the group G onto the group H. Prove: If W is a set of generators for G, then $W\phi$ is a set of generators for H.

**21. (Especially challenging)
    Let ϕ be a homomorphism from the group G onto the group H. Prove: If S is a set of generators for H, then $S\phi^{-1}$ is a set of generators for G.

§6. The Homomorphism and Isomorphism Theorems

In Chapters 4 and 5 we constructed factor groups G/N, where N is a normal subgroup of G, and factor rings R/A, where A is an ideal in R. In exercises and examples we saw that the function η defined by $x\eta = \bar{x}$ is, in each of the above cases, a homomorphism with the corresponding normal subgroup or ideal as its kernel. Thus, the factor groups and factor rings are images of the original groups and rings under homomorphisms. In the following two theorems we show that, *up to the equivalence relation of isomorphism, these are the only possible homomorphic images.*

Theorem 23 (The Fundamental Homomorphism Theorem for Groups)

If ϕ is a homomorphism from the group G onto the group H and K denotes the kernel of ϕ, then $G/K \cong H$.

Proof: Define $\bar{\phi}$ from G/K to H by $\bar{a}\,\bar{\phi} = a\phi$. We need first to show that this rule does, in fact, define a function.

If $\bar{a} = \bar{b}$, then $a \equiv b \pmod{K}$ and hence $ab^{-1} \in K$. Thus,

$$(ab^{-1})\phi = e$$
$$(a\phi)(b^{-1}\phi) = e$$
$$(a\phi)(b\phi)^{-1} = e$$

and therefore

$$a\phi = b\phi$$

Hence, if $\bar{a} = \bar{b}$, $a\phi = b\phi$, and therefore $\bar{\phi}$ is a well-defined function.

If $h \in H$, since ϕ is onto, for some x in G, $x\phi = h$, and therefore $\bar{x}\bar{\phi} = h$. Thus, $\bar{\phi}$ is onto.

Let $\bar{a}$ and $\bar{b}$ be elements of G/K. Then

$$(\overline{ab})\,\bar{\phi} = (\overline{ab})\,\bar{\phi} = (ab)\,\phi = (a\phi)(b\phi) = (\bar{a}\bar{\phi})(\bar{b}\bar{\phi})$$

and therefore $\bar{\phi}$ is a homomorphism.

If $\bar{a} \in \ker \bar{\phi}$, then $\bar{a}\bar{\phi} = e$, which implies $a\phi = e$, and therefore $a \in K$. Since $a \in K$, $\bar{a} = \bar{e}$. Thus, $\ker \bar{\phi} = \{\bar{e}\}$, and $\bar{\phi}$ is one-to-one. Thus, $\bar{\phi}$ is an isomorphism from G/K onto H. $\square$

Notice that $\eta \circ \bar{\phi} = \phi$, where η is the natural homomorphism from G onto G/K since $(x)(\eta \circ \bar{\phi}) = (x\eta)\,\bar{\phi} = \bar{x}\bar{\phi} = x\phi$. This is illustrated by the diagram in Figure 6.1.

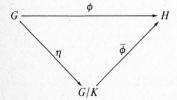

Figure 6-1

Theorem 24 (The Fundamental Homomorphism Theorem for Rings)

If ϕ is a homomorphism from the ring R onto the ring S and K denotes the kernel of ϕ, then $R/K \cong S$.

Proof: By Theorem 23 we know that the additive structure of R/K is isomorphic to that of S as groups and that the function $\bar{\phi}$ defined by $\bar{a}\bar{\phi} = a\phi$ is a group isomorphism. We need only show that $\bar{\phi}$ preserves multiplication.

If $\bar{a}$ and $\bar{b}$ are elements of R/K, then

$$(\overline{ab})\,\overline{\phi} \;=\; (\overline{ab})\,\overline{\phi} \;=\; (ab)\,\phi \;=\; (a\phi)(b\phi) \;=\; (\overline{a}\overline{\phi})(\overline{b}\overline{\phi})$$

Therefore, $\overline{\phi}$ is a ring isomorphism.   □

The following proposition follows directly from Theorem 23 and is a very useful result. It is a generalization of Proposition 5.

Proposition 25

If G is a cyclic group, then either $\mathbf{Z} \cong G$ *or, for some integer m,* $\mathbf{Z}_m \cong G$.

Proof: Since G is cyclic, $G = \langle a \rangle$ for some element a in G. Now define ϕ from $\mathbf{Z}$ to $\langle a \rangle$ by $x\phi = a^x$. Since $\langle a \rangle = \{a^n : n \in \mathbf{Z}\}$, we know that ϕ is onto. Let x and y be integers.

$$(x + y)\phi \;=\; a^{x+y} \;=\; a^x a^y \;=\; (x\phi)(y\phi)$$

Thus, ϕ is a homomorphism.

If ker $\phi = \{0\}$, then ϕ is one-to-one, and therefore ϕ is an isomorphism. If ker $\phi \neq \{0\}$, then we know ker $\phi = \langle m \rangle$ for some integer m. By Theorem 23, $\mathbf{Z}/\text{ker } \phi \cong G$ and thus, if $G \not\cong \mathbf{Z}$, $\mathbf{Z}/\langle m \rangle \cong G$, that is, $\mathbf{Z}_m \cong G$ for some integer m.   □

Proposition 26

If G is a finite group with p elements and p is prime, then $\mathbf{Z}_p \cong G$.

Proof: If a is an element of G and $a \neq e$, then since the order of a divides p, the order of a must be p and therefore $G = \langle a \rangle$. Then, by Proposition 25, $G \cong \mathbf{Z}_m$ for some positive integer m. But $\mathbf{Z}_m$ has p elements only if $m = p$. Therefore, $G \cong \mathbf{Z}_p$.   □

The above result provides a quick proof of our earlier assertions that all groups with two elements are isomorphic to $\mathbf{Z}_2$, and all groups with three elements are isomorphic to $\mathbf{Z}_3$.

By using Theorem 23 and Theorem 24 and analyzing the set of normal subgroups of a group or the set of ideals in a ring, we can study, up to isomorphism, all possible homomorphic images of a group or ring. For example, in Proposition 25 we used our knowledge of the possible subgroups of $\mathbf{Z}$ to determine all possible kernels and therefore all possible homomorphic images of $\mathbf{Z}$. Hidden in the proof of the proposition is the fact that any homomorphic image of $\mathbf{Z}$ is isomorphic to $\mathbf{Z}$ or to $\mathbf{Z}_m$ for some positive integer m. Using this and the fact that a cyclic group is a homomorphic image of $\mathbf{Z}$, we proved the proposition. In the exercises the student was asked to prove that a homomorphic image of a field is either a field or a ring with one element. This follows from the fact that the only possible kernels of the homomorphism are $\{0\}$ and the field itself, *because they are the only ideals.* Groups that possess no normal subgroups other than G and $\{e\}$ admit no homomorphic images other than those isomorphic to G and $\{e\}$. Such groups are called *simple* groups, and the problem of determining simple groups of finite order has been and continues to be a very important problem in the theory of groups.

The next two theorems are both very important in advanced studies and we include them here along with the other basic results concerning homomorphisms.

Theorem 27 (First Isomorphism Theorem)

If G is a group, M is a subgroup of G, and N is a normal subgroup of G, then $M/(N \cap M) \cong NM/N$.

Proof: Define ϕ from M to NM/N by

$$x\phi = Nx \qquad \text{for each } x \text{ in } M$$

Every element of NM/N is a coset of N and, since every element of NM is of the form nm for some n in N and m in M, the coset $Nnm = Nm$. Therefore each element of NM/N is a coset of the form Nm with m in M, and thus, ϕ is onto.

If x and y are elements of M, then

$$\begin{aligned}
(xy)\phi &= Nxy \\
&= NxNy \\
&= (x\phi)(y\phi)
\end{aligned}$$

Therefore, ϕ is a homomorphism.

If $x \in \ker \phi$, $x\phi = N$. But then $Nx = N$, and therefore $x \in N$.

Thus, $\ker \phi = \{x \in M : x \in N\} = N \cap M$.

By Theorem 23, $M/(N \cap M) \cong NM/N$. $\quad\square$

Theorem 28 (Second Isomorphism Theorem)

If A and B are normal subgroups of G and A is a subgroup of B, then $G/A \big/ B/A \cong G/B$.

Proof: Define ϕ from G/A to G/B by

$$(Ax)\phi = Bx \qquad \text{for each } Ax \text{ in } G/A$$

If $Ax = Ay$, then $x \equiv y \pmod{A}$, which implies $xy^{-1} \in A$. Since $A \subseteq B$, $xy^{-1} \in B$, so $x \equiv y \pmod{B}$, and therefore $Bx = By$. Thus the rule given does define a function.

Clearly, ϕ is onto.

$$(AxAy)\phi = (Axy)\phi = Bxy = BxBy = (x\phi)(y\phi)$$

Therefore, ϕ is a homomorphism.

If $(Ax)\phi = B$, $Bx = B$, and therefore $x \in B$.

Thus,

$$\ker \phi = \{Ax : x \in B\} = B/A$$

By Theorem 23,

$$G/A \Big/ B/A \cong G/B \quad \square$$

EXERCISE SET F

1. Show directly that, as groups, $\mathbf{Z}_4/\{\bar{0}, \bar{2}\} \cong \mathbf{Z}_2$ by defining a homomorphism from $\mathbf{Z}_4$ to $\mathbf{Z}_2$ with $\{\bar{0}, \bar{2}\}$ as the kernel of the homomorphism and applying Theorem 23.

2. Show that $\mathbf{Z}_8/\{\bar{0}, \bar{4}\} \cong \mathbf{Z}_4$ by defining a ring homomorphism from $\mathbf{Z}_8$ to $\mathbf{Z}_4$ with kernel $\{\bar{0}, \bar{4}\}$.

3. Define a suitable homomorphism from $\mathbf{R}[x]$ to $\mathbf{R}$ and use this homomorphism and Theorem 24 to show that $\mathbf{R}[x]/A \cong \mathbf{R}$, where A is the set of all polynomials with 5 as a root.

4. By determining all subgroups of $\mathbf{Z}_{12}$, prove that each homomorphic image of $\mathbf{Z}_{12}$ is isomorphic to one of the groups $\{0\}$, $\mathbf{Z}_2$, $\mathbf{Z}_3$, $\mathbf{Z}_4$, $\mathbf{Z}_6$, $\mathbf{Z}_{12}$. Show also that each of these is a homomorphic image of $\mathbf{Z}_{12}$.

5. Determine, up to isomorphism, all homomorphic images of $\mathbf{Z}_7$.

6. Determine, up to isomorphism, every homomorphic image of S_3.

**7. Determine, up to isomorphism, all homomorphic images of A_4.

8. Determine, up to isomorphism, all homomorphic images of the ring $\mathbf{Q}[\sqrt{2}]$.

**9. Determine, up to isomorphism, all homomorphic images of A_5; of S_5. (See Exercise Set G, Chapter 4.)

10. (a) Prove: If ϕ is a homomorphism of the group G onto the group H and N is a normal subgroup of G, then $N\phi$ is a normal subgroup of H.
    (b) State and prove an analogous assertion for rings.

**11. A normal subgroup N of a group G is said to be a *maximal* normal subgroup if N is not properly contained in any proper normal subgroup of G. Prove that for $N \neq G$, G/N is simple iff N is a maximal normal subgroup. (Hint: See Exercise 15, Set E.)

**12. An ideal A in a ring R is said to be *maximal* if A is not properly contained in any proper ideal of R. Prove that, for $A \neq R$, R/A has no proper ideals iff A is a maximal ideal in R. (Hint: Same as for Exercise 11.)

**13. Prove: If R is a commutative ring with $1 \neq 0$ and A is an ideal in R, then R/A is a field iff A is a maximal ideal in R and $A \neq R$.

§7. Automorphism Groups

If G is a group or ring we denote the set of all automorphisms of G by $A(G)$. Since the composition of automorphisms is an automorphism, and the inverse of an automorphism is an automorphism, $A(G)$ is a subgroup of $P(G)$, the group of all permutations of G.

If G is a group and a is an element of G, we define the function ϕ_a from G into G by

$$x\phi_a = a^{-1}xa \qquad \text{for each } x \text{ in } G$$

From the exercises we know that ϕ_a is one-to-one, onto, and operation preserving and therefore is an automorphism of G. This automorphism is called the *inner automorphism determined by a*. We denote the set of all inner automorphisms of G by $I(G)$.

If G is a group, we denote the center of G by $Z(G)$ and define it to be the set of all elements of G that commute with every element of G; that is,

$$Z(G) = \{x \in G : xg = gx \text{ for every element } g \text{ in } G\}$$

The following theorem is important in its own right but also serves as an example of the power of Theorem 23 and some of our earlier results. We shall combine three assertions into our theorem and prove them all at once.

Theorem 29

If G is a group, then

(a) $I(G)$ is a subgroup of $A(G)$,
(b) $Z(G)$ is a normal subgroup of G, and
(c) $G/Z(G) \cong I(G)$.

Proof: Define the function T from G into $A(G)$ by $aT = \phi_a$ for each element a in G. Thus, the image of an element under T is the inner automorphism determined by the element and therefore, by definition, $GT = I(G)$. If we can show that T is a homomorphism we will have shown that $I(G)$ is a subgroup of $A(G)$.

Let $a, b \in G$. Then

$$(ab)T = \phi_{ab} \quad \text{and} \quad (aT) \circ (bT) = \phi_a \circ \phi_b$$

If x is an element of G, then

$$\begin{aligned}
x\phi_{ab} &= (ab)^{-1}x(ab) \\
&= (b^{-1}a^{-1})x(ab) \\
&= b^{-1}(a^{-1}xa)b \\
&= (a^{-1}xa)\phi_b \\
&= (x\phi_a)\phi_b \\
&= (x)(\phi_a \circ \phi_b)
\end{aligned}$$

Since $\phi_{ab} = \phi_a \circ \phi_b$, $(ab)T = (aT) \circ (bT)$, and T is a homomorphism. Since $I(G) = GT$ is the image of G under a homomorphism, $I(G)$ is a subgroup of $A(G)$ and (a) is proved.

For each element w in G,

$$\begin{array}{llll}
w \in \ker T & \text{iff} & \phi_w = I & [I \text{ is the identity in } A(G)] \\
& \text{iff} & g\phi_w = g & \text{for every element } g \text{ in } G \\
& \text{iff} & w^{-1}gw = g & \text{for every element } g \text{ in } G
\end{array}$$

$$\text{iff} \qquad gw = wg \qquad \text{for every element } g \text{ in } G$$
$$\text{iff} \quad w \in Z(G)$$

Thus, $w \in \ker T$ iff $w \in Z(G)$ and $\ker T = Z(G)$. Since the kernel of a homomorphism is a normal subgroup, $Z(G) \lhd G$ and part (b) is proved.

Part (c) follows from Theorem 23, in that

$$G/\ker T \cong GT$$

and therefore

$$G/Z(G) \cong I(G) \qquad \square$$

For illustrative purposes, let us determine the automorphism groups of some elementary groups and rings.

Example 1. If ϕ is a group automorphism of $\mathbf{Z}_3$, since $\bar{1}\phi \neq \bar{0}$, either $\bar{1}\phi = \bar{1}$ or $\bar{1}\phi = \bar{2}$.

If $\bar{1}\phi = \bar{1}$, then the only possibility for $\bar{2}\phi$ is $\bar{2}$ and ϕ is the identity automorphism.

If we let $\bar{1}\mu = \bar{2}$ and $\bar{2}\mu = \bar{1}$, by direct computation we can verify that μ does preserve the operation and is, therefore, an automorphism. Since the only group automorphisms of $\mathbf{Z}_3$ are the identity and the function μ, $A(\mathbf{Z}_3) \cong \mathbf{Z}_2$. Since $\mathbf{Z}_3$ is an abelian group, $I(G)$ has only one element, the identity automorphism. The only ring automorphism of $\mathbf{Z}_3$ is the identity map, because $\bar{1}\phi = \bar{1}$ and $\bar{0}\phi = \bar{0}$ for any automorphism ϕ, and therefore the only possibility for $\bar{2}\phi$ is $\bar{2}$.

Example 2. In S_3, the only element that commutes with every element is i, and therefore $Z(S_3) = \{i\}$. By Theorem 29, then, $S_3/\{i\} \cong I(G)$. But $S_3/\{i\} \cong S_3$, and therefore $S_3 \cong I(G)$.

What are the other possibilities for automorphisms of S_3? The only elements of order 3 in S_3 are $(1\ \ 2\ \ 3)$ and $(1\ \ 3\ \ 2)$. The elements of order 2 are $(1\ \ 2)$, $(1\ \ 3)$, and $(2\ \ 3)$. If ϕ is an automorphism of S_3, since $\{(1\ \ 2), (1\ \ 2\ \ 3)\}$ generates S_3, once we know the images of $(1\ \ 2)$ and $(1\ \ 2\ \ 3)$, we shall have completely determined ϕ. The possibilities for $(1\ \ 2)\phi$ are $(1\ \ 2)$, $(1\ \ 3)$, and $(2\ \ 3)$. The possibilities for $(1\ \ 2\ \ 3)\phi$ are $(1\ \ 2\ \ 3)$ and $(1\ \ 3\ \ 2)$. Since there are three possibilities for $(1\ \ 2)\phi$ and two possibilities for $(1\ \ 2\ \ 3)\phi$, there are $3 \cdot 2$ possibilities for ϕ. Thus, there are no more than six possible automorphisms for S_3. Since $I(S_3) \cong S_3$ and $|S_3| = 6$, there are six inner automorphisms of S_3. Therefore, every automorphism of S_3 is inner, $I(S_3) = A(S_3)$, and $A(S_3) \cong S_3$.

Example 3. If ϕ is a group automorphism of $\mathbf{Z}_7$, there are six possibilities for $\bar{1}\phi$. Since $\bar{1}$ generates $\mathbf{Z}_7$, once the image of $\bar{1}$ under ϕ is determined, ϕ is completely determined. If $\bar{1}\phi = \bar{1}$, then ϕ is the identity automorphism. Suppose we define $\bar{1}\phi = \bar{2}$. This would determine a homomorphism, since it would be equivalent to multiplying each element by 2. Thus, $\bar{x}\phi = 2\bar{x}$ and

$$2(\bar{a} + \bar{b}) = 2\bar{a} + 2\bar{b} \qquad \text{for each } \bar{a}, \bar{b} \text{ in } \mathbf{Z}_7$$

Is ϕ one-to-one and onto? If $2\bar{a} = \bar{0}$, $2a \equiv 0$ (mod 7), which implies $7 \mid 2a$, and therefore $7 \mid a$. Thus, $a \equiv 0$ (mod 7) and $\bar{a} = \bar{0}$. Thus, ϕ is an automorphism. Furthermore, if n is any of the integers 1, 2, 3, 4, 5, or 6 the following argument will again hold; that is,

$$n(\bar{a} + \bar{b}) = n\bar{a} + n\bar{b} \quad \text{and} \quad \text{if } n\bar{a} = \bar{0}, \bar{a} = \bar{0}$$

Thus, each of these six integers determines an automorphism of $\mathbf{Z}_7$. If, for each element $\bar{a}$ in $\mathbf{Z}_7{}^*$, we define $\zeta_{(\bar{a})}$ by $\bar{x}\zeta_{(\bar{a})} = \bar{a}\bar{x}$, we see that

$$\bar{x}\zeta_{(\bar{a})} \circ \zeta_{(\bar{b})} = (\bar{x}\zeta_{(\bar{a})})\zeta_{(\bar{b})} = (\bar{a}\bar{x})\zeta_{(\bar{b})} = \bar{b}\bar{a}\bar{x} = \bar{a}\bar{b}\bar{x} = \bar{x}\zeta_{(\bar{a}\bar{b})}$$

Thus, if we define λ from $\mathbf{Z}_7{}^*$ to $A(\mathbf{Z}_7)$ by $\bar{a}\lambda = \zeta_{(\bar{a})}$, we see that λ is a homomorphism and, in fact, since $\zeta_{(\bar{a})} = \zeta_{(\bar{b})}$ implies $\bar{a}\bar{1} = \bar{b}\bar{1}$, which in turn implies $\bar{a} = \bar{b}$, λ is one-to-one. Since $\mathbf{Z}_7{}^*$ has six elements and $A(\mathbf{Z}_7)$ has at most six elements, λ is an isomorphism and $A(\mathbf{Z}_7) \cong \mathbf{Z}_7{}^*$.

Example 4. In examining $A(\mathbf{Z}_8)$, notice again that there are at most seven automorphisms, because there are at most seven possible images for $\bar{1}$ and once the image of $\bar{1}$ is determined, all images are determined. But $\bar{1}$ is of order 8 in $\mathbf{Z}_8$, whereas $\bar{4}$ is of order 2 and $\bar{2}$ and $\bar{6}$ are of order 4. Thus, there are at most four possible images for $\bar{1}$ under an automorphism: $\bar{1}, \bar{3}, \bar{5}$, and $\bar{7}$. If we proceed as in Example 3 we can show that the function $\zeta_{(\bar{a})}$ defined by $\bar{x}\zeta_{(\bar{a})} = \bar{a}\bar{x}$ is an isomorphism if $\bar{a} = \bar{1}, \bar{3}, \bar{5}$, or $\bar{7}$ and these functions are, in fact, distinct. The set $\{\bar{1}, \bar{3}, \bar{5}, \bar{7}\}$ is just the group of units of the ring $\mathbf{Z}_8$ and every element except $\bar{1}$ has order 2. By an argument similar to that in Example 3, it can be shown that $A(\mathbf{Z}_8)$ is isomorphic to this group and therefore $A(\mathbf{Z}_8)$ is a Klein 4-group.

Example 5. If we view $\mathbf{Z}_m$ as a ring, $\mathbf{Z}_m$ can have only one automorphism since, if ϕ is any ring automorphism, $\bar{1}\phi = \bar{1}$ and once the image of the generator $\bar{1}$ is determined, ϕ is completely determined.

Example 6. Viewing $\mathbf{Z}$ as a group, $A(\mathbf{Z}) \cong \mathbf{Z}_2$, because if ϕ is any automorphism of $\mathbf{Z}$ either $1\phi = 1$ or $1\phi = -1$. Otherwise, say $1\phi = n$ with $n \neq 1$ and $n \neq -1$, then $m\phi = nm$ for every integer m and $1 \notin \mathbf{Z}\phi$; that is, ϕ is not onto.

Example 7. In determining $A(\mathbf{Z}_2 \oplus \mathbf{Z}_2)$ notice first that there are at most six automorphisms, since there are six permutations of $\{(1, 0), (1, 1), (0, 1)\}$. It turns out that each of these permutations is an automorphism. We shall not verify this but leave it to the reader to examine some cases.

Notice that if

$$(1, 0)T = (1, 1), \quad (1, 1)T = (1, 0), \quad \text{and} \quad (0, 1)T = (0, 1)$$

and

$(1, 0)P = (0, 1), (1, 1)P = (1, 1)$, and $(0, 1)P = (1, 0)$. Hence $T \circ P \neq P \circ T$.

Thus, $A(\mathbf{Z}_2 \oplus \mathbf{Z}_2)$ is a noncommutative group with six elements, that is, $A(\mathbf{Z}_2 \oplus \mathbf{Z}_2) \cong S_3$.

Example 8. The only ring automorphism of $\mathbf{Q}$ is the identity. To see this recall again that if ϕ is any ring automorphism, then $1\phi = 1$. Since $1\phi = 1, n\phi = n$ for any integer n. Since the nonzero elements of $\mathbf{Q}$ form an abelian group, ϕ must be a group homomorphism, and therefore

$$\left(\frac{1}{p}\right)\phi = (p^{-1})\phi = (p\phi)^{-1} = \frac{1}{p\phi} = \frac{1}{p}$$

for any nonzero integer p. Thus,

$$\left(\frac{p}{q}\right)\phi = \left(p \cdot \frac{1}{q}\right)\phi = (p\phi)\left(\frac{1}{q}\right)\phi = p \cdot \frac{1}{q} = \frac{p}{q}$$

for any integer p and nonzero integer q. Thus $A(\mathbf{Q}) = \{i\}$.

EXERCISE SET G

1. Determine $A(\mathbf{Z}_4)$, viewing $\mathbf{Z}_4$ as a group.
2. Determine $A(\mathbf{Z}_6)$, viewing $\mathbf{Z}_6$ as a group.
**3. Combining Examples 1, 3, and 4 with your results from Exercises 1 and 2, can you determine $A(\mathbf{Z}_m)$ (viewing $\mathbf{Z}_m$ as a group) for every positive integer m? Prove your assertion.
4. If we view $\mathbf{Q}$ as an additive group, prove that $A(\mathbf{Q}) \cong \mathbf{Q}^*$, where $\mathbf{Q}^*$ denotes the multiplicative group of nonzero rational numbers.
5. (a) In $\mathbf{Q}(\sqrt{2})$, define ϕ by $(a + b\sqrt{2})\phi = a - b\sqrt{2}$. Prove that ϕ is a ring automorphism.
   *(b) Prove that I and ϕ are the only ring automorphisms of $\mathbf{Q}(\sqrt{2})$.
**6. Let $\mathcal{L}$ denote the group of linear functions from $\mathbf{R}$ to $\mathbf{R}$ that have nonzero slope. The group operation is composition. Prove that $I(\mathcal{L}) \cong \mathcal{L}$. (Hint: Prove $\mathbf{Z}(\mathcal{L}) = \{i\}$ and apply Theorem 29.)

§8. The Quotient Field of an Integral Domain

The integers $\mathbf{Z}$ is an example of an integral domain and is contained in the field $\mathbf{Q}$ of rational numbers, in fact, the subfield of $\mathbf{R}$ generated by $\mathbf{Z}$ is precisely $\mathbf{Q}$. In this section we shall show that every integral domain can be embedded in a field, that is, every integral domain is isomorphic to a subring of some field. In the process of proving this theorem we actually construct such a field using the elements of the given integral domain. The process used in the construction is motivated by the fact that rational numbers are fractions with integral numerators and denominators. By defining a suitable meaning for fractions with numerators and denominators from an integral domain, we can manufacture a system much like Q.

If D is an integral domain and D^* denotes the set of nonzero elements of D, we define the relation $\sim$ in $D \times D^*$ by

$$(a, b) \sim (c, d) \qquad \text{iff } ad = bc$$

Proposition 30

If D is an integral domain, the relation $\sim$ defined above is an equivalence relation on $D \times D^*$.

Proof: If $(x, y) \in D \times D^*$, since $xy = yx$, it follows that

$$(x, y) \sim (x, y)$$

If (x, y) and $(u, v) \in D \times D^*$ such that $(x, y) \sim (u, v)$, then $xv = yu$, which implies $uy = vx$, and therefore

$$(u, v) \sim (x, y)$$

If (x, y), (u, v), and (p, q) are in $D \times D^*$ such that $(x, y) \sim (u, v)$ and $(u, v) \sim (p, q)$, then $xv = yu$ and $uq = vp$. Multiplying yields $xvq = yuq$ and $yuq = yvp$. Thus, $xvq = yvp$ and $(xq)v = (yp)v$. Since D is an integral domain and $v \neq 0$, $xq = yp$, which implies $(x, y) \sim (p, q)$. $\square$

If D is an integral domain, let $Q(D)$ denote the set of all equivalence classes of $D \times D^*$ with respect to the equivalence relation $\sim$ defined above. Instead of denoting the classes by $\overline{(a, b)}$ we shall denote the class for (a, b) with respect to $\sim$ by a/b; that is, the class for (a, b) is a/b.

Proposition 31

If D is an integral domain and a/b and c/d are elements of $Q(D)$, then $a/b = c/d$ iff $ad = bc$.

Proof: Since a/b is the class for (a, b) and c/d is the class for (c, d),

$$a/b = c/d \qquad \text{iff } (a, b) \sim (c, d)$$
$$\text{iff } ad = bc \quad \square$$

Corollary 32

If a, b, and g are elements of the integral domain D and b and g are not zero, then

$$a/b = ag/bg$$

Proof: Exercise.

If D is an integral domain, we define operations $+$ and $\cdot$ in $Q(D)$ by

$$a/b + c/d = (ad + bc)/bd \quad \text{and} \quad (a/b) \cdot (c/d) = ac/bd$$

First we must show that these rules do, in fact, define operations. Since $b \neq 0$ and $d \neq 0$, $bd \neq 0$ and $(ad + bc)/bd$ and ac/bd are in $Q(D)$. Since the rules are given (as they were in $\mathbf{Z}_n$, factor groups, and factor rings) by using particular elements in the equivalence classes, we must make sure that these rules yield the same results even though different representations might be used in the computations.

Suppose $a/b = x/y$ and $c/d = u/v$, then

$$ay = bx \quad \text{and} \quad cv = du$$

Since $ay = bx$ and $cv = du$,

$$aycv = bxdu$$

that is, $(ac)(yv) = (bd)(xu)$, and therefore

$$ac/bd = xu/yv$$

Thus, the rule given for multiplication does define an operation on $Q(D)$. To show that addition is well defined, we need to show that

$$(ad + bc)/bd = (xv + yu)/yv$$

Since $ay = bx$ and $cv = du$ we have

$$(ay)(dv) = (bx)(dv) \quad \text{and} \quad (by)(cv) = (by)(du)$$

which implies

$$(ad)(yv) = (bd)(xv) \quad \text{and} \quad (bc)(yv) = (bd)(yu)$$

Adding yields

$$(ad)(yv) + (bc)(yv) = (bd)(xv) + (bd)(yu)$$

which implies

$$(ad + bc)(yv) = (bd)(xv + yu)$$

and therefore

$$(ad + bc)/bd = (xv + yu)/yv$$

Theorem 33

If D is an integral domain, then $Q(D)$ is a field with respect to the operations $+$ and $\cdot$ defined above.

Proof: If a/b, c/d, and x/y are elements of $Q(D)$, then

$$(a/b + c/d) + x/y = [(ad + bc)/bd] + x/y$$
$$= [(ad + bc)y + (bd)x]/bdy$$
$$= (ady + bcy + bdx)/bdy$$

and

$$a/b + (c/d + x/y) = a/b + [(cy + dx)/dy]$$
$$= [a(dy) + b(cy + dx)]/bdy$$
$$= (ady + bcy + bdx)/bdy$$

Therefore, addition is associative.

We leave the proofs of both commutative laws and the associative law of multiplication as exercises.

Since D is an integral domain, there is a 1 in D, $1 \neq 0$, and therefore $0/1$ is in $Q(D)$. If p is any nonzero element of D, $0/1 = 0/p$ since $0 \cdot p = 0 = 1 \cdot 0$. If $a/b \in Q(D)$, then

$$a/b + 0/1 = (a \cdot 1 + b \cdot 0)/(b \cdot 1) = (a + 0)/b = a/b$$

Therefore, $0/1$ is an additive identity.

If $a/b \in Q(D)$, then $a \in D$ and $-a \in D$.

$$a/b + (-a)/b = [ab + b(-a)]/bb$$
$$= [ab + (-ab)]/bb$$
$$= 0/bb$$
$$= 0/1$$

Thus, $-(a/b) = (-a)/b$ and each element of $Q(D)$ has an additive inverse.

Let a/b, c/d, u/v be elements of $Q(D)$.

$$(a/b + c/d)(u/v) = [(ad + bc)/bd](u/v)$$
$$= [(ad + bc)u]/bdv$$
$$= (adu + bcu)/bdv$$

and

$$(a/b)(u/v) + (c/d)(u/v) = au/bv + cu/dv$$
$$= [(au)(dv) + (bv)(cu)]/bvdv$$
$$= [(adu)v + (bcu)v]/(bdv)v$$
$$= (adu + bcu)v/(bdv)v$$
$$= adu + bcu/bdv$$

Therefore, one distributive law holds; the other follows once commutativity is established.

Since $(a/b)(1/1) = (a \cdot 1)/(b \cdot 1) = a/b$ for all a/b in $Q(D)$, $1/1$ is a multiplicative identity of $Q(D)$. Notice that if $a \neq 0$, $a/a = (1 \cdot a)/(1 \cdot a) = 1/1$.

We need yet to show that every nonzero element of $Q(D)$ has a multiplicative inverse.

If $a/b \neq 0/1$, then $a \cdot 1 \neq b \cdot 0$ and $a \neq 0$.

Thus, the nonzero elements of $Q(D)$ are just classes of the form a/b, where $a \neq 0$.

If $a \neq 0$, then, since $b \neq 0$, $(b, a) \in D \times D^*$ and b/a is an element of $Q(D)$.

Then $(a/b)(b/a) = ab/ba = ba/ba = 1/1$, and b/a is the multiplicative inverse of a/b. $\square$

Theorem 34

If D is an integral domain, then D is isomorphic to a subring of $Q(D)$.

Proof: Define ϕ from D into $Q(D)$ by $x\phi = x/1$ for each element x in D.
If a and b are in D, then

$$(a + b)\phi = (a + b)/1$$
$$= (a \cdot 1 + 1 \cdot b)/(1 \cdot 1)$$
$$= a/1 + b/1$$
$$= a\phi + b\phi$$

and

$$(ab)\phi = ab/1$$
$$= ab/(1 \cdot 1)$$
$$= (a/1)(b/1)$$
$$= (a\phi)(b\phi)$$

Thus, ϕ is a homomorphism.

If $x \in \ker \phi$, then $x\phi = 0/1$.
Since $x\phi = x/1$, $x/1 = 0/1$, which implies

$$x \cdot 1 = 1 \cdot 0$$

that is,

$$x = 0$$

Thus, $\ker \phi = \{0\}$ and ϕ is one-to-one.
Therefore, $D \cong D\phi$, which is a subring of $Q(D)$. $\square$

EXERCISE SET H

1. Prove Corollary 32.
2. Complete the proof of Theorem 33.
3. Prove that if D is a field, then $D \cong Q(D)$.

4. If $\mathbf{R}$ denotes the real numbers and $\mathbf{R}[x]$ the polynomials over $\mathbf{R}$, what is $Q(\mathbf{R}[x])$?

5. Prove: If F is a field and D is a subring of F containing 1, then $Q(D)$ is isomorphic to the subfield of F generated by D.

**6. Let R be a commutative ring with more than one element and with no proper zero divisors. Generalize the process for constructing $Q(D)$ to construct $Q(R)$, a field containing R. (Only the portions in which D has a 1 need be changed.)

§9. The Characteristic of a Ring

If R is a ring with 1 and $1 \neq 0$, then the subring generated by 1 is either finite or infinite. The following proposition shows that it is isomorphic to a familiar structure.

Proposition 35

If R is a ring with 1 and $1 \neq 0$, then the ring generated by 1 is either isomorphic to $\mathbf{Z}$ or to $\mathbf{Z}_m$ for some integer m.

Proof: We define ϕ from $\mathbf{Z}$ to R by $x\phi = x1$ where $x1$ is, as in Chapter 4, the sum of x 1's or the negative of the sum of $(-x)$ 1's. From previous arguments we know that ϕ is a group homomorphism. From the distributive law it also follows that, for integers x and y,

$$(xy)\phi = (xy)1 = x(y1) = (x1)(y1) = (x\phi)(y\phi)$$

Thus, ϕ is a homomorphism from $\mathbf{Z}$ into R. Since $\mathbf{Z}$ is generated by the integer 1, and $1\phi = 1$, $\mathbf{Z}\phi$ must be the ring generated by the ring element 1. If ker $\phi = \{0\}$, then $\mathbf{Z}\phi \cong \mathbf{Z}$ and if ker $\phi = \langle m \rangle$, then $\mathbf{Z}\phi \cong \mathbf{Z}_m$. $\square$

If R is a ring with 1, we say that R has *characteristic 0* if the ring generated by 1 is isomorphic to $\mathbf{Z}$ and has *characteristic m* if it is isomorphic to $\mathbf{Z}_m$. (Notice that the characteristic of R is the generator of the kernel of ϕ in the proof given above.) The ring is said to have *finite characteristic* if its characteristic is a positive integer.

Proposition 36

If R is a ring with 1 and $1 \neq 0$, then R has finite characteristic iff for some positive integer t, $tr = 0$ for each r in R.

Proof: If $m \neq 0$ and m is the characteristic of R, then $mr = m(1r) = (m1)r = 0r = 0$ for each r in R. Conversely, if $tr = 0$ for each r in R and t is a positive integer, then $t1 = 0$ and the ring generated by 1 has no more than t elements. $\square$

The notion of the characteristic of a ring can be extended to rings without 1 by redefining the characteristic of a ring R to be *the smallest positive integer m such that $mr = 0$ for all elements r in R if such an integer exists and to be zero otherwise.* By the arguments previously given we see that the two definitions are equivalent for rings with 1. If we examine Z_6 we see that the additive order of $\bar{2}$ is 3 and the additive order of $\bar{3}$ is 2. We know that Z_6 is not an integral domain, and the next assertion shows that such behavior cannot occur in an integral domain.

Proposition 37

If R is an integral domain, then all nonzero elements of R have the same additive order. Furthermore, if R has finite characteristic, then the characteristic is prime.

Proof: If every nonzero element has infinite order, then the theorem holds. Suppose $a \in R$ and the order of a is n for some $n \neq 0$. If $b \in R$ and $b \neq 0$, then $(nb)a = n(ba) = n(ab) = (na)b = 0b = 0$, and since R is an integral domain $nb = 0$. Thus, b has finite order. By a similar argument, if $mb = 0$ for some positive integer m, then $0 = (mb)a = m(ba) = m(ab) = (ma)b$, and therefore $ma = 0$. Thus, $n \mid m$ and since $nb = 0$, the order of b is n. Now suppose R has finite characteristic n and $n = pq$ for some positive integers p and q. Then $0 = n1 = (pq)1 = p(q1) = (p1)(q1)$ and, since R is an integral domain, either $p1 = 0$ or $q1 = 0$. Thus, either $p = n$ or $q = n$ and n must be prime. $\square$

Corollary 38

If F is a field with finite characteristic, then the characteristic of F is prime.

Proof: Immediate from Proposition 37.

Corollary 39

If R is an integral domain with finite characteristic, then at least one of the subrings of R is a field.

Proof: The subring generated by 1 is isomorphic to Z_p for some prime integer p. $\square$

EXERCISE SET I

1. What is the characteristic of $Z_2 \oplus Z_3$? Of $Z_2 \oplus Z_4$? Of $Z_3 \oplus Z_5$? Of $Z_6 \oplus Z_9$? In general, of $Z_n \oplus Z_m$?
2. What is the characteristic of the ring of even integers? Of the subring $\{\bar{0}, \bar{2}, \bar{4}, \bar{6}\}$ in Z_8? Of the subring $\{\bar{0}, \bar{6}, \overline{12}\}$ in Z_{18}?
3. Let R and S be rings with 1, let the characteristic of R be finite and let ϕ be a homomorphism from R onto S. Prove that S has finite characteristic and that the characteristic of S divides the characteristic of R.
4. Under the hypotheses of Exercise 3, if R and S are integral domains, what can be said about their characteristics?

5. Under the hypotheses of Exercise 3, what can be said about the characteristic of the kernel of ϕ?
6. Under the hypotheses of Exercise 4, what can be said about the characteristic of the kernel of ϕ? Now rethink your answer to Exercise 5.
**7. What is the characteristic of

$$\bigoplus_{p \in \mathbf{N}} Z_p$$

§10. The Ring of Endomorphisms of an Abelian Group

If M is an abelian group, which we shall denote additively, then the set of all functions from M into M is an abelian group with the operation of pointwise addition; that is

$$(x)(f + g) = xf + xg$$

In fact, if W is any nonempty set, the set of all functions from W into M is an abelian group with respect to pointwise addition for

$$(x)(f + g) = xf + xg = xg + xf = (x)(g + f)$$

for each x in W.

The identity in this group is the constant function 0, that is, the function that assigns each element of W to 0,

$$x0 = 0 \qquad \text{for each } x \text{ in } W,$$

since $(x)(f + 0) = xf + x0 = xf + 0 = xf$.
The inverse of such a function f is the function $-f$, defined by

$$(x)(-f) = -(xf) \qquad \text{for each element } x \text{ in } W$$

because

$$
\begin{aligned}
(x)[f + (-f)] &= xf + (x)(-f) \\
&= xf + [-(xf)] \\
&= 0 \\
&= x0
\end{aligned}
$$

and therefore $f + (-f) = 0$, the zero constant function.

The above argument is part of the justification of the general process discussed in §8 of Chapter 4 and can be expanded and modified to provide a solution for Exercise 11 in Set H of Chapter 4.

If M is an abelian group, we denote the set of all endomorphisms of M by $E(M)$. The set $E(M)$ is a subset of the group of all functions from M into M and is, in fact, endowed with a great deal of structure.

Theorem 40

If M is an abelian group, then $E(M)$ is a ring with respect to the operations of functional addition and composition.

Proof: First we need to show that $E(M)$ is a group with respect to addition. We can do so by showing that $E(M)$ is a subgroup of the group of all functions from M into M. Since the constant function 0 and the identity function I are in $E(M)$, $E(M) \neq \emptyset$.

Let $\phi, \lambda \in E(M)$, and let x and y be elements of M. Then

$$(x + y)(\phi + \lambda) = (x + y)\phi + (x + y)\lambda$$

by definition of functional addition

$$= x\phi + y\phi + x\lambda + y\lambda$$

because ϕ and λ are endomorphisms

$$= x\phi + x\lambda + y\phi + y\lambda$$

because M is abelian

$$= (x)(\phi + \lambda) + y(\phi + \lambda)$$

by definition of functional addition

Therefore, $\phi + \lambda$ is an endomorphism of M; that is,

$$\phi + \lambda \in E(M)$$

Let $\phi \in E(M)$ and let x and y be elements of M. Then

$$(x + y)(-\phi) = -[(x + y)\phi]$$

by definition of $-\phi$

$$= -[x\phi + y\phi]$$

because $\phi \in E(M)$

$$= [-(x\phi)] + [-(y\phi)]$$

because M is abelian

$$= x(-\phi) + y(-\phi)$$

by definition of $-\phi$

Therefore, $-\phi \in E(M)$.

By Theorem 16 of Chapter 4, $E(M)$ is a subgroup of the group of all functions from M into M.

Since the composition of homomorphisms is a homomorphism, the composition of endomorphisms is an endomorphism, and therefore composition is an operation on $E(M)$. Since the composition of functions in general is associative, the composition of endomorphisms is associative.

We need yet to establish the two distributive laws. Let ϕ, λ, and μ be in $E(M)$, and let $x \in M$.

$$(x)[(\phi + \lambda) \circ \mu] = [x(\phi + \lambda)]\mu$$

by definition of composition

$$= [x\phi + x\lambda]\mu$$

by definition of functional addition

$$= (x\phi)\mu + (x\lambda)\mu$$

because $\mu \in E(M)$

$$= (x)(\phi \circ \mu) + x(\lambda \circ \mu)$$

by definition of composition

$$= (x)[\phi \circ \mu + \lambda \circ \mu]$$

by definition of functional addition

Therefore, $(\phi + \lambda) \circ \mu = \phi \circ \mu + \lambda \circ \mu$.

Also

$$(x)[\mu \circ (\phi + \lambda)] = (x\mu)[\phi + \lambda]$$

by definition of composition

$$= (x\mu)\phi + (x\mu)\lambda$$

by definition of functional addition

$$= (x)(\mu \circ \phi) + (x)(\mu \circ \lambda) \qquad \text{by definition of composition}$$

$$= (x)[\mu \circ \phi + \mu \circ \lambda] \qquad \text{by definition of functional addition}$$

Therefore, $\mu \circ (\phi + \lambda) = \mu \circ \phi + \mu \circ \lambda$.

Since $E(M)$ satisfies Axioms (i)–(vi), $E(M)$ is a ring. $\square$

Notice that in the proof of the distributive laws the fact that ϕ, λ, and μ are endomorphisms was needed only for the first law and not for the second. It is not too hard to show that the set of all functions from an abelian group to itself satisfies all the ring axioms except one distributive law. The law that fails to hold depends on whether one writes functions on the right or left.

Theorem 41

If R is a ring with 1, then R is isomorphic to a subring of $E(R)$, the ring of endomorphisms of the additive structure of R.

Proof: For each element a in R, define f_a by $xf_a = xa$ for each element x in R.

If u and v are elements of R, then

$$(u + v)f_a = (u + v)a = ua + va = uf_a + vf_a$$

and therefore each f_a is an endomorphism of the abelian group R.

Define ϕ from R to $E(R)$ by $a\phi = f_a$ for each element a in R. Then

$$(a + b)\phi = f_{a+b} \quad \text{and} \quad a\phi + b\phi = f_a + f_b$$

If u is in R,

$$uf_{a+b} = u(a + b) = ua + ub = uf_a + uf_b = u(f_a + f_b)$$

Therefore, $(a + b)\phi = a\phi + b\phi$.

Also if u is in R,

$$uf_{ab} = uab = (ua)f_b = (uf_a)f_b = u(f_a \circ f_b)$$

Therefore,

$$f_{ab} = f_a \circ f_b \quad \text{and} \quad (ab)\phi = (a\phi) \circ (b\phi)$$

Thus, ϕ is a homomorphism.

If $a \in \ker \phi$, then $f_a = 0$, that is, f_a is the constant function zero. But $1f_a = 1a = a$. Thus, if $f_a = 0$, $a = 0$. Therefore, $\ker \phi = \{0\}$ and ϕ is one-to-one.

Since $R\phi$ is a subring of $E(R)$, ϕ is an isomorphism onto $R\phi$. $\square$

EXERCISE SET J

1. Determine all endomorphisms of the group Z_4. Make tables for the ring $E(Z_4)$.

2. Determine all endomorphisms of the group Z_6. Make tables for the ring $E(Z_6)$.

3. Prove that the ring of integers is isomorphic to the ring of endomorphisms of the additive group Z.

4. Does an analogous assertion to that from Exercise 3 hold for Z_m for every integer m? Justify your answer.

5. Prove that every endomorphism of the group $Z \oplus Z$ can be expressed in the form $(x, y)\phi = (ax + by, cx + dy)$, where a, b, c, and d are integers. Show, furthermore, that every function so expressed is an endomorphism.

6. (a) If $(x, y)\phi = (ax + by, cx + dy)$ and $(x, y)\lambda = (Ax + By, Cx + Dy)$ compute $\phi \circ \lambda$.

   (b) Can you find a more convenient notation for the elements of $E(Z \oplus Z)$?

7. (a) Show that $E(Z \oplus Z)$ is not a commutative ring.

   (b) Exhibit a zero divisor in $E(Z \oplus Z)$.

   (c) What are the units in $E(Z \oplus Z)$?

8. Let M be an abelian group and $E(M)$ denote the ring of endomorphisms of M. Show that $A(M)$ is the group of units of $E(M)$.

7
VECTOR SPACES

In many calculus courses a portion of the study is concerned with vectors in two- and three-dimensional space. These vectors are frequently characterized by arrows and said to be objects having two characteristics: direction and magnitude. These "arrows" are added by the "parallelogram law." (See Figure 7-1.)

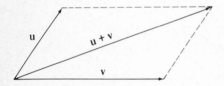

Figure 7-1

If one multiplies a vector v by a number r, the resultant is a vector $|r|$ times as long as v and in the same direction as v if $r > 0$ and in the opposite direction if $r < 0$. If v is a vector, $-v$ is the vector with the same magnitude but the opposite direction, and when one adds v and $-v$, the result is the zero vector 0. It is not too hard to show that vector addition satisfies both the Associative and Commutative laws, and therefore the set of all vectors in the plane is an abelian group and the set of all vectors in three-dimensional Euclidean space is an abelian group. If v is a vector in two- or three-dimensional space and r is a real number, then rv is also a vector and we can "multiply" vectors by real numbers.

Do we know of any other abelian groups in which the elements can be multiplied by real numbers? If we let $\mathbf{R}[x]$ denote the abelian group of polynomials over $\mathbf{R}$, then if

$$p(x) = a_0 + a_1 x + a_2 x^2 + \cdots + a_n x^n$$

and r is a real number,

184

$$rp(x) = ra_0 + ra_1 x + ra_2 x^2 + \cdots + ra_n x^n$$

is also a polynomial. If we let $\mathcal{C}$ denote the abelian group of continuous functions and let $\mathcal{D}$ denote the abelian group of differentiable functions we see that the elements in these groups also can be multiplied by real numbers; that is, if r is a real number and f is a continuous function then rf is a continuous function, and if g is a differentiable function, rg is also a differentiable function.

Consider now the group of polynomials $\mathbf{Q}[x]$, where $\mathbf{Q}$ denotes the field of rational numbers. For any polynomial $p(x)$ in $\mathbf{Q}[x]$ and any rational number b, $bp(x)$ will also be a polynomial in $\mathbf{Q}[x]$. If, in the group $\mathbf{Q} \oplus \mathbf{Q}$, we define multiplication by a rational number b by $b(m, n) = (bm, bn)$, we have a way of multiplying each element of this group by each rational number. Similarly, if, in the group $\mathbf{Z}_7 \oplus \mathbf{Z}_7 \oplus \mathbf{Z}_7$, we define multiplication by an element of $\mathbf{Z}_7$ by

$$\overline{m}(\overline{a}, \overline{b}, \overline{c}) = (\overline{ma}, \overline{mb}, \overline{mc})$$

we have a way of multiplying each element of the group by each element of the field $\mathbf{Z}_7$.

If we let α and β represent elements of the field under consideration and u and v represent elements of the group under consideration, it is easy to verify that the following equations hold in each of the examples discussed above.

(i) $1u = 1$, where 1 is the multiplicative identity in the field.
(ii) $\alpha(\beta u) = (\alpha\beta)u$.
(iii) $\alpha(u + v) = \alpha u + \alpha v$.
(iv) $(\alpha + \beta)u = \alpha u + \beta u$.

In each of the examples we had an abelian group, a field, and a way of multiplying group elements by field elements so that the product was again a group element and, furthermore, equations (i)–(iv) held in each example. In this chapter we shall abstract from these examples and study the class of systems known as vector spaces.

§1. Axioms and Elementary Properties of Vector Spaces

Axioms for a Vector Space

A set V is said to be a *vector space* over the field F if V is an abelian group (we shall denote the operation in V additively) and for each pair of elements α in F and v in V, there is an element in V, called the *product* of α and v and denoted by αv, such that the following conditions hold:

(i) If 1 is the multiplicative identity of F, and u is in V, then $1u = u$.
(ii) If α and β are elements of F and u is an element of V, then $\alpha(\beta u) = (\alpha\beta)u$.

(iii) If α is an element of F and u and v are elements of V, then $\alpha(u + v) = \alpha u + \alpha v$.

(iv) If α and β are elements of F and u is an element of V, then $(\alpha + \beta)u = \alpha u + \beta u$.

The elements of the vector space V are called *vectors*, and the elements of the field F are called *scalars*. The process of forming the product αv, where $\alpha \in F$ and $v \in V$, is called *scalar multiplication*. A vector space over the field F will frequently be called an F-space; for example, $Q[x]$ is a Q-space and C is an R-space. By the phrase "V is an F-space" we mean that F is a field and that V is a vector space over F.

Other examples of vector spaces are the following:

(i) The additive structure of any field is a vector space over the field itself.

(ii) The additive group of all functions from R into R is an R-space.

(iii) The additive structure of R is a Q-space with the usual multiplication.

(iv) The additive structure of the complex numbers is an R-space.

(v) If K is a field and F is a subfield of K, then K is an F-space with the usual multiplication.

(vi) If S is a ring, F is a subring of S and F is, in fact, a field, then S is an F-space with the usual multiplication.

(vii) If F is a field, then $F \oplus F \oplus F \oplus F$ is an F-space with multiplication defined by $a(b, c, d, f) = (ab, ac, ad, af)$ for any elements a, b, c, d, and f in F.

(viii) The group of all integrable functions from $[0, 1]$ into R is an R-space.

In discussing vector spaces we shall use the same symbol 0 for the zero of the vector space and for the zero of the field. The context will make clear which element is being discussed.

Since a vector space is an abelian group, all assertions previously established for groups will hold for vector spaces. Before proceeding further, we list some of those that we shall use frequently.

Proposition 1

If V is an F-space and u is an element of V such that $u + u = u$, then $u = 0$.

Proposition 2

If V is an F-space and u, v, and w are elements of V such that $u + v = w + v$, then $u = w$.

Corollary 3

If V is an F-space and u and v are elements of V such that $u + v = 0$, then $u = -v$.

Corollary 4

If u is an element of the F-space V, then $-(-u) = u$.

Proposition 5

If V is an F-space and u and v are elements of V, then $-(u + v) = (-u) + (-v)$.

The vector space results we obtain that are not purely group theoretic in nature *must* be related to the scalar multiplication. Some of the elementary properties that one might expect are established by the following propositions.

Proposition 6

If V is an F-space and u is an element of V, then $0u = 0$.

Proof: Exercise (Hint: Adapt the proof of Proposition 5, Chapter 2.)

Proposition 7

If u is an element of the F-space V, then $(-1)u = -u$.

Proof:

$$
\begin{aligned}
(-1)u + u &= (-1)u + 1u && \text{by axiom (i)} \\
&= [(-1) + 1]u && \text{by axiom (iv)} \\
&= 0u \\
&= 0 && \text{by Proposition 6}
\end{aligned}
$$

Therefore,

$$(-1)u = -u \qquad \text{by Corollary 3} \quad \square$$

Corollary 8

If u is an element of the F-space V and α is an element of F, then $-(\alpha u) = (-\alpha)u = \alpha(-u)$.

Proof: Exercise.

As in all additively denoted abelian groups, subtraction in vector spaces is defined by

$$u - v = u + (-v)$$

for any vectors u and v in the F-space V.

Proposition 9

If V is an F-space, u and v are elements of V, and α and β are elements of F, then

$$(\alpha - \beta)u = \alpha u - \beta u \quad and \quad \alpha(u - v) = \alpha u - \alpha v$$

Proof: Exercise.

EXERCISE SET A

1. Which of the following sets are vector spaces over $\mathbf{R}$? If not, why not?
   (a) all polynomials over $\mathbf{R}$ of degree 8 or less
   (b) all functions from $\mathbf{R}$ into $\mathbf{R}$ that have at most finitely many discontinuities
   (c) all bounded functions from $\mathbf{R}$ into $\mathbf{R}$
   (d) all periodic functions from $\mathbf{R}$ into $\mathbf{R}$
   (e) all functions f from $\mathbf{R}$ into $\mathbf{R}$ such that $f = f \cdot f$
   (f) all differentiable functions g such that $g' = g$
   (g) all differentiable functions f such that $f'' + f = 0$
   (h) all integrable functions g such that $\displaystyle\int_0^1 g(x)\, dx \leq 1$

2. Prove Proposition 6.
3. Prove Corollary 8.
4. Prove Proposition 9.
5. Prove: If 0 denotes the zero of the F-space V and $\alpha \in F$, then $\alpha 0 = 0$.
6. Prove: If V is an F-space, α is an element of F, and u is a vector in V such that $\alpha u = 0$, then either $\alpha = 0$ or $u = 0$.
7. Prove: If α and β are elements of the field F and u is a vector in an F-space such that $\alpha u = \beta u$, then either $\alpha = \beta$ or $u = 0$.
8. Prove: If u and v are vectors in the F-space V and α and β are nonzero scalars such that $\alpha u + \beta v = 0$, then for some scalar γ, $u = \gamma v$.

§2. Subspaces

The concept of a subspace of a vector space is analogous to the ideas of a subgroup of a group, a subring of a ring, and a subfield of a field.

If V is a vector space over the field F and W is a subset of V, W is a *subspace* of V if W is a vector space over F with the same addition and scalar multiplication as defined for V.

According to the definition, W is a subspace of V if W is a subgroup of V with respect to addition and satisfies the axioms for the scalar multiplication. Since Axioms (i)–(iv) hold for all vectors in V, they must hold for those in W and we might be led to believe that there is nothing to worry about with respect to scalar multiplication. This is, however, not the case. In order to be an F-space, scalar multiplication must be defined in W; that is, if $\alpha \in F$ and $w \in W$, we must have αw in W. At this point it is convenient to introduce a new meaning for a familiar word. If V is an F-space and U is a subset of V, we say that U is *closed under scalar multiplication* if for each α in F and u in U, αu is in U.

Now we are ready to establish an assertion that provides a simple criterion for determining if a subset of a vector space is a subspace.

Proposition 10

If V is an F-space and W is a nonempty subset of V, then W is a subspace of V iff W is closed under vector addition and scalar multiplication.

Proof: Clearly every subspace is closed under vector addition and scalar multiplication.

Suppose W is a nonempty subset of V closed under vector addition and scalar multiplication. If u is an element of W, then $(-1)u$ is in W. Since $(-1)u = -u$, $-u$ is in W. Since W is closed under addition and the inverse of each element of W is also in W, W is a subgroup of V. Since W is closed under scalar multiplication and Axioms (i)–(iv) are hereditary, W is a subspace of V. □

The following assertion provides another method of showing that a subset of a vector space is a subspace and is probably the most commonly used criterion.

Proposition 11

If V is an F-space and W is a nonempty subset of V, then W is a subspace of V if for each pair of elements α and β in F and each pair of vectors u and v in W, $\alpha u + \beta v$ is in W.

Proof: If u and v are elements of W, then $1u + 1v$ is in W; that is, $u + v$ is in W, and therefore W is closed under vector addition. If $\alpha \in F$ and $v \in W$, then $\alpha v + 0v$ is in W; that is, αv is in W, and therefore W is closed under scalar multiplication. □

From previous studies we know that the intersection of a collection of subgroups of a group is a subgroup, the intersection of a collection of subrings of a ring is a subring, and the intersection of a collection of subfields of a field is a subfield. An analogous assertion holds for subspaces and we leave the proof as an exercise.

Proposition 12

If V is an F-space and S is a nonempty collection of subspaces of V, then $\bigcap S$ is a subspace of V.

Proof: Exercise.

If S is any subset of the F-space V, the set of all subspaces of V containing S is not empty, because V itself is such a subspace. By Proposition 12 the intersection of all subspaces containing S is a subspace. The following definition is analogous to those given in the study of groups, rings, and fields.

If S is a subset of the F-space V, the intersection of all subspaces containing S is called the *subspace generated by* S and will be denoted by $\langle S \rangle$.

If W is the subspace generated by S, S is said to be a set of *generators* for W. Another, probably more commonly used phrasing, is to say that S is a *spanning set* for W or that W is the subspace *spanned* by S. A subspace generated by one vector is, as expected, said to be *cyclic*. The subspace generated by a single vector u is denoted by $\langle u \rangle$.

Proposition 13

If V is an F-space and u is a vector in V, then $\langle u \rangle = \{\alpha u : \alpha \in F\}$.

Proof: Exercise.

If V is an F-space and T is a subset of V, a finite sum of the form

$$\alpha_1 t_1 + \alpha_2 t_2 + \cdots + \alpha_n t_n$$

where each α_i is in F and each t_i is in T is called a *linear combination* of vectors in T. In general, not all linear combinations of vectors in T will again be in T; in fact this happens only if T is a subspace.

Proposition 14

If V is an F-space and S is a nonempty subset of V, then the subspace of V spanned by S is the set of all linear combinations of vectors in S.

Proof: Let

$$u = \alpha_1 t_1 + \alpha_2 t_2 + \cdots + \alpha_n t_n \quad \text{and} \quad v = \beta_1 s_1 + \beta_2 s_2 + \cdots + \beta_m s_m$$

where each α_i and β_j is in F and each t_i and s_j is in S. If γ and δ are elements of F, then

$$\gamma u + \delta v = \gamma(\alpha_1 t_1 + \cdots + \alpha_n t_n) + \delta(\beta_1 s_1 + \cdots + \beta_m s_m)$$
$$= \gamma \alpha_1 t_1 + \cdots + \gamma \alpha_n t_n + \delta \beta_1 s_1 + \cdots + \delta \beta_m s_m$$

Thus, $\gamma u + \delta v$ is a linear combination of vectors in S. By Proposition 11 the set of all linear combinations of vectors in S is a subspace, and this subspace contains S, because $s = 1s$ is a linear combination for each vector s in S. Hence, the subspace spanned by S is contained in the set of all linear combinations of vectors in S.

Also, however, if W is any subspace containing S, then W must contain all linear combinations of vectors in S, because W is closed under vector addition and scalar multiplication. Since the subspace spanned by S must contain all linear combinations of vectors in S and the set of all such linear combinations must contain the subspace spanned by S, the subspace spanned by S is the set of all linear combinations of vectors in S. $\square$

EXERCISE SET B

1. Which of the following are subspaces of $\mathbf{R} \oplus \mathbf{R} \oplus \mathbf{R}$, the usual three-dimensional Euclidean space? If not, why not?
   (a) The XY plane

(b) The Y axis

(c) $\{(x, y, z):z = 3, x \in \mathbf{R}, y \in \mathbf{R}\}$

(d) $\{(x, y, z):x + y = 0 \text{ and } z = 0\}$

(e) $\{(x, y, z):x + y + z = 0\}$

(f) $\{(x, y, z):x + y + z = 2\}$

(g) $\{(x, y, z):x = y = 2z\}$

(h) $\{(x, y, z):x = y + 1 = z - 1\}$

(i) $\{(x, y, z):x = y^2 = z\}$

(j) $\{(x, y, z):x + 2y + 3z = 0 \quad \text{and} \quad 5x + 2y + 3z = 0\}$

(k) $\{(x, y, z):x + y + z \geq 0\}$

(l) $\{(x, y, z): |x| = |y| = |z|\}$

2. Which of the following are subspaces of the **R**-space of all functions from **R** into **R**? If not, why not?

(a) the set of increasing functions

(b) the set of differentiable functions f such that $2f' + f = 0$

(c) the set of all differentiable functions g such that $g' + g = \sin x$

(d) the set of all twice-differentiable functions

(e) the set of all twice-differentiable functions f such that $4f'' + f = 0$

(f) the set of all five-times differentiable functions g such that

$$7g^{[5]} - 4g^{[4]} + 8g^{[3]} + 19g^{[2]} - 0.3g' + 1.8g = 0$$

**(g) Generalize part (f).

3. Prove Proposition 12.

4. Prove Proposition 13.

5. In $\mathbf{R} \oplus \mathbf{R} \oplus \mathbf{R}$, describe geometrically the subspaces given below.

(a) $\langle (1, 2, 3) \rangle$

(b) $\langle \{(1, 0, 0), (0, 1, 0)\} \rangle$

(c) $\langle \{(1, 0, 0), (0, 1, 0), (-1, -1, 0)\} \rangle$

(d) $\langle \{(x, y, z):x > y \geq 0 \text{ and } z = 0\} \rangle$

(e) $\langle \{(1, 0, 0), (1, 1, 0), (1, 1, 1)\} \rangle$

(f) $\langle \{(x, y, z): |x| = |y| \text{ and } z = 0\} \rangle$

(g) $\langle \{(x, y, z):x = 0 \text{ and } y^2 + z^2 = 1\} \rangle$

(h) $\langle \{(x, y, z):x^2 + y^2 + z^2 = 1\} \rangle$

(i) $\langle \{(0.1, 0, 0), (0, 17, 0), (0, 0, \sqrt{3})\} \rangle$

(j) $\langle \{(1, 2, 3), (2, -1, 0), (3, 6, -5)\} \rangle$

6. If V is an F-space and S and T are subsets of V, we define $S + T$ by

$$S + T = \{x + y:x \in S \text{ and } y \in T\}$$

(This is just the complex addition from group theory.) Prove: If V is an F-space and S and T are subspaces of V, then $S + T$ is a subspace of V.

§3. Vector Space Constructions

In the study of groups and rings we constructed factor groups and factor rings, and we now perform an analogous construction for *factor spaces*. The proofs are very much like those made previously and we leave most of them as exercises.

If W is a subspace of the F-space V, and u and v are vectors in V, we say that *u is congruent to v* modulo W iff $u - v \in W$. We denote this relation by $u \equiv v \,(\text{mod } W)$.

Since subspaces are, in fact, subgroups, we know from group theory that congruence modulo a subspace is an equivalence relation and that the congruence classes are just the right cosets. If W is a subspace of the F-space V, every right coset is a left coset because V is abelian. Thus, the congruence class for the vector u can be written as

$$\bar{u} = W + u = u + W$$

Furthermore, since F-spaces are abelian groups, all subspaces are normal, and therefore congruence modulo a subspace is compatible with vector addition; that is, if W is a subspace of the vector space V and u, v, w, and z are vectors in V such that $u \equiv v \,(\text{mod } W)$ and $w \equiv z \,(\text{mod } W)$, then $u + w \equiv v + z \,(\text{mod } W)$.

In the following assertion we see that congruence modulo a subspace is also compatible with scalar multiplication.

Proposition 15

If W is a subspace of the F-space V and u and v are vectors in V such that $u \equiv v \,(mod \, W)$ and α is an element of F, then $\alpha u \equiv \alpha v \,(mod \, W)$.

Proof: If $u \equiv v \,(\text{mod } W)$, then $u - v \in W$.
If $\alpha \in F$, then $\alpha(u - v) \in W$, because W is a subspace.
Then $\alpha u - \alpha v \in W$ and, by definition, $\alpha u \equiv \alpha v \,(\text{mod } W)$.   □

If W is a subspace of the F-space V, we know from group theory that V/W is a group with addition of classes defined by $\bar{u} + \bar{v} = \overline{u + v}$ for each pair of classes $\bar{u}$ and $\bar{v}$ in V/W. If α is in F and u is in V, we define $\alpha\bar{u}$ by

$$\alpha\bar{u} = \overline{\alpha u}$$

We need to show that this rule does adequately define a scalar multiplication on V/W. If $\bar{u} = \bar{v}$, then $u \equiv v \,(\text{mod } W)$ and, by Proposition 15, $\alpha u \equiv \alpha v \,(\text{mod } W)$. Thus, $\overline{\alpha u} = \overline{\alpha v}$ and the rule does, in fact, define a scalar multiplication on V/W.

Theorem 16

If W is a subspace of the F-space V, then V/W is an F-space with respect to the operation of addition defined by $\bar{u} + \bar{v} = \overline{u + v}$ for each $\bar{u}$ and $\bar{v}$ in V/W and the scalar multiplication defined by $\alpha\bar{u} = \overline{\alpha u}$ for each α in F and each $\bar{u}$ in V/W.

Proof: From group theory we know that V/W is a group with respect to the addition defined above. We need to verify that Axioms (i)–(iv) are satisfied by the scalar multiplication.

If $\bar{u} \in V/W$, $1\bar{u} = \overline{1u} = \bar{u}$. Therefore, Axiom (i) is satisfied.
If $\bar{u} \in V/W$ and α and β are elements of F, then

$$\alpha(\beta\bar{u}) = \alpha(\overline{\beta u}) = \overline{\alpha(\beta u)} = \overline{(\alpha\beta)u} = (\alpha\beta)\bar{u}$$

and Axiom (ii) is satisfied.
We leave the establishment of Axioms (iii) and (iv) as exercises. □

If H and K are F-spaces, we know that $H \oplus K$ is an abelian group. In order to extend the notion of direct sums to F-spaces, we need a scalar multiplication. For each scalar α in F and each element (h, k) in $H \times K$, we define scalar multiplication by

$$\alpha(h, k) = (\alpha h, \alpha k)$$

Proposition 17

If H and K are F-spaces, then $H \times K$ is an F-space with respect to addition defined by $(a, b) + (c, d) = (a + c, b + d)$ for each (a, b) and (c, d) in $H \times K$ and scalar multiplication defined by $\alpha(h, k) = (\alpha h, \alpha k)$ for each α in F and (h, k) in $H \times K$.

Proof: As noted, $H \times K$ is known to be a group with respect to component-wise addition. We shall verify Axiom (iii) and leave the verifications of the other axioms as exercises.

Let $\alpha \in F$ and (a, b) and (c, d) be elements of $H \times K$. Then

$$\begin{aligned}
\alpha[(a, b) + (c, d)] &= \alpha(a + c, b + d) \\
&= (\alpha[a + c], \alpha[b + d]) \\
&= (\alpha a + \alpha c, \alpha b + \alpha d) \\
&= (\alpha a, \alpha b) + (\alpha c, \alpha d) \\
&= \alpha(a, b) + \alpha(c, d) \quad \square
\end{aligned}$$

If H and K are F-spaces, the set $H \times K$ with the addition and scalar multiplication defined above is called the *direct sum* of H and K and denoted by $H \oplus K$.

More generally, if $V_1, V_2, \ldots, V_n$ is any collection of F-spaces, we can form the abelian group $V_1 \oplus V_2 \oplus \cdots \oplus V_n$, where the elements are ordered n-tuples and the operation is component-wise addition. In this group we define scalar multiplication by $\alpha(x_1, x_2, \ldots, x_n) = (\alpha x_1, \alpha x_2, \ldots, \alpha x_n)$ for each $(x_1, x_2, \ldots, x_n)$ in $V_1 \oplus V_2 \oplus \cdots \oplus V_n$ and each α in F. It is easy to verify that this scalar multiplication satisfies Axioms (i)–(iv) and we leave the verification to the reader.

Theorem 18

If $V_1, \ldots, V_n$ is a collection of F-spaces, then $V_1 \oplus \cdots \oplus V_n$ is an F-space with respect to the vector addition defined by $(a_1, \ldots, a_n) + (b_1, \ldots, b_n) = (a_1 + b_1, \ldots, a_n + b_n)$ for each pair $(a_1, \ldots, a_n)$ and $(b_1, \ldots, b_n)$ in

$V_1 \oplus \cdots \oplus V_n$ and scalar multiplication defined by $\alpha(x_1, \ldots, x_n) = (\alpha x_1, \ldots, \alpha x_n)$ for each α in F and $(x_1, \ldots, x_n)$ in $V_1 \oplus \cdots \oplus V_n$.

Proof: Exercise (Hint: Just adapt the proof of Proposition 17.)

If S is a nonempty set and V is an F-space, the set of all functions from S into V is known to be an abelian group with respect to pointwise addition of functions. If α is in F and g is a function from S into V, we define a new function αg by

$$(x)(\alpha g) = \alpha(xg)$$

Thus, the image of an element x under αg is defined to be the product of α and the image of x under g. Again, it is easy to verify that this scalar multiplication satisfies Axioms (i)–(iv) and we leave it as an exercise.

Proposition 19

If S is a nonempty set and V is an F-space, then the set of all functions from S into V is an F-space with respect to pointwise addition and scalar multiplication defined above.

Proof: Exercise.

EXERCISE SET C

1. Complete the proof of Theorem 16.
2. In $\mathbf{R} \oplus \mathbf{R} \oplus \mathbf{R}$, consider the subspace $W = \{(x, y, z):x = y = z\}$. Describe W geometrically. Now describe geometrically $(\mathbf{R} \oplus \mathbf{R} \oplus \mathbf{R})/W$.
3. In $\mathbf{R} \oplus \mathbf{R} \oplus \mathbf{R}$, let $M = \{(x, y, z):z = 0\}$. Describe M geometrically. Now describe geometrically $(\mathbf{R} \oplus \mathbf{R} \oplus \mathbf{R})/M$.
4. In $\mathbf{R} \oplus \mathbf{R} \oplus \mathbf{R}$, let $P = \{(x, y, z):x + 2y + 3z = 0\}$. Describe P geometrically. Now describe geometrically $(\mathbf{R} \oplus \mathbf{R} \oplus \mathbf{R})/P$.
5. Complete the proof of Proposition 17.
6. Prove Theorem 18.
7. Prove Proposition 19.
8. Let S be the set of all sequences of real numbers with the usual addition and scalar multiplication of sequences. By Proposition 19 we know S is an $\mathbf{R}$-space, because sequences are just functions that have the set $\mathbf{N}$ of positive integers as a domain. Let W be all sequences that are 0 from some point on; that is,

$$W = \{\{a_n\}:\text{for some integer } k, \text{if } m > k, a_m = 0\}$$

   Show that W is a subspace of S.
9. Let V be an F-space, T a nonempty set, and W the F-space of all functions from T into V. Define H by

$$H = \{f \in W:xf \neq 0 \text{ for at most finitely many elements } x \text{ in } T\}$$

   Show that H is a subspace of W.

**10. Let V be an F-space and let $\sim$ be an equivalence relation on V that is compatible with vector addition and scalar multiplication; that is, if u, v, y, and w are vectors such that $u \sim v$ and $y \sim w$, then $u + y \sim v + w$ and if α is a scalar and u and v are vectors such that $u \sim v$, then $\alpha u \sim \alpha v$. Prove that $\sim$ is congruence modulo some subspace of V. [Hint: Let $\overline{0}$ denote the class for 0 with respect to $\sim$ and show first that $\overline{0}$ is a subspace. Then show that $u \sim v$ iff $u \equiv v \pmod{\overline{0}}$.]

**11. Let $\{V_i : i \in I\}$ be a nonempty collection of F-spaces, one space V_i associated with each element i of the set I. Let $S = \bigcup V_i$ and now define ΠV_i to be the set of all functions f from I into S such that $(i)f \in V_i$ for each i in I. Show that ΠV_i is an F-space under pointwise addition of functions and scalar multiplication defined by $(i)(\alpha f) = \alpha[(i)f]$ for each function f, scalar α, and element i in I. The F-space ΠV_i is the *complete direct sum* of the family $\{V_i : i \in I\}$.

**12. In the vector space ΠV_i above, define $\oplus V_i$ by

$$\oplus V_i = \{f \in \Pi V_i : (i)f = 0 \text{ for all but finitely many elements } i \text{ in } I\}$$

Show that $\oplus V_i$ is an F-space. The space $\oplus V_i$ is usually called the *restricted direct sum* or just the *direct sum* of the family $\{V_i : i \in I\}$.

§4. Isomorphisms and Linear Transformations

In Chapter 6 we studied operation-preserving functions from groups into groups and from rings into rings. We now extend these ideas to vector spaces.

If V and W are F-spaces and ϕ is a function from V into W, ϕ is a *linear transformation* from V into W if, for each pair of elements u and v in V and each pair of elements α and β in F,

$$(\alpha u + \beta v)\phi = \alpha(u\phi) + \beta(v\phi)$$

Notice that, in the case where $\alpha = \beta = 1$, the above equation becomes

$$(u + v)\phi = u\phi + v\phi$$

and therefore ϕ is a group homomorphism. If we let $\beta = 0$, the above equation becomes

$$(\alpha u)\phi = \alpha(u\phi)$$

which means that ϕ preserves scalar multiplication. Thus, a linear transformation is a function that preserves vector addition and scalar multiplication. Conversely one can show that a function which preserves addition and scalar multiplication is a linear transformation.

If V and W are F-spaces and ϕ is a function from V into W, ϕ is an *isomorphism* if ϕ is a one-to-one, onto, linear transformation.

If there is an isomorphism from the F-space V onto the F-space W, we say that V is isomorphic to W and write $V \cong W$. As with homomorphisms, a one-to-one linear transformation is called an *injection*, an *embedding*, a *monomorphism*, or an *isomorphism into*. An onto linear transformation is called a *surjection* or an *epimorphism*.

The following assertions concerning linear transformations are analogous to those concerning homomorphisms. We leave the proofs as exercises.

Proposition 20

If V, W, and Y are F-spaces, ϕ is a linear transformation from V into W, and λ is a linear transformation from W into Y, then $\phi \circ \lambda$ is a linear transformation from V into Y.

Proof: Exercise.

Proposition 21

If ϕ is an isomorphism from the F-space V onto the F-space W, then ϕ^{-1} is an isomorphism from W onto V.

Proof: Exercise.

Proposition 22

If ϕ is an isomorphism from the F-space V onto the F-space W and λ is an isomorphism from W onto the F-space Y, then $\phi \circ \lambda$ is an isomorphism from V onto Y.

Proof: Exercise.

Since the identity function from an F-space to itself is clearly an isomorphism, we see again that the relation of "is isomorphic to" is an equivalence relation on any set of F-spaces.

If ϕ is a linear transformation from the F-space V into the F-space W, the *kernel* of ϕ is defined as it was for all groups; that is,

$$\text{kernel of } \phi = \{v \in V : v\phi = 0\} = 0\phi^{-1}$$

Proposition 23

If ϕ is a linear transformation from the F-space V into the F-space W, then the kernel of ϕ is a subspace of V.

Proof: Exercise.

Recall that in Chapter 6 we saw that a group homomorphism ϕ is one-to-one if and only if ker ϕ is the identity. Since linear transformations are group homomorphisms, a linear transformation ϕ is one-to-one iff ker $\phi = \{0\}$.

Theorem 24 (Fundamental Homomorphism Theorem for Vector Spaces)

If ϕ is a linear transformation from the F-space V onto the F-space W and K is the kernel of ϕ, then $V/K \cong W$.

Proof: From group theory we know that the function $\bar{\phi}$ defined by $\bar{u}\bar{\phi} = u\phi$ is a group isomorphism from V/K onto W. We need yet to show that $\bar{\phi}$ preserves scalar multiplication. Let $\alpha \in F$ and $\bar{u} \in V/K$. Then

$$
\begin{aligned}
(\alpha\bar{u})\bar{\phi} &= (\overline{\alpha u})\bar{\phi} && \text{by definition of scalar multiplication in } V/K \\
&= (\alpha u)\phi && \text{by definition of } \bar{\phi} \\
&= \alpha(u\phi) && \text{because } \phi \text{ preserves scalar multiplication} \\
&= \alpha(\overline{u\phi}) && \text{by definition of } \bar{\phi} \quad \square
\end{aligned}
$$

If H and K are subsets of the F-space V, our definition of complex addition from Chapter 4 defines $H + K$ by

$$H + K = \{x + y : x \in H \text{ and } y \in K\}$$

If H and K are subspaces of the F-space V, it is easy to show that $H + K$ is a subspace of V and in fact, $H + K$ is the subspace generated by $H \cup K$. The following theorems are analogues to Theorems 27 and 28 from Chapter 6, and we leave the proofs as exercises. In each case the proof given previously needs to be translated to additive notation and the given isomorphism must be shown to preserve scalar multiplication.

Theorem 25

If M and N are subspaces of the F-space V, then $M/(M \cap N) \cong (M + N)/N$

Proof: Exercise.

Theorem 26

If V is an F-space, B is a subspace of V, and A is a subspace of B, then $V/A/B/A \cong V/B$.

Proof: Exercise (Hint: In our definition of scalar multiplication in a factor space V/W, $\alpha\bar{u} = \overline{\alpha u}$ means that $\alpha(W + u) = W + \alpha u$.)

If V and W are F-spaces and ϕ and λ are linear transformations from V into W, then the pointwise sum of ϕ and λ defined by

$$(v)(\phi + \lambda) = v\phi + v\lambda \qquad \text{for each } v \text{ in } V$$

is a function from V into W. From our discussion in §3 we know that the set of all functions from V into W is an F-space. We shall let $\mathcal{L}(V, W)$ denote the set of all linear transformations from V into W and, in the case where $V = W$, we shall abbreviate $\mathcal{L}(V, V)$ to $\mathcal{L}(V)$.

Theorem 27

If V and W are F-spaces, then $\mathcal{L}(V, W)$ is a subspace of the F-space of all functions from V into W.

Proof: $\mathcal{L}(V, W) \neq \phi$, because the constant function 0 is a linear transformation. We need to show that $\mathcal{L}(V, W)$ is closed under addition and scalar multiplication.

Let ϕ and λ be functions in $\mathcal{L}(V, W)$ and suppose u and v are vectors in V. Then

$$
\begin{aligned}
(u + v)(\phi + \lambda) &= (u + v)\phi + (u + v)\lambda && \text{by definition of} \\
&&& \text{addition of functions} \\
&= (u\phi + v\phi) + (u\lambda + v\lambda) && \text{because } \phi \text{ and } \lambda \text{ are} \\
&&& \text{linear transformations} \\
&= (u\phi + u\lambda) + (v\phi + v\lambda) && \text{by associativity and} \\
&&& \text{commutativity of} \\
&&& \text{vector addition} \\
&= u(\phi + \lambda) + v(\phi + \lambda) && \text{by definition of} \\
&&& \text{functional addition}
\end{aligned}
$$

If $\beta \in F$ and $u \in V$, then

$$
\begin{aligned}
(\beta u)(\phi + \lambda) &= (\beta u)\phi + (\beta u)\lambda && \text{by definition of functional} \\
&&& \text{addition} \\
&= \beta(u\phi) + \beta(u\lambda) && \text{because } \phi \text{ and } \lambda \text{ are linear} \\
&= \beta[u\phi + u\lambda] && \text{by Axiom (iii) for vector spaces} \\
&= \beta[u(\phi + \lambda)] && \text{by definition of functional} \\
&&& \text{addition}
\end{aligned}
$$

Thus, $\phi + \lambda$ preserves addition and scalar multiplication and $\mathcal{L}(V, W)$ is closed under addition.

We need yet to show that $\mathcal{L}(V, W)$ is closed under scalar multiplication. Let $\alpha \in F$ and $\phi \in \mathcal{L}(V, W)$ and suppose u and v are vectors in V. Then

$$
\begin{aligned}
(u + v)(\alpha\phi) &= \alpha[(u + v)\phi] && \text{by definition of scalar} \\
&&& \text{multiplication of functions} \\
&= \alpha(u\phi + v\phi) && \text{because } \phi \text{ is a linear} \\
&&& \text{transformation} \\
&= \alpha(u\phi) + \alpha(v\phi) && \text{by Axiom (iii) for vector spaces} \\
&= u(\alpha\phi) + v(\alpha\phi) && \text{by definition of scalar} \\
&&& \text{multiplication of functions}
\end{aligned}
$$

If $\beta \in F$ and $u \in V$, then

$$
(\beta u)(\alpha\phi) = \alpha[(\beta u)\phi] \qquad \text{by definition of scalar multiplication of} \\
\text{functions}
$$

$$\begin{aligned}
&= \alpha[\beta(u\phi)] &&\text{because } \phi \text{ is linear}\\
&= (\alpha\beta)(u\phi) &&\text{by Axiom (ii) for vector spaces}\\
&= (\beta\alpha)(u\phi) &&\\
&= \beta[\alpha(u\phi)] &&\text{by Axiom (ii) again}\\
&= \beta[u(\alpha\phi)] &&\text{by definition of scalar multiplication of}\\
&&&\text{functions}
\end{aligned}$$

Thus, $\alpha\phi$ preserves addition and scalar multiplication and $\mathcal{L}(V, W)$ is closed under scalar multiplication. □

Proposition 28

If V, W and Y are F-spaces, ϕ and μ are in $\mathcal{L}(V, W)$, and θ and λ are in $\mathcal{L}(W, Y)$, then

$$(\phi + \mu)\circ\lambda = \phi\circ\lambda + \mu\circ\lambda$$

and

$$\phi\circ(\theta + \lambda) = \phi\circ\theta + \phi\circ\lambda$$

Proof: For each vector x in V,

$$\begin{aligned}
(x)[(\phi + \mu)\circ\lambda] &= [(x)(\phi + \mu)]\lambda &&\text{by definition of}\\
&&&\text{composition}\\
&= (x\phi + x\mu)\lambda &&\text{by definition of}\\
&&&\text{functional addition}\\
&= (x\phi)\lambda + (x\mu)\lambda &&\text{because } \lambda \text{ is linear}\\
&= (x)(\phi\circ\lambda) + (x)(\mu\circ\lambda) &&\text{by definition of}\\
&&&\text{composition}\\
&= (x)[\mu\circ\lambda + \mu\circ\lambda] &&\text{by definition of}\\
&&&\text{functional addition}
\end{aligned}$$

Therefore, $(\phi + \mu)\circ\lambda = \phi\circ\lambda + \mu\circ\lambda$.

We leave the rest of the proof as an exercise. □

Theorem 29

If V is an F-space, then $\mathcal{L}(V)$ is a ring with respect to the operations of pointwise addition of functions and composition of functions.

Proof: From Theorem 27 we know that $\mathcal{L}(V)$ is an F-space with respect to pointwise addition of functions, and therefore the first four axioms for a ring are satisfied. By Proposition 20 we know that $\mathcal{L}(V)$ is closed under composition of functions, and therefore composition is an associative operation on $\mathcal{L}(V)$. The two Distributive laws follow from Proposition 28. □

As we did with permutations, we shall find it convenient to write the composition of linear transformations using just multiplicative notation.

Theorem 30

If V, W, M, and N are F-spaces such that $V \cong M$ and $W \cong N$, then $\mathcal{L}(V, W) \cong \mathcal{L}(M, N)$ as F-spaces and $\mathcal{L}(V) \cong \mathcal{L}(M)$ as rings.

Proof: Let μ be an isomorphism from V onto M, and let λ be an isomorphism from W onto N. Define the function ϕ by $T\phi = \mu^{-1}T\lambda$ for each T in $\mathcal{L}(V, W)$. Since μ^{-1} is a function from M to V, T is a function from V to W, and λ is a function from W to N, $\mu^{-1}T\lambda$ is a function from M to N. Since μ^{-1}, T, and λ are all linear, $\mu^{-1}T\lambda$ is linear. Thus ϕ maps $\mathcal{L}(V, W)$ into $\mathcal{L}(M, N)$.

If T and P are in $\mathcal{L}(V, W)$, then

$$(T + P)\phi = \mu^{-1}(T + P)\lambda$$
$$= \mu^{-1}(T\lambda + P\lambda)$$
$$= \mu^{-1}T\lambda + \mu^{-1}P\lambda$$
$$= T\phi + P\phi$$

Also, if $\alpha \in F$ and $T \in \mathcal{L}(V, W)$, then

$$(\alpha T)\phi = \mu^{-1}(\alpha T)\lambda$$
$$= \mu^{-1}(\alpha T\lambda)$$
$$= \alpha(\mu^{-1}T\lambda)$$

Thus, ϕ is a linear transformation.
If $T\phi = 0$, then $\mu^{-1}T\lambda = 0$.
Thus, $\mu(\mu^{-1}T\lambda)\lambda^{-1} = \mu 0\lambda^{-1} = 0$;
that is, $I_V T I_W = 0$ and $T = 0$.
Therefore, ker $\phi = \{0\}$ and ϕ is one-to-one.
If $P \in \mathcal{L}(M, N)$, then $\mu P\lambda^{-1} \in \mathcal{L}(V, W)$ and

$$(\mu P\lambda^{-1})\phi = \mu^{-1}\mu P\lambda^{-1}\lambda = P$$

Therefore, ϕ is onto and we have shown that ϕ is a vector space isomorphism.

If, in the above arguments, we let $V = W$, $M = N$, and $\mu = \lambda$, we have $T\phi = \mu^{-1}T\mu$ and we have shown that $\mathcal{L}(V) \cong \mathcal{L}(M)$ as F-spaces. We need only show that ϕ preserves multiplication to show that $\mathcal{L}(V) \cong \mathcal{L}(M)$ as rings.

If T and P are in $\mathcal{L}(V)$, then

$$(TP)\phi = \mu^{-1}(TP)\mu$$
$$= (\mu^{-1}T)(P\mu)$$
$$= (\mu^{-1}T)(\mu\mu^{-1})(P\mu)$$
$$= (\mu^{-1}T\mu)(\mu^{-1}P\mu)$$
$$= (T\phi)(P\phi) \quad \square$$

EXERCISE SET D

1. Let V and W be F-spaces and let ϕ be a function from V into W such that $(u + v)\phi = u\phi + v\phi$ and $(\alpha u)\phi = \alpha(u\phi)$ for each u and v in V and each α in F. Prove that ϕ is a linear transformation.
2. Which of the following functions are linear transformations? If not, why not?
   (a) the derivative operator as a function from $\mathbf{R}[x]$ to $\mathbf{R}[x]$
   (b) the operator I_0 from $\mathbf{R}[x]$ to $\mathbf{R}[x]$ defined by

   $$fI_0 = \int_0^x f(t)\, dt$$

   (c) the function λ from $\mathbf{R} \oplus \mathbf{R}$ to $\mathbf{R} \oplus \mathbf{R}$ defined by $(x, y)\lambda = (x + y, x - y)$
   (d) Let $\mathbf{C}$ denote the complex numbers and define ϕ from $\mathbf{C} \oplus \mathbf{C}$ to $\mathbf{C} \oplus \mathbf{C}$ by $(a, b)\phi = (\bar{a}, \bar{b})$, where $\bar{a}$ is the complex conjugate of a, that is, $\overline{x + iy} = x - iy$ for real numbers x and y.
   (e) the function T from $\mathbf{R} \oplus \mathbf{R}$ to $\mathbf{R} \oplus \mathbf{R}$ defined by $(x, y)T = (xy, y)$
   (f) the function ϕ from $\mathbf{R} \oplus \mathbf{R} \oplus \mathbf{R}$ to $\mathbf{R} \oplus \mathbf{R} \oplus \mathbf{R}$ defined by

   $$(x, y, z)\phi = (x + y + z, 2x - y - z, 3x + 4y - 5z)$$

3. Which of the linear transformations above are one-to-one? Which ones are onto?
4. Prove Proposition 20.
5. Prove Proposition 21.
6. Prove Proposition 22.
7. Prove Proposition 23.
**8. Prove Theorem 25.
**9. Prove Theorem 26.
10. Complete the proof of Proposition 28.

§5. Basis and Dimension

The functions $\sin x$ and $\cos x$ are elements of the $\mathbf{R}$-space of continuous functions from $\mathbf{R}$ into $\mathbf{R}$. If we try to find real numbers α and β such that

$$\alpha \sin x + \beta \cos x = 0$$

α and β would be such that

$$\alpha \sin 0 + \beta \cos 0 = 0 \quad \text{and} \quad \alpha \sin \frac{\pi}{2} + \beta \cos \frac{\pi}{2} = 0$$

that is

$$\alpha \cdot 0 + \beta \cdot 1 = 0 \quad \text{and} \quad \alpha \cdot 1 + \beta \cdot 0 = 0$$

which implies that

$$\beta = 0 \quad \text{and} \quad \alpha = 0$$

Thus, if α and β are real numbers such that $\alpha \sin x + \beta \cos x = 0$, then $\alpha = \beta = 0$.

Similarly, if in $\mathbf{R} \oplus \mathbf{R} \oplus \mathbf{R}$ we attempt to find real numbers α, β, and γ such that

$$\alpha(1, 0, 0) + \beta(0, 1, 0) + \gamma(0, 0, 1) = (0, 0, 0)$$

we would have

$$(\alpha, 0, 0) + (0, \beta, 0) + (0, 0, \gamma) = (0, 0, 0)$$

which implies

$$(\alpha, \beta, \gamma) = (0, 0, 0) \quad \text{and therefore} \quad \alpha = \beta = \gamma = 0$$

On the other hand if, in $\mathbf{R} \oplus \mathbf{R} \oplus \mathbf{R}$, we try to find real numbers α, β, and γ such that

$$\alpha(1, 1, 0) + \beta(1, 0, 1) + \gamma(3, 2, 1) = (0, 0, 0)$$

we would have

$$(\alpha, \alpha, 0) + (\beta, 0, \beta) + (3\gamma, 2\gamma, \gamma) = (0, 0, 0)$$

which implies

$$\alpha + \beta + 3\gamma = 0, \quad \alpha + 2\gamma = 0, \quad \text{and} \quad \beta + \gamma = 0$$

If we let

$$\gamma = 1, \quad \beta = -1, \quad \text{and} \quad \alpha = -2$$

we see that there are nonzero real numbers α, β, and γ satisfying the equation above.

If S is a nonempty set of vectors in the F-space V, then S is said to be a *linearly independent* set if, for each linear combination of distinct elements in S,

$$\alpha_1 s_1 + \alpha_2 s_2 + \cdots + \alpha_n s_n$$

whenever

$$\alpha_1 s_1 + \alpha_2 s_2 + \cdots + \alpha_n s_n = 0$$

then

$$\alpha_1 = \alpha_2 = \cdots = \alpha_n = 0$$

Thus, S is linearly independent iff the only linear combinations of distinct vectors in S that are the zero vector are those in which every scalar is the 0 field element. A set that is not linearly independent is said to be *linearly dependent*. Notice that any set of vectors containing the zero vector is linearly dependent.

In the examples above, $\{\sin x, \cos x\}$ is a linearly independent subset of the $\mathbf{R}$-space of continuous functions from $\mathbf{R}$ into $\mathbf{R}$ and

$$\{(1, 0, 0), (0, 1, 0), (0, 0, 1)\}$$

is a linearly independent subset of $\mathbf{R} \oplus \mathbf{R} \oplus \mathbf{R}$, whereas

$\{(1, 1, 0), (1, 0, 1), (3, 2, 1)\}$

is a linearly dependent subset of this space.

Suppose u is a vector and $v_1, \ldots, v_n$ are distinct elements of some set S of linearly independent vectors. If $\alpha_1, \ldots, \alpha_n, \beta_1, \ldots, \beta_n$ are scalars such that

$$u = \alpha_1 v_1 + \alpha_2 v_2 + \cdots + \alpha_n v_n \quad \text{and}$$
$$u = \beta_1 v_1 + \beta_2 v_2 + \cdots + \beta_n v_n$$

then

$$(\alpha_1 v_1 + \alpha_2 v_2 + \cdots + \alpha_n v_n) - (\beta_1 v_1 + \cdots + \beta_n v_n) = u - u = 0$$

Thus,

$$\alpha_1 v_1 - \beta_1 v_1 + \alpha_2 v_2 - \beta_2 v_2 + \cdots + \alpha_n v_n - \beta_n v_n = 0$$

which implies

$$(\alpha_1 - \beta_1)v_1 + (\alpha_2 - \beta_2)v_2 + \cdots + (\alpha_n - \beta_n)v_n = 0$$

Since $\{v_1, \ldots, v_n\}$ is a linearly independent set of vectors it follows that

$$\alpha_1 - \beta_1 = 0, \quad \alpha_2 - \beta_2 = 0, \ldots, \quad \alpha_n - \beta_n = 0$$

and therefore

$$\alpha_1 = \beta_1, \quad \alpha_2 = \beta_2, \ldots, \quad \alpha_n = \beta_n$$

Thus, there is only one way to express the vector u as a linear combination of distinct vectors in S.

Proposition 31

If V is an F-space, S is a spanning set for V, and S is a linearly independent set of vectors, then every vector in V can be uniquely expressed as a linear combination of distinct vectors in S.

Proof: From Proposition 14 we know that every vector in V is a linear combination of vectors in S. From our discussion above we know that this representation is unique. □

If V is an F-space and S is a subset of V, then S is said to be a *basis* for V if S spans V and S is a linearly independent set of vectors.

Examples

(i) $\{(1, 0, 0), (0, 1, 0), (0, 0, 1)\}$ is a basis for $\mathbf{R} \oplus \mathbf{R} \oplus \mathbf{R}$.
   $\{(1, 0, 0), (1, 1, 0), (1, 1, 1)\}$ is also.
(ii) $\{1, x, x^2, \ldots, x^n, \ldots\}$ is a basis for the $\mathbf{R}$-space $\mathbf{R}[x]$.
(iii) $\{1, i\}$ is a basis for the complex numbers viewed as an $\mathbf{R}$-space.

The concept of a basis is very important in the study of vector spaces. By our arguments above we know that if B is a basis for the F-space V, then every vector in V can be expressed as a linear combination of vectors in B and, furthermore, such a representation is unique. Thus, for example, because of the uniqueness of the representation we can define functions whose domain is V by defining the image of a vector using its representation as linear combinations of vectors in B. This is, in fact, the way we usually define functions from Euclidean 3-space into $\mathbf{R}$ or any set. For example, the function defined by

$$(x, y, z)f = x^2 + y - z^3$$

is so expressed since every point has a unique expression in the form $x(1, 0, 0) + y(0, 1, 0) + z(0, 0, 1)$. In general, functions of three variables can be viewed as functions from $\mathbf{R} \oplus \mathbf{R} \oplus \mathbf{R}$ into the range where x, y, and z are the coefficients of the basis vectors $(1, 0, 0)$, $(0, 1, 0)$, and $(0, 0, 1)$.

In the next few assertions we shall show that vector spaces with finite bases can be completely described. Analogous assertions can be proved for all vector spaces, but to do so one needs more powerful set-theoretic tools than we have available to us in this text. Using such tools we could show that every vector space has a basis. We shall concern ourselves with the case where a finite spanning set is known to exist and shall, after developing the necessary theory, show that such a space does, in fact, have a basis.

Proposition 32

If V is an F-space and S is a nonempty collection of nonzero vectors in V, then either S is linearly independent or one of the vectors in S can be expressed as a linear combination of the other vectors in S.

Proof: If S is not linearly independent, then there exist distinct vectors $t_1, \ldots, t_n$ in S and scalars $\alpha_1, \ldots, \alpha_n$ in F such that $\alpha_1 t_1 + \cdots + \alpha_n t_n = 0$ and some $\alpha_i \neq 0$. Then

$$(-\alpha_i)t_i = \alpha_1 t_1 + \cdots + \alpha_{i-1}t_{i-1} + \alpha_{i+1}t_{i+1} + \cdots + \alpha_n t_n$$

Since α_i is a nonzero element in a field, α_i^{-1} exists in F. Thus,

$$t_i = (-\alpha_i^{-1})(-\alpha_i t_i) = (-\alpha_i^{-1}\alpha_1)t_1 + \cdots + (-\alpha_i^{-1}\alpha_{i-1})t_{i-1}$$
$$+ (-\alpha_i^{-1}\alpha_{i+1})t_{i+1} + \cdots + (-\alpha_i^{-1}\alpha_n)t_n$$

and t_i can be expressed as a linear combination of other vectors in S. □

Proposition 33

If V is an F-space, n is a positive integer greater than 1, and $B = \{v_1, \ldots, v_n\}$ is a basis for V, then $\{\bar{v}_1, \ldots, \bar{v}_{n-1}\}$ is a basis for $V/\langle v_n \rangle$.

Proof: First we show that $\{\bar{v}_1, \ldots, \bar{v}_{n-1}\}$ is linearly independent.

Suppose $\alpha_1, \ldots, \alpha_{n-1}$ are scalars such that $\alpha_1 \bar{v}_1 + \cdots + \alpha_{n-1} \bar{v}_{n-1} = \bar{0}$. We need to show that each $\alpha_i = 0$.

Since $\alpha_1 \bar{v}_1 + \cdots + \alpha_{n-1} \bar{v}_{n-1} = \bar{0}$, $\overline{\alpha_1 v_1 + \cdots + \alpha_{n-1} v_{n-1}} = \bar{0}$.

Thus, $\alpha_1 v_1 + \cdots + \alpha_{n-1} v_{n-1} \equiv 0 \pmod{\langle v_n \rangle}$ and therefore

$$\alpha_1 v_1 + \cdots + \alpha_{n-1} v_{n-1} \in \langle v_n \rangle$$

Since $\langle v_n \rangle = \{ \beta v_n : \beta \in F \}$,

$$\alpha_1 v_1 + \alpha_2 v_2 + \cdots + \alpha_{n-1} v_{n-1} = \beta v_n$$

for some scalar β in F and therefore

$$\alpha_1 v_1 + \cdots + \alpha_{n-1} v_{n-1} - \beta v_n = 0$$

Since B is a linearly independent set, $\alpha_1 = \alpha_2 = \cdots = \alpha_{n-1} = \beta = 0$. Thus, each $\alpha_i = 0$ and $\{ \bar{v}_1, \ldots, \bar{v}_{n-1} \}$ is a linearly independent set.

If $\bar{u}$ is a vector in $V / \langle v_n \rangle$, then since $u \in V$, $u = \alpha_1 v_1 + \cdots + \alpha_n v_n$ for some scalars $\alpha_1, \ldots, \alpha_n$ in F. Then

$$
\begin{aligned}
\bar{u} &= \overline{\alpha_1 v_1 + \cdots + \alpha_n v_n} \\
&= \alpha_1 \bar{v}_1 + \cdots + \alpha_{n-1} \bar{v}_{n-1} + \alpha_n \bar{v}_n \\
&= \alpha_1 \bar{v}_1 + \cdots + \alpha_{n-1} \bar{v}_{n-1} + \alpha_n \bar{0} \qquad \text{because } \bar{v}_n = \bar{0} \\
&= \alpha_1 \bar{v}_1 + \cdots + \alpha_{n-1} \bar{v}_{n-1}
\end{aligned}
$$

Therefore, $\{ \bar{v}_1, \ldots, \bar{v}_{n-1} \}$ spans $V / \langle v_n \rangle$.

By the above arguments $\{ \bar{v}_1, \ldots, \bar{v}_{n-1} \}$ is a basis for $V / \langle v_n \rangle$. □

Proposition 34

If V is an F-space and $B = \{ v_1, \ldots, v_k \}$ is a basis for V, then any subset of V containing more than k vectors is linearly dependent.

Proof: By induction on k.

If $B = \{ v_1 \}$ and u and y are distinct vectors in V, then $u = \alpha v_1$ and $y = \beta v_1$ for some scalars α and β. We may assume $\alpha \neq 0$ and $\beta \neq 0$ because otherwise $\{ u, y \}$ is trivially dependent.

Then, since α and β are nonzero elements of a field, there exist scalars α^{-1} and β^{-1} in F. Then

$$
\begin{aligned}
\alpha^{-1} u + (-\beta^{-1}) y &= \alpha^{-1} \alpha v_1 + (-\beta^{-1} \beta) v_1 \\
&= 1 v_1 + (-1) v_1 \\
&= v_1 - v_1 \\
&= 0
\end{aligned}
$$

Thus, $\{ u, y \}$ is linearly dependent, and therefore any set containing more than one vector is linearly dependent.

Now assume that the assertion is true for every space with a basis containing exactly n elements and suppose $B = \{v_1, \ldots, v_n, v_{n+1}\}$ is a basis for a vector space V. Let $u_1, \ldots, u_{n+1}, u_{n+2}$ be any collection of vectors in V and now consider the vectors $\overline{u}_1, \ldots, \overline{u}_{n+1}, \overline{u}_{n+2}$ in $V/\langle v_{n+1} \rangle$. If the vectors $\overline{u}_1, \ldots, \overline{u}_{n+2}$ are not distinct, then $\overline{u}_i = \overline{u}_j$ for some $i \neq j$. We know that $\{\overline{v}_1, \ldots, \overline{v}_n\}$ is a basis for $V/\langle v_{n+1} \rangle$ and by the induction hypothesis we know that if the vectors are distinct, then $\overline{u}_1, \overline{u}_2, \ldots, \overline{u}_{n+1}, \overline{u}_{n+2}$ is linearly dependent in $V/\langle v_{n+1} \rangle$. In either case, one of these vectors can be expressed as a linear combination of the others. Without loss of generality we can let $\overline{u}_{n+2}$ be such a vector. Thus, there exist scalars $\alpha_1, \ldots, \alpha_{n+1}$ such that

$$\overline{u}_{n+2} = \alpha_1 \overline{u}_1 + \cdots + \alpha_{n+1} \overline{u}_{n+1}$$
$$= \overline{\alpha_1 u_1 + \cdots + \alpha_{n+1} u_{n+1}}$$

Then

$$u_{n+2} \equiv \alpha_1 u_1 + \cdots + \alpha_{n+1} u_{n+1} \qquad (\text{mod } \langle v_{n+1} \rangle)$$

and therefore $u_{n+2} = \alpha_1 u_1 + \cdots + \alpha_{n+1} u_{n+1} + \beta v_{n+1}$ for some β in F.

If $\beta = 0$, then $\alpha_1 u_1 + \cdots + \alpha_{n+1} u_{n+1} + (-1) u_{n+2} = 0$, $u_1, \ldots, u_{n+2}$ is linearly dependent and the proof is complete.

Now assume $\beta \neq 0$. Either $\overline{u}_i = \overline{u}_j$ for some $i \neq j$; $i, j < n + 2$ or, by our induction hypothesis, $\overline{u}_1, \ldots, \overline{u}_{n+1}$ is a linearly dependent set in $V/\langle v_{n+1} \rangle$. In either case, there exist scalars $\gamma_1, \ldots, \gamma_{n+1}$, not all zero, such that

$$\gamma_1 \overline{u}_1 + \gamma_2 \overline{u}_2 + \cdots + \gamma_{n+1} \overline{u}_{n+1} = \overline{0}$$

Then

$$\overline{\gamma_1 u_1 + \gamma_2 u_2 + \cdots + \gamma_{n+1} u_{n+1}} = \overline{0}$$

and therefore

$$\gamma_1 u_1 + \gamma_2 u_2 + \cdots + \gamma_{n+1} u_{n+1} \in \langle v_{n+1} \rangle$$

Thus, for some scalar δ,

$$\gamma_1 u_1 + \gamma_2 u_2 + \cdots + \gamma_{n+1} u_{n+1} = \delta v_{n+1}$$

If $\delta = 0$, then, since some $\gamma_i \neq 0$, $\{u_1, \ldots, u_{n+1}\}$ is a dependent set and the proof is complete.

Now assume $\delta \neq 0$.

$$u_{n+2} = \alpha_1 u_1 + \cdots + \alpha_{n+1} u_{n+1} + \beta v_{n+1}$$
$$= \alpha_1 u_1 + \cdots + \alpha_{n+1} u_{n+1} + (\beta \delta^{-1})(\delta v_{n+1})$$
$$= \alpha_1 u_1 + \cdots + \alpha_{n+1} u_{n+1} + (\beta \delta^{-1})(\gamma_1 u_1 + \cdots + \gamma_{n+1} u_{n+1})$$

Therefore,

$$(\alpha_1 + \beta\delta^{-1}\gamma_1)u_1 + (\alpha_2 + \beta\delta^{-1}\gamma_2)u_2 + \cdots$$
$$+ (\alpha_{n+1} + \beta\delta^{-1}\gamma_{n+1})u_{n+1} + (-1)u_{n+2} = 0$$

and $\{u_1, \ldots, u_{n+2}\}$ is linearly dependent.

The proposition now follows by the Principle of Mathematical Induction. $\square$

Theorem 35

If V is an F-space and B is a finite basis for V, then every basis of V has the same number of vectors as B.

Proof: If $B = \{v_1, \ldots, v_n\}$ and D is a basis for V, D cannot contain more than n vectors for then, by Proposition 34, D would be linearly dependent. On the other hand, if D contains less than n vectors and is a basis, B would be a linearly dependent set by the same proposition. Therefore, D and B must have the same number of elements. $\square$

If V is an F-space and V has a finite basis, then the *dimension* of V is the number of vectors in a basis for V.

Proposition 36

If V is an F-space, B is a basis for V, B contains n elements, and D is any set of n linearly independent vectors, then D is a basis for V.

Proof: We need to show that D spans V. Clearly, every vector in D is in the space spanned by D. Let $D = \{u_1, \ldots, u_n\}$ and let v be any vector in the space V other than those in D. Then the set $\{u_1, \ldots, u_n, v\}$ contains $n + 1$ vectors and must, therefore, be linearly dependent. Thus, there exist scalars $\alpha_1, \ldots, \alpha_n, \beta$, not all zero, such that

$$\alpha_1 u_1 + \alpha_2 u_2 + \cdots + \alpha_n u_n + \beta v = 0$$

If $\beta = 0$, then

$$\alpha_1 u_1 + \alpha_2 u_2 + \cdots + \alpha_n u_n = 0 \qquad \text{and some} \quad \alpha_i \neq 0$$

But this is impossible, because D is a linearly independent set. Thus, $\beta \neq 0$,

$$v = (-\beta^{-1}\alpha_1)u_1 + \cdots + (-\beta^{-1}\alpha_n)u_n$$

and D spans V. $\square$

Proposition 37

If U is a finite set of nonzero vectors that spans the F-space V, then U contains a basis of V.

Proof: By induction on the number of vectors in U.
If $U = \{u_1\}$, $\{u_1\}$ is independent since $\alpha_1 u_1 = 0$ implies $\alpha_1 = 0$.

Now assume the proposition is true for all sets of nonzero vectors containing less than m elements. If U is linearly independent, there is nothing to prove. If U is not linearly independent, then one of the vectors in U can be expressed as a linear combination of the others. Without loss of generality we may assume that u_m can be so expressed. Then every vector that can be expressed as a linear combination of $\{u_1, \ldots, u_m\}$ can be expressed as a linear combination of $\{u_1, \ldots, u_{m-1}\}$. That is, since

$$u_m = \beta_1 u_1 + \cdots + \beta_{m-1} u_{m-1}$$

if

$$v = \alpha_1 u_1 + \cdots + \alpha_{m-1} u_{m-1} + \alpha_m u_m$$

then

$$v = \alpha_1 u_1 + \cdots + \alpha_{m-1} u_{m-1} + \alpha_m(\beta_1 u_1 + \cdots + \beta_{m-1} u_{m-1})$$

and

$$v = (\alpha_1 + \alpha_m \beta_1) u_1 + \cdots + (\alpha_{m-1} + \alpha_m \beta_{m-1}) u_{m-1}$$

Thus, $\{u_1, \ldots, u_{m-1}\}$ spans V and by our induction hypothesis $\{u_1, \ldots, u_{m-1}\}$ contains a basis for V. Since $U \supseteq \{u_1, \ldots, u_{m-1}\}$, U contains a basis for V. The proposition now follows by the Principle of Mathematical Induction (course of values form). □

Proposition 38

If V is an F-space and S is a linearly independent subset of V, then either S is a basis of V or S is properly contained in a linearly independent subset of V.

Proof: If S is a linearly independent subset of V and S is not a basis for V, then S does not span V. Let y be a vector in V not in $\langle S \rangle$, and let $T = \{y\} \cup S$. To see that T is linearly independent, suppose

$$\alpha_1 s_1 + \alpha_2 s_2 + \cdots + \alpha_n s_n + \beta y = 0 \quad \text{and} \quad s_1, \ldots, s_n \text{ are in } S$$

If $\beta \neq 0$, then $y = -\beta^{-1}(\alpha_1 u_1 + \cdots + \alpha_n u_n)$
and $y \in \langle S \rangle$, which is impossible.
Thus, $\beta = 0$ and $\alpha_1 s_1 + \cdots + \alpha_n s_n = 0$. Since S is linearly independent, $\alpha_1 = \alpha_2 = \cdots = \alpha_n = 0$, and therefore T is linearly independent. □

Proposition 39

If V is an n-dimensional F-space and U is a linearly independent subset of V, then U is contained in a basis of V.

Proof: Let $\mathcal{L}$ denote all linearly independent subsets of V containing U. No set in $\mathcal{L}$ contains more than n vectors. However, by Proposition 38, every set in $\mathcal{L}$ is either a basis or is contained in a larger set in $\mathcal{L}$. Therefore, the largest sets in $\mathcal{L}$ contain n elements. Let S be a maximal set in $\mathcal{L}$. Then S must contain n vectors and, since S is linearly independent, S is a basis of V containing U.   □

Proposition 40

If V is an n-dimensional F-Space, every subspace of V is of finite dimension less than or equal to n.

Proof: If W is a subspace of V, any subset of W containing more than n vectors is linearly dependent. Let k be the largest number of vectors in a linearly independent subset of W. Then any subset of W with k linearly independent vectors is a basis of W because otherwise, by Proposition 38, it would be properly contained in a linearly independent subset of W. Therefore, dim $W = k$ for some $k \leq n$.   □

Proposition 41

If V is an n-dimensional F-space, W is an F-space, and T is a linear transformation from V into W, then $n = dim\ ker\ T + dim\ image\ T$.

Proof: The image of a set of generators is always a set of generators, and since V has a finite basis, $VT =$ image T is finite dimensional. Since ker T is a subspace of V, ker T is finite dimensional. Let $u_1, \ldots, u_k$ be a basis of ker T, and let $y_1, \ldots, y_p$ be a basis of VT. Then for some $v_1, \ldots, v_p$ in V, $v_1 T = y_1, \ldots, v_p T = y_p$.   Consider the set $\{u_1, \ldots, u_k, v_1, \ldots, v_p\}$. If $\alpha_1 u_1 + \cdots + \alpha_k u_k + \beta_1 v_1 + \cdots + \beta_p v_p = 0$, then

$$(\alpha_1 u_1 + \cdots + \alpha_k u_k + \beta_1 v_1 + \cdots + \beta_p v_p)T = 0$$
$$\alpha_1 u_1 T + \cdots + \alpha_k u_k T + \beta_1 v_1 T + \cdots + \beta_p v_p T = 0$$
$$0 + \cdots + 0 + \beta_1 y_1 + \cdots + \beta_p y_p = 0$$
$$\beta_1 y_1 + \cdots + \beta_p y_p = 0$$

Since $\{y_1, \ldots, y_p\}$ is linearly independent, $\beta_1 = \beta_2 = \cdots = \beta_p = 0$. Thus

$$\alpha_1 u_1 + \cdots + \alpha_k u_k = 0$$

and since $\{u_1, \ldots, u_k\}$ is linearly independent,

$$\alpha_1 = \alpha_2 = \cdots = \alpha_k = 0$$

Thus, $\{u_1, \ldots, u_k, v_1, \ldots, v_p\}$ is linearly independent. If $v \in V$, then $vT = \beta_1 y_1 + \cdots + \beta_p y_p$ for some $\beta_1, \ldots, \beta_p$. Thus,

$$[v - (\beta_1 v_1 + \cdots + \beta_p v_p)]T = 0 \quad \text{and}$$
$$v - (\beta_1 v_1 + \cdots + \beta_p v_p) \in \text{ker } T$$

Hence, $v - (\beta_1 v_1 + \cdots + \beta_p v_p) = \alpha_1 u_1 + \cdots + \alpha_k u_k$ for some $\alpha_1, \ldots, \alpha_k$.

Therefore,

$$v = \alpha_1 u_1 + \cdots + \alpha_k u_k + \beta_1 v_1 + \cdots + \beta_p v_p$$

$\{u_1, \ldots, u_k, v_1, \ldots, v_p\}$ is a spanning set for V, and

$$n = k + p$$

Since dim ker $T = k$ and dim $VT = p$,

$$n = \dim V = \dim \ker T + \dim \text{image } T \quad \square$$

If T is a linear transformation from the finite dimensional F-space V into the F-space W, then the *rank of* T is defined to be the dimension of VT, the image of V under T, and the *nullity of* T is defined to be the dimension of the kernel of T. Proposition 41 can thus be restated as dim V = (nullity of T) + (rank of T).

If F is a field, we let F^n denote the F-space of all ordered n-tuples of elements of F.

Theorem 42

If V is an F-space and V has dimension n, then $F^n \cong V$.

Proof: Let $B = \{v_1, \ldots, v_n\}$ be a basis for V.
Define ϕ from F^n to V by

$$(\alpha_1, \alpha_2, \ldots, \alpha_n)\phi = \alpha_1 v_1 + \alpha_2 v_2 + \cdots + \alpha_n v_n$$

If $(\alpha_1, \alpha_2, \ldots, \alpha_n)$ and $(\beta_1, \ldots, \beta_n)$ are vectors in F^n, then

$$
\begin{aligned}
[(\alpha_1, \ldots, \alpha_n) + (\beta_1, \ldots, \beta_n)]\phi &= (\alpha_1 + \beta_1, \ldots, \alpha_n + \beta_n)\phi \\
&= (\alpha_1 + \beta_1)v_1 + \cdots + (\alpha_n + \beta_n)v_n \\
&= \alpha_1 v_1 + \beta_1 v_1 + \cdots + \alpha_n v_n + \beta_n v_n \\
&= (\alpha_1 v_1 + \cdots + \alpha_n v_n) \\
&\quad + (\beta_1 v_1 + \cdots + \beta_n v_n) \\
&= (\alpha_1, \ldots, \alpha_n)\phi + (\beta_1, \ldots, \beta_n)\phi
\end{aligned}
$$

Therefore, ϕ preserves addition.
If $\gamma \in F$ and $(\alpha_1, \ldots, \alpha_n) \in F^n$, then

$$
\begin{aligned}
[\gamma(\alpha_1, \ldots, \alpha_n)]\phi &= (\gamma\alpha_1, \ldots, \gamma\alpha_n)\phi \\
&= \gamma\alpha_1 v_1 + \cdots + \gamma\alpha_n v_n \\
&= \gamma(\alpha_1 v_1 + \cdots + \alpha_n v_n) \\
&= \gamma[(\alpha_1, \ldots, \alpha_n)\phi]
\end{aligned}
$$

Therefore, ϕ preserves scalar multiplication.
Thus far we have shown that ϕ is a linear transformation.
If $u \in V$, then

$$u = \alpha_1 v_1 + \cdots + \alpha_n v_n \qquad \text{for some scalars } \alpha_1, \ldots, \alpha_n$$

Thus

$$(\alpha_1, \ldots, \alpha_n)\phi = \alpha_1 v_1 + \cdots + \alpha_n v_n = u$$

and ϕ is onto.

If $(\alpha_1, \ldots, \alpha_n) \in \ker \phi$,

then $(\alpha_1, \ldots, \alpha_n)\phi = 0$ and since

$$(\alpha_1, \ldots, \alpha_n)\phi = \alpha_1 v_1 + \cdots + \alpha_n v_n$$

$$\alpha_1 v_1 + \cdots + \alpha_n v_n = 0$$

Since $\{v_1, \ldots, v_n\}$ is a linearly independent set,

$$\alpha_1 = \alpha_2 = \cdots = \alpha_n = 0 \quad \text{and} \quad (\alpha_1, \ldots, \alpha_n) = (0, \ldots, 0)$$

Thus $\ker \phi = \{(0, \ldots, 0)\}$ and ϕ is one-to-one.

Since ϕ is a one-to-one, onto, linear transformation, ϕ is an isomorphism. $\square$

Corollary 43

If V and W are two F-spaces and both have dimension n, then $V \cong W$.

By Theorem 42 and Corollary 43 we see that the structure of any F-space with a finite basis is completely determined. These results, combined with Proposition 37, provide the strongest structure theorem in this text.

Theorem 44

If V is an F-space and V has a nonempty finite set of generators, then $V \cong F^n$ for some positive integer n.

For example, in the space of continuous functions from **R** to **R**, the subspace spanned by any finite collection of functions will be isomorphic to $\mathbf{R}^n$ for some positive integer n; for example,

$$\langle \{\sin x, \cos x\} \rangle \cong \mathbf{R}^2$$

and

$$\langle \{\sin x, \cos x, x, e^x\} \rangle \cong \mathbf{R}^4$$

Notice, however, that

$$\langle \{x, x^2, 3x^2 - 4x\} \rangle \not\cong \mathbf{R}^3$$

but, in fact,

$$\langle \{x, x^2, 3x^2 - 4x\} \rangle \cong \mathbf{R}^2$$

because the set $\{x, x^2, 3x^2 - 4x\}$ is not linearly independent but does contain the linearly independent set $\{x, x^2\}$.

The proof of the following assertion is contained in the proof of Theorem 42. We leave finding it and adapting it as an exercise.

Proposition 45

If V and W are F-spaces, $B = \{v_1, \ldots, v_n\}$ is a basis for V, and $D = \{u_1, \ldots, u_n\}$ is any collection of vectors in W, then the function ϕ defined by

$$(\alpha_1 v_1 + \cdots + \alpha_n v_n)\phi = \alpha_1 u_1 + \cdots + \alpha_n u_n \qquad \text{for all scalars } \alpha_i$$

is a linear transformation. Furthermore ϕ is one-to-one iff D is a linearly independent set. Also, if W has dimension n, ϕ is one-to-one iff ϕ is onto.

Proof: Exercise.

EXERCISE SET E

1. Determine which of the following sets are linearly independent.
   (a) $\{(1, 1, 0), (0, 1, 0), (3, 1, 0)\}$ in $\mathbf{R}^3$
   (b) $\{(2, 1, 4), (3, 1, 2), (1, 1, 1)\}$ in $\mathbf{R}^3$
   (c) $\{3, \sqrt{2}\}$ viewing $\mathbf{R}$ as a $\mathbf{Q}$-space
   (d) $\{\pi, \pi^2\}$ viewing $\mathbf{R}$ as a $\mathbf{Q}$-space
   (e) $\{4, \cos^2 x, \sin^2 x\}$ in the space of continuous functions over $\mathbf{R}$
   (f) $\{\cos x, \sec x\}$ in the $\mathbf{R}$-space of continuous functions from $[0, 1]$ into $\mathbf{R}$
   (g) $\{1, \tan x, \sec x\}$ in the same $\mathbf{R}$-space
   (h) $\{3, \tan^2 x, \sec^2 x\}$ in the same $\mathbf{R}$-space.
2. Show that the following sets span $\mathbf{R}^3$. Find a basis for $\mathbf{R}^3$ in each set.
   (a) $\{(1, 1, -1), (2, 4, 3), (4, 6, 1), (3, 5, 2)\}$
   (b) $\{(1, 0, 1), (1, 1, 0), (0, 1, 1), (1, 1, 1)\}$
   (c) $\{(\pi, \sqrt{2}, e), (\sqrt{17}, \sqrt{5}, 0), (-4.3, 0, 0)\}$
3. (a) Express $(4, 1, 3)$ as a linear combination of $\{(1, 1, 0), (1, 1, 1),$ and $(1, 0, 0)\}$.
   (b) Express $(4, 1, 3)$ as a linear combination of $\{(1, 0, 0), (0, 0, 1),$ and $(1, 1, 1)\}$.
   (c) Express $(4, 1, 3)$ as a linear combination of $\{(2, 1, 4), (3, -1, 5),$ and $(7, 1, 1)\}$.
4. Show that $\{(1, 2, 3, 4), (1, 1, 1, 1), (2, 4, 4, 5), (3, 3, 8, 8)\}$ is a basis for $\mathbf{Q} \oplus \mathbf{Q} \oplus \mathbf{Q} \oplus \mathbf{Q}$
5. Determine a basis for $\mathbf{R}^3$ that contains:
   (a) $\{(2, 3, 1), (1, 4, 7)\}$
   (b) $\{(1, 2, 3), (1, 1, 1)\}$
   (c) $\{(1, 1, 1), (1, 1, 2)\}$
6. Prove: If V is an F-space and R is a *finite* field, then V is finite dimensional iff V is finite.
7. Let F be a field and in F^n define e_i to be the n-tuple that has a 1 in the ith place and 0 elsewhere, for example,

   $$e_2 = (0, 1, \ldots, 0)$$

   Show that $\{e_1, \ldots, e_n\}$ is a basis for F^n.
8. Prove Proposition 45.
**9. Prove: If V is a finite-dimensional vector space, then every element of $\mathcal{L}(V)$ is either 0, a unit, or a proper zero divisor.

10. Prove: If V and W are finite dimensional F-spaces, T is a linear transformation from T into W and Q is an automorphism of W, then rank T = rank TQ and nullity T = nullity TQ.

**11. (Especially challenging at this point.)
Prove: If V and W are finite dimensional F-spaces, then dim $\mathcal{L}(V, W)$ = (dim V) · (dim W).

§6. Modules

The idea of a vector space is abstracted from the usual Euclidean spaces and spaces of functions. The basic notion is the idea of being able to multiply elements in an abelian group by elements of a field and obtain an element of the group as the product. If we let the system of scalars be a ring instead of a field, we obtain the concept of a *module*.

A set M is said to be a *module* over the ring R if M is an abelian group (we shall denote the operation on M additively) and for each pair of elements α in R and v in M, there is an element in M, called the *product* of α and v and denoted by αv (or by $v\alpha$) such that the following conditions hold.

(i) If α and β are elements of R and u is an element of M, then $\alpha(\beta u) = (\alpha\beta)u$.

(ii) If α is an element of R and u and v are elements of M, then $\alpha(u + v) = \alpha u + \alpha v$.

(iii) If α and β are elements of R and u is an element of M, then $(\alpha + \beta)u = \alpha u + \beta u$.

If R is a ring with 1, M is said to be *unitary* if $1u = u$ for each u in M.

The process of forming the product αv, where $\alpha \in R$ and $v \in M$ is, as for vector spaces, called *scalar multiplication*. A module over the ring R will frequently be called an R-module and by the phrase "M is an R-module" we mean that R is a ring and that M is a module over R.

Examples

(i) All F-spaces are unitary F-modules.

(ii) If V is an F-space, V is an $\mathcal{L}(V)$ module where the scalar multiplication is just the action of the function; that is, $uT = (u)T$ if $u \in V$ and $T \in \mathcal{L}(V)$.

(iii) If M is any additively denoted abelian group, then M is a Z-module under the notation adopted in Chapter 4.

(iv) If R is a ring, the additive structure of R can be viewed as an R-module with the multiplication in R as the scalar multiplication.

(v) If V is an F-space and R is a subring of F, then V is an R-module under the scalar multiplication defined for elements of F.

(vi) Let M be an abelian group, let R be any ring, and define $\alpha v = 0$ for each α in R and v in M.

(vii) For each element y in $\mathbf{Z}$ and each pair (a, b) in $\mathbf{Z} \oplus \mathbf{Z}$, define $y(a, b) = (0, yb)$. Then $\mathbf{Z} \oplus \mathbf{Z}$ is a $\mathbf{Z}$-module but is not unitary.

(viii) The group of functions from a ring R into itself is an R-module under the natural way of defining scalar multiplication.

The concept of a submodule is defined analogously to the way subgroups, subrings, subfields, and subspaces are defined, and the corresponding theorems also hold. In fact, all the assertions proved in §§1–4 can, after making the expected appropriate definitions, be proved for unitary R-modules. Not until the study of linear independence and bases in §5 did we need to use the fact that nonzero scalars had inverses. We shall not bother to restate or reprove these results but shall leave the translations to the reader.

For R-modules, the concept of linear independence is defined somewhat differently in that $v_1, \ldots, v_n$ are said to be linearly independent if

$$\alpha_1 v_1 + \cdots + \alpha_n v_n = 0$$

implies

$$\alpha_1 v_1 = \alpha_2 v_2 = \cdots = \alpha_n v_n = 0$$

The change is necessary since it is possible in R-modules for $\alpha v = 0$ even though $\alpha \neq 0$ and $v \neq 0$. Because we cannot expect α^{-1} to exist in R when $\alpha \neq 0$, the results from §5 cannot be extended to R-modules.

EXERCISE SET F

1. Develop suitable definitions of submodule, linear transformation, and isomorphism for R-modules.

2. Prove: If V is an additively denoted abelian group and F is a field, then V can be viewed as an F-space iff there is an isomorphism from F into the ring $E(V)$ such that the image of 1 is the identity function in $E(V)$. (Hint: Denote scalar multiplication on the right.)

3. Prove: If M is an additively denoted abelian group and R is a ring, then with each homomorphism from R into $E(M)$ there is a scalar multiplication that makes M an R-module. If R has a 1 and the image of 1 is the identity endomorphism, then M is a unitary R-module.

**4. Prove: In the case of either Exercise 2 or Exercise 3, the linear transformations are just those endomorphisms that commute with the images of each of the elements of F or R.

**5. Let R be a ring and let S be the set of all sequences $\{a_n\}$ in R, that is, all functions from $\mathbf{N}$ into R. From earlier work we know that S, with pointwise addition, is an abelian group. Show that S is an R-module with respect to scalar multiplication defined by

$$r\{a_n\} = \{ra_n\} \qquad \text{for each } r \text{ in } R$$

8
POLYNOMIAL RINGS AND FIELD EXTENSIONS

The first type of algebraic objects with which a student learns to operate, after numbers, are expressions such as $x^2 + 2x$, $4x^3 - 3x^2 + 1$, $\frac{1}{2}x^4 - \frac{2}{3}x^3 + 17x$, and so on. These are called polynomials and, in elementary courses in algebra, a student learns the arithmetic of polynomials and then finds, in calculus, that the functions defined by polynomials are the simplest to differentiate and to integrate. If we examine our usual techniques for handling polynomials, we find that all we really need is a way of adding and multiplying coefficients such that the usual arithmetic laws hold. This leads us to study polynomials in which the coefficients need not be real numbers but are elements of some ring. In this chapter we shall study such polynomials, with special attention to the case where the coefficients are elements of some field.

§1. Definitions and Elementary Properties

Intuitively, a polynomial in the one variable x over the ring R is a formal sum of the form

$$a_0 + a_1 x + a_2 x^2 + \cdots + a_n x^n$$

where each of the a_i's are elements of R. The sum is formal in the sense that we can, for example, only indicate the sum of x and x^2 by $x + x^2$; we have no simpler form for this element. If we have two polynomials

$$a_0 + a_1 x + a_2 x^2 + \cdots + a_n x^n$$

and

$$b_0 + b_1 x + b_2 x^2 + \cdots + b_n x^n$$

we add them by adding the corresponding coefficients and obtain

$$(a_0 + b_0) + (a_1 + b_1) x + (a_2 + b_2) x^2 + \cdots + (a_n + b_n) x^n$$

215

Multiplying polynomials is not as simple in that the coefficients of the product are obtained by adding together several products of coefficients; that is, the coefficient of x^k in the product is obtained by adding

$$a_0 b_k + a_1 b_{k-1} + a_2 b_{k-2} + \cdots + a_{k-1} b_1 + a_k b_0$$

since when we perform the multiplication in the usual intuitive sense,

$$(a_0)(b_k x^k) = (a_0 b_k) x^k, \quad (a_1 x)(b_{k-1} x^{k-1}) = a_1 b_{k-1} x^k, \text{ etc.}$$

Summation notation is very convenient for polynomials if we adopt the convention that $x^0 = 1$. Then we can denote

$$a_0 + a_1 x + a_2 x^2 + \cdots + a_n x^n \quad \text{by} \quad \sum_{i=0}^{n} a_i x^i$$

The rule for addition becomes

$$\sum_{i=0}^{n} a_i x^i + \sum_{i=0}^{n} b_i x^i = \sum_{i=0}^{n} (a_i + b_i) x^i$$

and the rule for multiplication becomes

$$\left(\sum_{i=0}^{n} a_i x^i \right) \left(\sum_{j=0}^{m} b_j x^j \right) = \sum_{k=0}^{n+m} c_k x^k$$

where

$$c_k = \sum_{i+j=k} a_i b_j$$

The phrase "a formal sum of the form $\sum_{i=0}^{n} a_i x^i$" is not really a good definition of a polynomial. Before going further into the study of polynomials we shall formalize and make our intuitive notions precise by using sequences.

Let R be a ring and let $\mathbf{N}_0$ denote the set of all nonnegative integers. A *sequence in R* is a function from $\mathbf{N}_0$ into R. Sequences are usually described differently than other functions; instead of writing $(i)a$ or $a(i)$ we write a_i for the image of i under the function. We usually denote such a sequence by $\{a_i\}$. Intuitively we usually think of sequences in a form such as

$$\{a_0, a_1, a_2, a_3, \ldots\}$$

The ring elements a_i are called the *terms* of the sequence. A sequence is said to be *finite* if it has, at most, finitely many nonzero terms, that is, if for some integer p, $a_i = 0$ for all $i > p$.

From group theory we know that the set of all sequences in a ring R is an abelian group under the usual functional addition; that is,

$$\{a_i\} + \{b_i\} = \{a_i + b_i\}$$

Notice the similarity of this addition and the addition of polynomials. If we were to view polynomials as sequences that are zero from some point on, the addition would be the same. We use this idea to define polynomials formally.

Let R be a ring and let S denote the set of all sequences of elements of R. We define the sum and product of two sequences $\{a_i\}$ and $\{b_i\}$ by

$$\{a_i\} + \{b_i\} = \{a_i + b_i\} \quad \text{and} \quad \{a_i\} \cdot \{b_i\} = \{c_i\}$$

where

$$c_i = \sum_{i=j+k} a_j b_k$$

Theorem 1

If R is a ring and S is the set of all sequences of elements in R, then S is a ring with respect to the operations $+$ and $\cdot$ defined above. If R has a 1, then S has a 1. If R is commutative, then S is commutative.

Proof: From group theory we know that S is an abelian group with respect to functional addition. The proofs of the Associative and Distributive laws are straightforward but messy because of the number of summations involved.

Let $\{d_i\}$, $\{p_i\}$, and $\{q_i\}$ be any elements of S. Then

$$\{d_i\}(\{p_i\}\{q_i\}) = \{d_i\}\left\{ \sum_{i=j+k} p_j q_k \right\}$$

$$= \left\{ \sum_{i=h+m} \left(d_h \sum_{m=j+k} p_j q_k \right) \right\}$$

$$= \left\{ \sum_{i=h+m} \sum_{m=j+k} d_h p_j q_k \right\}$$

$$= \left\{ \sum_{i=h+j+k} d_h p_j q_k \right\}$$

Now

$$(\{d_i\}\{p_i\})\{q_i\} = \left\{ \sum_{i=h+j} d_h p_j \right\}\{q_i\}$$

$$= \left\{ \sum_{i=m+k} \left(\sum_{m=h+j} d_h p_j \right) q_k \right\}$$

$$= \left\{ \sum_{i=m+k} \sum_{m=h+j} d_h p_j q_k \right\}$$

$$= \left\{ \sum_{i=h+j+k} d_h p_j q_k \right\}$$

Therefore,

$$\{d_i\}(\{p_i\}\{q_i\}) = (\{d_i\}\{p_i\})\{q_i\}$$

Again suppose $\{d_i\}$, $\{p_i\}$, and $\{q_i\}$ are elements of S. Then

$$\{d_i\}(\{p_i\} + \{q_i\}) = \{d_i\}\{p_i + q_i\}$$

$$= \left\{\sum_{i=j+k} d_j(p_k + q_k)\right\}$$

$$= \left\{\sum_{i=j+k} (d_j p_k + d_j q_k)\right\}$$

$$= \left\{\sum_{i=j+k} d_j p_k + \sum_{i=j+k} d_j q_k\right\}$$

$$= \left\{\sum_{i=j+k} d_j p_k\right\} + \left\{\sum_{i=j+k} d_j q_k\right\}$$

$$= \{d_i\}\{p_i\} + \{d_i\}\{q_i\}$$

We leave the proof of the right-hand distributive law as an exercise. We also leave it to the reader to show that S is commutative if R is and if R has a 1, the sequence $\{1, 0, 0, \ldots\}$ is a multiplicative identity for S. □

Proposition 2

If R is a ring and S is the ring of all sequences in R, then the set of all finite sequences in R is a subring of S.

Proof: If $\{a_i\}$ and $\{b_i\}$ are finite sequences, then for some integer p, $a_i = 0$ and $b_i = 0$ for $i > p$. Then $\{a_i\} - \{b_i\} = \{a_i - b_i\}$ is a finite sequence, because for $i > p$, $a_i - b_i = 0 - 0 = 0$. Furthermore, $\{a_i\}\{b_i\}$ is a finite sequence for $\{a_i\}\{b_i\} = \{c_i\}$, where

$$c_i = \sum_{i=j+k} a_j b_k$$

If $i > 2p$, then if $j + k = i$, either $j > p$ or $k > p$, and therefore $a_j = 0$ or $b_k = 0$ and $a_j b_j = 0$. Thus, for $i > 2p$, $c_i = 0$ and $\{c_i\}$ is a finite sequence. Since the set of finite sequences is closed under subtraction and multiplication, it is a subring of the ring of all sequences. □

We now define a *polynomial over the ring R* to be a finite sequence in R; polynomials are added and multiplied as sequences. Thus, the set of all polynomials in R is the set of all finite sequences in R and, by Proposition 2, is a subring of the ring of all sequences.

This definition is all well and good from a formal standpoint, but one of the nice intuitive properties of polynomials was being able to express them in the form

$$\sum_{i=0}^{n} a_i x^i \quad \text{or} \quad a_0 + a_1 x + a_2 x^2 + \cdots + a_n x^n$$

where each a_i is in R. This is done rather easily if R is a ring with 1 and we first make an R-module out of the set of sequences by defining

$$r\{a_i\} = \{ra_i\}$$

for each element r in R and each sequence $\{a_i\}$ in R. From exercises in Chapter 7 we know that this definition does, in fact, make S into an R-module. Furthermore, if $\{a_i\}$ is finite, clearly $r\{a_i\}$ is finite, and therefore the finite sequences form a submodule of S. Now define X to be the sequence $\{0, 1, 0, \ldots\}$; that is, X is the sequence defined by

$$X = \{x_i\}$$

where $x_1 = 1$ and $x_i = 0$ for $i \neq 1$.

Proposition 3

If n is a nonnegative integer, then $X^n = \{0, 0, \ldots, 1, 0, 0, \ldots\}$; that is, $X^n = \{y_i\}$, where $y_n = 1$ and $y_i = 0$ for $i \neq n$.

Proof: If $n = 0$, $X^0 = 1 = \{1, 0, 0, \ldots\}$
since $\{1, 0, 0, \ldots\}$ is the identity of S.
If $n = 1$, $X^1 = X = \{0, 1, 0, \ldots\}$.
Assume that the assertion holds for m.
Then

$$\begin{aligned}
X^{m+1} &= X^m X \\
&= \{0, 0, \ldots, 1, 0, \ldots\} \cdot \{0, 1, \ldots\} \\
&= \{y_i\}\{x_i\}
\end{aligned}$$

where

$$\begin{aligned}
y_m &= 1 \quad \text{and} \quad y_i = 0 \qquad \text{for } i \neq m \\
x_1 &= 1 \quad \text{and} \quad x_i = 0 \qquad \text{for } i \neq 1
\end{aligned}$$

Thus,

$$X^{m+1} = \left\{ \sum_{i=j+k} y_j x_k \right\}$$

But if $i \neq m + 1$, then either $j \neq m$ or $k \neq 1$.
Therefore,

$$\sum_{i=j+k} y_j x_k = 0 \qquad \text{if } i \neq m + 1$$

and

$$\sum_{m+1=j+k} y_j x_k = y_m x_1 = 1 \cdot 1 = 1$$

Therefore, $X^{m+1} = \{z_i\}$, where $z_{m+1} = 1$ and $z_i = 0$ for $i \neq 1$ and the proposition follows by mathematical induction. $\square$

Let $\{a_i\}$ be a finite sequence with no more than n nonzero terms. For each i such that $0 \leq i \leq n$, define the sequence $\{b_{ij}\}$ by $b_{ii} = a_i$ and $b_{ij} = 0$ for $i \neq j$. Then

$$\{a_i\} = \sum_{j=0}^{n} \{b_{ij}\}$$

Since, for each i, $\{b_{ij}\} = a_i X^i$, it follows that

$$\{a_i\} = \sum_{j=0}^{n} a_j X^j$$

Thus, we have proved the following assertion.

Proposition 4

If R is a ring with 1, every polynomial over R can be expressed in the form

$$\sum_{i=0}^{n} a_i X^i$$

We shall also use the common notation of $p(X)$ or just the letter f to denote a polynomial. The ring of all polynomials over the ring R is denoted by $R[X]$.

The above may seem like quite a bit of work to get back to where our intuition started, but now we have the same type of foundation underlying polynomial theory as we have underlying the rest of our study.

A polynomial $p(X)$ over the ring R is said to have *degree n* if n is a nonnegative integer such that $p(X) = \Sigma_{i=0}^{n} a_i X^i$ and $a_n \neq 0$. The degree of a polynomial p is frequently denoted by δp. The zero polynomial can be considered to have degree $-\infty$, where $-\infty$ obeys the rules $(-\infty)n = -\infty$, $(-\infty) + m = -\infty$ for any positive integers n and m and also $(-\infty) + (-\infty) = -\infty$. We shall also consider $-\infty$ to be less than any nonnegative integer. This may seem like an artificial definition, but it enables us to state many propositions more simply by giving the zero polynomial a degree. For example, the polynomial $\Sigma_{i=0}^{n} a_i X^i$ is assured of having degree less than or equal to n.

Clearly, R is isomorphic to the set of all polynomials of degree 0 or $-\infty$, and we can make the natural identification and view R as a subring of $R[X]$.

Proposition 5

If p and q are polynomials of degree less than or equal to n over the ring R, then $p + q$ has degree less than or equal to n.

Proof: Let $p(X) = \Sigma_{i=0}^n a_i X^i$ and $q(X) = \Sigma_{i=0}^n b_i X^i$. Then

$$(p + q)(X) = \sum_{i=0}^n (a_i + b_i)X^i$$

which has degree less than or equal to n. $\square$

Proposition 6

If p and q are polynomials of degree n and m, respectively, then the degree of pq is less than or equal to $n + m$.

Proof: If $p(X) = \Sigma_{i=0}^n a_i X^i$ and $q(X) = \Sigma_{i=0}^m b_i X^i$, then $pq = \{c_i\}$, where $c_i = \Sigma_{i=j+k} a_j b_k$.
If $i > n + m$, either $j > n$ or $k > m$ and then $a_j b_k = 0$.
Thus, $\{c_i\} = \Sigma_{i=0}^{n+m} c_i X^i$ and the degree of pq is less than or equal to $n + m$. $\square$

The above proof can be adapted to obtain an important result for polynomials over an integral domain.

Proposition 7

If R is an integral domain and p and q are nonzero polynomials over R of degree n and m, respectively, then the degree of pq is $n + m$.

Proof: Let $p(X) = \Sigma_{i=0}^n a_i X^i$ and $q(X) = \Sigma_{i=0}^m b_i X^i$, with $a_n \neq 0$ and $b_m \neq 0$.
Then

$$pq = \sum_{i=0}^{n+m} c_i X^i$$

where

$$c_i = \sum_{i=j+k} a_j b_k$$

Then

$$c_{n+m} = \sum_{n+m=j+k} a_j b_k$$

If $j < n$, then $k > m$, which implies $b_k = 0$, and therefore $a_j b_k = 0$.
Also, if $j > n$, $a_j = 0$ and $a_j b_k = 0$.
Thus, $c_{n+m} = a_n b_m$.
Since $a_n \neq 0$ and $b_m \neq 0$, $c_{n+m} \neq 0$ and the degree of pq is $n + m$. $\square$

If p is a polynomial over the ring R, p determines a function from R into R defined by $(t)p = p(t) = \Sigma_{i=0}^n a_i t^i$, where $p = \Sigma_{i=0}^n a_i X^i$ and t represents an element in R. This is the usual evaluation process as done in elementary mathematics courses. If b is an element of the ring R such that $(b)p = p(b) = 0$, we call b a *root* or a *zero* of $p(X)$. The evaluation process determines functions from $R[X]$ to R if, for each element b in R we define E_b by $pE_b = p(b)$.

Proposition 8

If R is a commutative ring and b is an element of R, the function E_b defined by $pE_b = p(b)$ is a homomorphism from $R[X]$ onto R.

Proof: Exercise.

The kernel of E_b in the assertion above is just the set of all polynomials that have b for a root. Since the kernel of a homomorphism is an ideal, we have the following assertion.

Corollary 9

If b is an element of the commutative ring R, the set of all polynomials that have b for a root is an ideal in the ring of all polynomials over R.

Since the homomorphism E_b is onto R, if we let K_b denote the kernel of E_b or the set of all polynomials that have b for a root, we have $R[X]/K_b \cong R$.

EXERCISE SET A

1. Complete the proof of Theorem 1.
2. Prove Proposition 8.
3. If $\mathbf{R}$ denotes the set of real numbers and E_5 is the evaluating map from $\mathbf{R}[X]$ to $\mathbf{R}$, describe ker E_5.
4. Let $\mathbf{Q}$ denote the rational numbers and define $E_{\sqrt{2}}$ from $\mathbf{Q}[X]$ to $\mathbf{R}$ by $pE_{\sqrt{2}} = p(\sqrt{2})$. Prove that $E_{\sqrt{2}}$ is a homomorphism. Describe ker $E_{\sqrt{2}}$.
**5. Let F be a field, and let K be a subfield of F. For each element b in F, define E_b from $K[X]$ to F by $pE_b = p(b)$. The proof of Proposition 8 can easily be adapted to show that E_b is a homomorphism. We say that b is *algebraic* over K if b is a root of some polynomial over K, otherwise b is, said to be *transcendental*. Prove that the image of $K[X]$ under E_b is a subfield of F iff b is algebraic over K. (Hint: First argue that if $p(X) = \Sigma_{i=0}^{n} a_i X^i$ is the non-zero polynomial of smallest degree such that $p(b) = 0$, then $a_0 \neq 0$.)

§2. Divisibility and Factorization in $F[X]$

Much of the divisibility theory for $\mathbf{Z}$ can be generalized to the ring of polynomials over a field and, in fact, some of it can be extended to polynomials over commutative rings. In this section we shall concern ourselves only with the ring of polynomials over a field and leave the generalizations as exercises.

If $p(X) = \sum_{i=0}^{n} a_i X^i$ is a polynomial over the ring R, the ring elements $a_0, \ldots, a_n$ are called *coefficients* and if the degree of p is n, a_n is called the *leading* coefficient. A polynomial over a ring with 1 is said to be *monic* if its leading coefficient is 1. Polynomials of degree 0 or of degree $-\infty$ are called *constant* polynomials. If p and q are polynomials over the field (or commutative ring) F, we say that p divides q if $pf = q$ for some polynomial f in $F[X]$. As before we write $p \mid q$. The following two assertions are easily proved using the same ideas used in Chapter 2.

Proposition 10

If p, g, f, h, and w are polynomials over the field F such that $p \mid g$ and $p \mid f$, the $p \mid (gh + fw)$.

Proof: Exercise.

Proposition 11

If p, q, and f are polynomials over the field F such that $p \mid q$ and $q \mid f$, then $p \mid f$.

Proof: Exercise.

Proposition 12

If p and q are polynomials over the field F such that $p \mid q$ and $q \neq 0$, then $\delta p \leq \delta q$.

Proof: If $p \mid q$, then $pg = q$ for some polynomial g. Thus, $\delta(pg) = \delta q$. Since F is an integral domain, from Proposition 7 we know that $\delta q = \delta(pg) = \delta p + \delta g$. Since $q \neq 0, g \neq 0$ and $\delta q \geq 0$ and $\delta g \geq 0$. Therefore, $\delta p \leq \delta p + \delta g = \delta q$, which implies $\delta p \leq \delta q$. $\square$

Corollary 13

If p and q are monic polynomials over the field F such that $p \mid q$ and $q \mid p$, then $p = q$.

Proof: Since $p \mid q$ and $q \mid p$, $\delta p \leq \delta q$ and $\delta q \leq \delta p$, which implies $\delta p = \delta q$. Thus, $pg = q$ for some polynomial g and, since $\delta(pg) = \delta p + \delta g = \delta p$, it follows that $\delta g = 0$ and g is just some element of F. Now if

$$p = \sum_{i=0}^{n} a_i X^i \qquad q = \sum_{i=0}^{n} b_i X^i$$

then

$$pg = \sum_{i=0}^{n} a_i g X^i = \sum_{i=0}^{n} b_i X^i \quad \text{and} \quad a_n g = b_n$$

Since both p and q are monic, $a_n = 1$ and $b_n = 1$, which implies $1g = 1$ and g must be 1. Since $pg = q$, $p = p1 = q$. $\square$

Theorem 14

If f and g are polynomials over the field F, $y \neq 0$, then there exist unique polynomials q and r such that $f = gq + r$ and $\delta r < \delta g$.

Proof: We adapt the proof of Theorem 20 in Chapter 2 but incorporate induction on the degree of f. If $\delta f < \delta g$ then we have $f = g \cdot 0 + f$, where $r = f$ and $\delta f < \delta g$. Now assume $\delta f \geq \delta g$. If $\delta f = 0$, then $\delta g = 0$ and $f = g(g^{-1}f) + 0$, where $g \in F$ and $r = 0$. Now assume that the theorem is true for any polynomial of degree less than $n = \delta f$. Let $M = \{f - gh : h \in F[X]\}$, and let r be a polynomial in M of smallest degree. Let $f = \sum_{i=0}^{n} a_i X^i$ and $g = \sum_{i=0}^{m} b_i X^i$ with $\delta g = m$. Since F is a field, there is an element b_m^{-1} in F.

Let $h = a_n b_m^{-1} X^{n-m}$. Then

$$f - gh = \sum_{i=0}^{n} a_i X^i - \left[\left(\sum_{i=0}^{m} b_i X^i \right) a_n b_m^{-1} X^{n-m} \right]$$

$$= \sum_{i=0}^{n} a_i X^i - \left(\sum_{i=0}^{m} b_i a_n b_m^{-1} X^{i+n-m} \right)$$

$$= \sum_{i=0}^{n-1} a_i X^i + a_n X^n - \left(\sum_{i=0}^{m-1} b_i a_n b_m^{-1} X^{i+n-m} + b_m a_n b_m^{-1} X^n \right)$$

$$= \sum_{i=0}^{n-1} a_i X^i - \sum_{i=0}^{m-1} b_i a_n b_m^{-1} X^{i+n-m}$$

and therefore M contains polynomials of degree less than δf, in particular $f - gh$. If we let p denote $f - gh$, then p is a polynomial of degree less than n and, by the induction hypothesis, there are polynomials w and r such that

$$p = gw + r \quad \text{and} \quad \delta r < \delta g$$

Then

$$f - gh = p = gw + r$$

or

$$f = gh + gw + r$$
$$= g(h + w) + r$$

and if we let $h + w = q$ we have

$$f = gq + r \quad \text{and} \quad \delta r < \delta g$$

Now suppose

$$f = gq + r \quad \text{and} \quad f = gp + t \qquad \text{where } \delta r < \delta g \quad \text{and} \quad \delta t < \delta g$$

Then

$$gq + r = gp + t$$
$$gq - gp = t - r$$
$$g(q - p) = t - r \quad \text{and} \quad g \,|\, (t - r)$$

Now $\delta g > \delta t$ and $\delta g > \delta r$ so $\delta g > \delta(t - r)$.
But since $g \,|\, (t - r)$, either $\delta g \leq \delta(t - r)$ or $t - r = 0$.
Since $\delta g > \delta(t - r)$, $t - r = 0$ and $t = r$.
Since $t = r$, $gq = gp$ and $q = p$.
Thus, there are unique polynomials q and r such that $f = gq + r$ and $\delta r < \delta g$. $\square$

Theorem 14 is just the common division algorithm for polynomials rather than for integers. Since this analogous theorem holds, one might expect other theorems for $\mathbf{Z}$ to have analogues in $F[X]$. This is, in fact, the case as we now proceed to establish in the following sequence of assertions.

The notion of a greatest common divisor of two polynomials is just a generalization of the concept of the greatest common divisor of two integers.

If f and g are two polynomials over the field F, the monic polynomial d is said to be the *greatest common divisor* of f and g if

(i) $d \,|\, f$ and $d \,|\, g$.
(ii) If c is any polynomial such that $c \,|\, f$ and $c \,|\, g$, then $c \,|\, d$.

We shall again abbreviate the words *greatest common divisor* to GCD. As before, the word "the" in the above definition is justified by the following proposition.

Proposition 15

If the GCD of two polynomials exists, it is unique.

Proof: Exercise.

Theorem 16

If f and g are polynomials over the field F such that $f \neq 0$ or $g \neq 0$, then the GCD of f and g exists.

Proof: This is almost a word-by-word translation of the proof of Theorem 22 in Chapter 2. Let $M = \{fp + gq : p$ and q are polynomials over $F\}$. Since $f \neq 0$ or $g \neq 0$, M contains polynomials of nonnegative degree. Let h be an element of M of smallest nonnegative degree. Then, for some polynomials u and v, $h = fu + gv$.

If h is not monic, by multiplying both sides of the equation by the proper constant we get a monic polynomial on the left; that is, if

$$h = \sum_{i=0}^{n} b_i X^i \quad \text{and} \quad b_n \neq 0$$

then

$$hb_n^{-1} = f[ub_n^{-1}] + g[vb_n^{-1}]$$

If we let $d = hb_n^{-1}$, $y = ub_n^{-1}$, and $z = vb_n^{-1}$ we have

$$d = fy + gz$$

Thus d is monic, $d \in M$, and no nonzero polynomial in M has a smaller degree.

Now let us show that d is, in fact, the GCD of f and g. There exist polynomials q and r such that

$$f = dq + r \quad \text{and} \quad \delta r < \delta d$$

Then $f - dq = r$ and, since $d = fy + gz$,
we have $f - (fy + gz)q = r$
or $f(1 - yq) + g(-zq) = r$ and therefore $r \in M$.
Since $\delta r < \delta d$ and d is an element of smallest nonnegative degree in M, $r = 0$ and $f = dq$; that is $d \,|\, f$. By a similar argument $d \,|\, g$.

Now, if c is a polynomial such that $c \,|\, f$ and $c \,|\, g$, then $c \,|\, (fy + gz)$ and therefore $c \,|\, d$.
Thus, d is the GCD of f and g. $\square$

A polynomial f over the field F is said to be *irreducible* if $\delta f > 0$ and whenever g and h are polynomials such that $gh = f$, either $\delta g = 0$ or $\delta h = 0$; that is, either g or h is a constant. A monic irreducible polynomial will be called a *prime* polynomial. Two polynomials are said to be *relatively prime* if their GCD is 1.

Proposition 17

All polynomials of degree 1 over the field F are irreducible. Furthermore, if $a \in F$, then $a + X$ is a prime polynomial.

Proof: Exercise.

Proposition 18

If p and g are polynomials over the field F, p is irreducible, and p does not divide g, then p and g are relatively prime.

Proof: Exercise.

Proposition 19

If p is an irreducible polynomial and f and g are polynomials such that $p \,|\, fg$, then $p \,|\, f$ or $p \,|\, g$.

Proof: Exercise (Hint: See Proposition 23, Chapter 2.)

Theorem 20 (Unique Factorization Theorem)

Every nonzero polynomial over the field F is either a constant or can be expressed uniquely as the product of an element of F and prime polynomials.

Proof: Assume that the theorem holds for all polynomials of degree less than m, and let f be a polynomial of degree m.
If $m = 0$, then $\delta f = 0$ and the theorem holds for f.
If $m \neq 0$, then

$$f = \sum_{i=0}^{m} a_i X^i \qquad \text{with } a_m \neq 0$$

and

$$f = a_m \sum_{i=0}^{m} b_i X^i$$

where $b_i = a_m^{-1} a_i$ for $i = 0, 1, \ldots, m$.   Notice that $b_m = 1$ since $b_m = a_m^{-1} a_m = 1$.

If we let $g = \sum_{i=0}^{m} b_i X^i$, then f can be expressed as the product of a_n and the monic polynomial g. Suppose g is prime. If q is a prime polynomial that divides f, then $q \mid a_n g$ and, since q does not divide a_n, q must divide g. Since q and g are both prime, $q = g$. Thus, if g is prime, the representation of f is unique. Now suppose g is not prime. Then $g = pq$ for some polynomials p and q such that $\delta p > 0$ and $\delta q > 0$. Since $\delta g = \delta p + \delta q$, $\delta p < \delta g$ and $\delta q < \delta g$. By the induction hypothesis the conditions of the theorem hold for p and q because $\delta p < m$ and $\delta q < m$. Thus, $p = ch_1 h_2 \cdots h_z$ and $q = dk_1 k_2 \cdots k_y$, where $c \in F$, $d \in F$, and each h_i and k_j is prime. Thus, $g = cdh_1 h_2 \cdots h_z k_1 \cdots k_y$ and, since g is monic and so is each h_i and k_j, $cd = 1$ and $g = h_1 h_2 \cdots h_z k_1 \cdots k_y$. Therefore, $f = a_m h_1 h_2 \cdots h_z k_1 \cdots k_y$, where $a_m \in F$ and each h_i and k_j is prime. Now suppose $f = w s_1 \cdots s_u$, where each s_i is prime.
Since each s_i is monic, $w = a_m$.
Thus, $h_1 h_2 \cdots h_z k_1 \cdots k_y = s_1 \cdots s_u$ and $h_1 \mid s_1 (s_2 \cdots s_u)$.
Since h_1 is prime, either $h_1 \mid s_1$ or $h_1 \mid s_2 \cdots s_u$.
If $h_1 \mid s_1$, then $h_1 = s_1$.
If h_1 does not divide s_1, then $h_1 \mid s_2 \cdots s_u$ and, since $\delta s_1 > 0$, $\delta(s_2 \cdots s_u) < m$.
By the induction hypothesis, the prime factorization of $s_2 \cdots s_u$ is unique and $h_1 = s_j$ for some j.
Thus, h_1 is one of the prime polynomials $s_1, \ldots, s_u$.
Without loss of generality we may assume $h_1 = s_1$.
Then $h_1 h_2 \cdots h_z k_1 \cdots k_y = s_1 s_2 \cdots s_u$ and $h_2 \cdots h_z k_1 \cdots k_y = s_2 \cdots s_u$.
Since $\delta(h_2 \cdots h_z k_1 \cdots k_y) < m$, the prime factorization of $h_2 \cdots h_z k_1 \cdots k_y$ is unique; that is, each of the s_i's is one of the h_i's or k_j's and the factorization of f is unique. $\square$

EXERCISE SET B

1. Prove Proposition 10.
2. Prove Proposition 11.
3. Prove Proposition 15.
4. Prove Proposition 17.
5. Prove Proposition 18.
6. Prove Proposition 19.

§3. Ideals in $F[X]$

What must the ideals in $F[X]$ look like? We know that the ideals in $\mathbf{Z}$ are all generated by single integers; that is, all ideals in $\mathbf{Z}$ are of the form $\langle m \rangle$ or $m\mathbf{Z}$. The key in the proof is the division algorithm, and we shall use the analogue to obtain a similar result in $F[X]$.

Theorem 21

If F is a field, then every ideal in $F[X]$ is generated by a single polynomial; that is, every ideal is of the form $pF[X]$.

Proof: Let A be an ideal in $F[X]$.
If $A = \{0\}$, then $A = \langle 0 \rangle = 0F[X]$.
If $A \neq \{0\}$, then A contains nonzero polynomials.

Let $p(X)$ be a nonzero polynomial of smallest degree in A. Since A is an ideal, $pg \in A$ for all polynomials g in $F[X]$, and therefore $pF[X] \subseteq A$.
   If f is any polynomial in A, then there exist polynomials q and r such that

$$f = pq + r \quad \text{and} \quad \delta r < \delta p$$

Since $p \in A$ and A is an ideal, $pq \in A$.
Thus, $f - pq \in A$, which imples $r \in A$.
Since $\delta r < \delta p$ and p is a nonzero polynomial of smallest degree in A, r must be 0 and $f = pq$; that is, $f \in pF[X]$ and thus $A \subseteq pF[X]$.
Since $pF[X] \subseteq A$ and $A \subseteq pF[X]$, $A = pF[X]$.   □

   From Chapter 6 we know that if R is a commutative ring with 1 and A is a proper ideal in R then R/A is a field iff A is maximal, that is, A is not contained in any other proper ideal.

Proposition 22

If F is a field and A is an ideal in $F[X]$, then $F[X]/A$ is a field iff $A = pF[X]$ for some prime polynomial p in $F[X]$.

Proof: We need only show that A is maximal iff $A = pF[X]$ for some prime polynomial p in $F[X]$.
   Since A is an ideal in $F[X]$, $A = qF[X]$ for some polynomial q in

$F[X]$ and we might as well assume that q is monic. We need to show that A is maximal iff q is prime.

If q is prime and B is an ideal containing $qF[X]$, then $B = fF[X]$ for some polynomial f. Since $A \subseteq B$, then $qF[X] \subseteq fF[X]$ and $q \in fF[X]$. But if $q \in fF[X]$, $q = fg$ for some polynomial g in $F[X]$. Since q is prime, q is irreducible, and therefore either $f \in F$ or $g \in F$. If $f \in F$, then $fF[X] = F[X]$ and B is not a proper ideal. If $g \in F$, then, since $q \neq 0, g \neq 0$ and $qg^{-1} = f$. Thus, $f \in qF[X]$ and therefore $B = fF[X] \subseteq qF[X] = A$, which implies $B = A$. Thus, if q is prime, then $A = qf[X]$ is a maximal proper ideal.

If q is not prime then, since q is monic, q is not irreducible. Thus, there exist polynomials f and g such that $q = fg$ and $\delta f < \delta q$ and $\delta g < \delta q$. Furthermore, $q \in fF[X]$, and therefore $qF[X] \subseteq fF[X]$. Since every nonzero polynomial in $qF[X]$ has degree at least as great as δq and $fF[X]$ contains f, which has degree less than δq, $qF[X]$ is properly contained in $fF[X]$. Furthermore, $fF[X]$ is a proper ideal because $fF[X]$ contains no polynomials of degree 0, and therefore $qF[X] = A$ is not maximal. □

Proposition 23

If F is a field and A is a proper ideal in $F[X]$, then F is isomorphic to a subring of $F[X]/A$.

Proof: The natural function η from $F[X]$ to $F[X]/A$ is known to be a homomorphism, and therefore $\eta \mid_F$ (the restriction of η to F) is a homomorphism from F into $F[X]/A$. Since A is a proper ideal in $F[X]$, $A = qF[X]$ for some polynomial q of positive degree. Thus, $F \cap A = F \cap qF[X] = \{0\}$, $\eta \mid_F$ is one-to-one, and the image of F under η is isomorphic to F. □

Since $F[X]/A$ contains an isomorphic copy of F, we can make the natural identification and view F as a subring of $F[X]/A$. In the case where A is a maximal proper ideal, F can be viewed as a subfield of $F[X]/A$.

EXERCISE SET C

1. Use Theorem 21 to develop another proof for Theorem 16 by showing that the ideal generated by two polynomials f and g is generated by the GCD of f and g.
**2. Make a suitable definition of LCM for polynomials over a field F and show that the intersection of the ideals generated by f and g is generated by the LCM of f and g.

§4. Field Extensions

If K is a field and F is a subfield of K, then K is called an *extension* of F. If a is an element of K, then a is said to be *algebraic* over F if $p(a) = 0$

for some nonzero polynomial p in $F[X]$, otherwise a is said to be *transcendental*. Since a is a root of $X - a$, every element of F is algebraic over F.

Examples. Although the real number $\sqrt{2}$ is not in $\mathbf{Q}$, $\sqrt{2}$ is algebraic over $\mathbf{Q}$ because $\sqrt{2}$ is a root of $X^2 - 2$.

The complex number $i = \sqrt{-1}$ is algebraic over $\mathbf{R}$ because i is a root of $X^2 + 1$.

The number π is transcendental over $\mathbf{Q}$.

Lemma 24

If p is a polynomial over the field F and $a \in F$, then the remainder after dividing $p(X)$ by $(X - a)$ is $p(a)$, that is, for the unique polynomials q and r such that

$$p(X) = (X - a)q(X) + r(X) \text{ where } \delta r < \delta(X - a),$$

r is constant and $r = p(a)$.

Proof: Since $p(X) = (X - a)q(X) + r(X)$ and $\delta r < \delta(X - a)$, $\delta r < 1$, and therefore r is in F. Then

$$p(a) = (a - a)q(a) + r$$

or

$$p(a) = 0 \cdot q(a) + r$$

that is,

$$p(a) = r \quad \square$$

Proposition 25

If p is a polynomial over the field F and $a \in F$, then a is a root of p iff $(X - a) \mid p(X)$.

Proof: If $(X - a) \mid p(X)$, $p(X) = (X - a)q(X)$ and $p(a) = 0 \cdot q(a) = 0$.

Conversely, by Lemma 24, if $p(a) = 0$, the remainder after dividing $p(X)$ by $X - a$ is 0, and therefore $(X - a) \mid p(X)$. $\quad \square$

If p is a polynomial over the field F and a is a root of p, then, since $X - a$ is prime, $X - a$ must be one of the prime factors of p. We now use this idea to show that the number of roots is limited by the degree of the polynomial.

Proposition 26

If p is a polynomial over the field F and $\delta p = n$, then p has no more than n roots in F.

Proof: Suppose $a_1, \ldots, a_n$ are roots of p. Then, since for each i, $X - a_i$ is a prime factor of p, the unique factorization theorem implies that

$$p(X) = c(X - a_1)(X - a_2)\cdots(X - a_n)$$

There can be no other prime factors, since the polynomial $(X - a_1)$ $(X - a_2)\cdots(X - a_n)$ has degree n and any other prime factor would be of degree 1 or more and then the product would have degree greater than $n = \delta p$.

If $b \in F$ and b is a root of p, then $(X - b)\,|\,p$. By the unique factorization theorem, $X - b = X - a_i$ for some a_i, and therefore $b = a_i$ for some a_i. Thus, every root is one of $a_1, \ldots, a_n$ and p has no more than n roots. $\square$

If F is a subfield of the field K, every polynomial over F can be viewed as a polynomial over K. The above proposition then also proves the following assertion.

Corollary 27

If p is a polynomial of degree n over the field F, then p has no more than n roots in any extension of F.

We essentially defined an irreducible polynomial to be a polynomial divisible only by constants and constant multiples of itself. This seems somewhat backwards, since we have not yet defined what we mean by a reducible polynomial. A *reducible* polynomial over the field F is, as expected, a polynomial that is not irreducible. A polynomial p over the field is said to be *completely* reducible if all of its prime factors have degree 1, that is, if for some elements $c, a_1, \ldots, a_n$ in F,

$$p(X) = c(X - a_1)(X - a_2)\cdots(X - a_n)$$

Examples. The polynomial $2X^2 - 6X + 4$ is completely reducible over **Q** because $2X^2 - 6X + 4 = 2(X - 2)(X - 1)$.

The polynomial $X^2 - 2$ is irreducible over **Q** but is completely reducible over **R** because $X^2 - 2 = (X - \sqrt{2})(X + \sqrt{2})$.

The polynomial $X^2 + 1$ is irreducible over **R** but is completely reducible over **C** because $X^2 + 1 = (X - i)(X + i)$.

The fundamental theorem of algebra says that every polynomial of positive degree over the complex numbers has a complex root and by using induction, one can show that every polynomial of positive degree over **C** is completely reducible. The complex numbers are an extension of **Q** and of **R**, and therefore every polynomial over **Q** or **R**, is completely reducible over some extension of **Q** or **R**, namely **C**. We need not use the entire field of complex numbers to obtain the reducibility. In fact, if $p(X)$ is a polynomial over **Q** and $c, a_1, \ldots, a_n$ are complex numbers such that $p(X) = c(X - a_1)(X - a_2)\cdots(X - a_n)$, then p is completely reducible over the subfield of **C** generated by $\{a_1, \ldots, a_n\}$. (In the above case, $c \in$ **Q**.)

It seems natural to ask if this is, in fact, the case for polynomials over arbitrary fields. The answer is yes and is assured by the following theorem.

Theorem 28

If p is a polynomial of positive degree over the field F, there is an extension K of F over which p is completely reducible.

Proof: By induction on δp. Without loss of generality we may assume that p is monic. If $\delta p = 1$, then $p = X - a$ for some a in F and p is completely reducible over F. Now assume the theorem is true for all polynomials of degree less than n and $\delta p = n$. Let $p = q_1 q_2 \cdots q_m$, where each q_i is a prime polynomial. If each q_i has degree 1, then p is completely reducible over F. If not, then some q_j is of degree greater than 1, without loss of generality let $\delta q_1 = s > 1$.

Now consider the ideal $q_1 F[X] = \langle q_1 \rangle$ and the factor ring $F[X]/\langle q_1 \rangle$. Since q_1 is prime, by Proposition 22, $F[X]/\langle q_1 \rangle$ is a field. Furthermore, the natural function η from $F[X]$ to $F[X]/\langle q_1 \rangle$ embeds F into $F[X]/\langle q_1 \rangle$, and therefore F can be viewed as a subfield of $F[X]/\langle q_1 \rangle$ and p can be viewed as a polynomial over $F[X]/\langle q_1 \rangle$. If we denote $X\eta$ in $F[X]/\langle q_1 \rangle$ by the element α and, since $\eta \mid_F$ is one-to-one, we identify the elements of F with their images, then since $q_1 = \Sigma_{i=0}^m a_i X^i$, it follows that $q_1\eta = \Sigma_{i=0}^m a_i \alpha^i$. Since $\bar{0} = \bar{0} = q_1\eta$, $\Sigma_{i=0}^m a_i \alpha^i = \bar{0}$. Since α is a root of q_1 when viewed as a polynomial over $F[X]/\langle q_1 \rangle$, $(X - \alpha) \mid q_1$ and $q_1(X) = (X - \alpha)f$ for some polynomial f over $F[X]/\langle q_1 \rangle$.

Now, if we view p as a polynomial over $F[X]/\langle q_1 \rangle$, $p(X) = (X - \alpha)fq_2 \cdots q_s$ and the polynomial $fq_2 \cdots q_s$ has degree $n - 1$. By our induction hypothesis there is an extension K of $F[X]/\langle q_1 \rangle$ over which $fq_2 \cdots q_s$ is completely reducible. Then K can be viewed as an extension of F, and if we let $\alpha = \alpha_1$ and let $fq_2 \cdots q_s = (X - \alpha_2) \cdots (X - \alpha_n)$ where $\alpha_2, \ldots, \alpha_n$ are in K, then

$$p(X) = (X - \alpha_1)(X - \alpha_2) \cdots (X - \alpha_n)$$

and p is completely reducible over K. □

Example 1. Consider the polynomial $X^2 - \bar{2}$ over $\mathbf{Z}_3$. Since $\bar{1}^2 = \bar{1}$,

$$\bar{1}^2 - \bar{2} = -\bar{1} \neq \bar{0}$$

and since $\bar{2}^2 = \bar{1}$,

$$\bar{2}^2 - \bar{2} = -\bar{1} \neq \bar{0}$$

Thus, $X^2 - \bar{2}$ has no roots over $\mathbf{Z}_3$. In the factor ring $\mathbf{Z}_3[X]/\langle X^2 - \bar{2} \rangle$, if we let $\alpha = X\eta$, then

$$\alpha^2 - \bar{2} = (X\eta)^2 - \bar{2}$$
$$= (X^2)\eta - \bar{2}\eta \qquad \text{(since we identify 2 with } 2\eta\text{)}$$
$$= (X^2 - \bar{2})\eta$$
$$= \bar{0}$$

Thus, α is a root and $X^2 - \bar{2}$ is reducible over $\mathbf{Z}_3[X]/\langle X^2 - \bar{2}\rangle$ since

$$(X - \alpha)(X + \alpha) = X^2 - \alpha^2$$
$$= X^2 - \bar{2} \qquad (\alpha^2 = \bar{2})$$

In this case we can define α to be $\sqrt{\bar{2}}$ and can denote $\mathbf{Z}_3[X]/\langle X^2 - \bar{2}\rangle$ by $\mathbf{Z}_3[\sqrt{\bar{2}}]$, where we mean all polynomials in $\sqrt{\bar{2}}$ and require that $\sqrt{\bar{2}}$ satisfies $(\sqrt{\bar{2}})^2 = \bar{2}$. This means essentially to consider all expressions of the form $a + b\sqrt{\bar{2}}$, where $a, b \in \mathbf{Z}_3$ with addition and multiplication as usual in $\mathbf{Z}_3$ with $(\sqrt{\bar{2}})^2 = \bar{2}$.

Example 2. Consider the polynomial $X^2 + X + \bar{2}$ over $\mathbf{Z}_3$. Since $\bar{1}^2 + \bar{1} + \bar{2} = \bar{1}$ and $\bar{2}^2 + \bar{2} + \bar{2} = \bar{1} + \bar{2} + \bar{2} = \bar{2}$, $X^2 + X + \bar{2}$ has no roots in $\mathbf{Z}_3$. In the factor ring $\mathbf{Z}_3[X]/\langle X^2 + X + \bar{2}\rangle$, if we let $\alpha = X\eta$, then

$$\alpha^2 + \alpha + \bar{2} = (X\eta)^2 + (X\eta) + \bar{2}$$
$$= (X^2)\eta + X\eta + \bar{2}\eta \qquad \text{we identify } \bar{2} \text{ and } \bar{2}\eta$$
$$= (X^2 + X + \bar{2})\eta$$
$$= \bar{0}$$

Thus, over $\mathbf{Z}_3[X]/\langle X^2 + X + \bar{2}\rangle$, $X^2 + X + \bar{2}$ is reducible, since

$$(X - \alpha)(X + \alpha + \bar{1}) = X^2 + X\alpha + X - \alpha X - \alpha^2 - \alpha$$
$$= X^2 + X + (-\alpha^2 - \alpha)$$
$$= X^2 + X + \bar{2} \qquad (\text{since } -\alpha^2 - \alpha = \bar{2})$$

We can, in this case, denote $\mathbf{Z}_3[X]/\langle X^2 + X + \bar{2}\rangle$ by $\mathbf{Z}_3[\alpha]$, where we mean all polynomials in α and require that α satisfies $\alpha^2 + \alpha + \bar{2} = 0$. This means essentially to consider expressions of the form $a + b\alpha$, where a and b are in $\mathbf{Z}_3$ with addition and multiplication as usual in $\mathbf{Z}_3$ and $\alpha^2 = -\alpha - \bar{2}$.

Notice that, if we view $X^2 + X + \bar{2}$ as a polynomial over $\mathbf{Z}_3[\sqrt{\bar{2}}]$, it is completely reducible. If we apply the quadratic formula we obtain

$$X = \frac{-\bar{1} \pm \sqrt{\bar{1} - \bar{2}}}{2} = \frac{-\bar{1} \pm \sqrt{-\bar{1}}}{2} = \frac{-\bar{1} \pm \sqrt{\bar{2}}}{2} = -\bar{2} \pm \bar{2}\sqrt{\bar{2}}$$

If we let $\alpha = -\bar{2} + \bar{2}\sqrt{\bar{2}}$, then

$$X^2 + X + \bar{2} = (X - \alpha)(X + \alpha + \bar{1})$$
$$= (X + \bar{2} - \bar{2}\sqrt{\bar{2}})(X - \bar{1} + \bar{2}\sqrt{\bar{2}})$$

Similarly, $X^2 - \bar{2}$ is reducible over $\mathbf{Z}_3[\alpha]$ since

$$
\begin{aligned}
[X - (\bar{2}\alpha + \bar{1})][X + (\bar{2}\alpha + \bar{1})] &= X^2 - (\bar{2}\alpha + \bar{1})^2 \\
&= X^2 - (\alpha^2 + \alpha + \bar{1}) \\
&= X^2 - (\bar{1} + \bar{1}) \qquad (\alpha^2 + \alpha = \bar{1}) \\
&= X^2 - \bar{2}
\end{aligned}
$$

By the examples above we see that we cannot expect to have a unique method of obtaining an extension over which a polynomial is reducible.

EXERCISE SET D

1. Let F be a subfield of the field K, and let ϕ be an automorphism of K that fixes each element of F. Prove that if $a \in K$ and a is a root of the polynomial $p(X)$, then $a\phi$ is also a root of $p(X)$.

2. Prove: If F is a subfield of the field K and K is a finite-dimensional F-space, then every element of K is algebraic over F.

3. Prove: If F is a subfield of the field K and t is an element of K that is transcendental over F, then the subring of K generated by F and t is isomorphic to $F[X]$.

**4. Let F be a subfield of the field K, and let p be an irreducible polynomial over F that is completely reducible over K. Let T be the subfield of K generated by the roots of p. Prove that the group of automorphisms of T that fix the elements of F is a finite group.

9
MATRICES AND DETERMINANTS

Throughout the text we have assumed that the reader is familiar with 2×2 and 3×3 matrices with entries from $\mathbf{R}$ but for convenience we have provided, in an appendix, a synopsis of the arithmetic of these matrices. In this chapter we study matrices as notational devices for describing linear transformations from F^n to F^m and, in fact, as devices for describing linear transformations from finite-dimensional F-spaces into finite-dimensional F-spaces.

§1. Definitions and Elementary Properties

If V and W are F-spaces of dimensions n and m, respectively, then we know that $V \cong F^n$ and $W \cong F^m$ and also that $\mathcal{L}(V, W) \cong \mathcal{L}(F^n, F^m)$. Therefore, we can study linear transformations from V to W by studying the linear transformations from F^n to F^m. In proving Proposition 45 in Chapter 7 one actually shows that if B is a basis for the F-space V, then every function ϕ from B into an F-space W can be extended to a linear transformation from V into W. In fact, only one linear extension exists, because if $v_1\phi, \ldots, v_n\phi$ are known and ϕ is linear, then

$$(\alpha_1 v_1 + \cdots + \alpha_n v_n)\phi = \alpha_1(v_1\phi) + \cdots + \alpha_n(v_n\phi)$$

Thus, if B is a basis for the F-space V, we can determine or describe all linear transformations from V into the F-space W by determining or describing all functions from B into W.

In F^n we know that the n-tuples that have 1 as one component and 0 as the other components constitute a basis for the space. Thus, every function from this basis to F^m will determine a linear transformation from F^n to F^m, and all linear transformations can be described by describing all of these functions. One of the most common ways of describing such a function is by means of a matrix. A matrix is a rectangular array of elements and, for our purposes, these elements will always be elements of the field F.

Thus

$$\begin{bmatrix} 1 & 2 & 4 & 7 \\ 3 & 9 & 2 & -8 \end{bmatrix} \quad \text{and} \quad \begin{bmatrix} -3 & 0 & 17 \\ \pi & \sqrt{2} & 0.9 \\ 80 & 73 & 9.2 \\ 0.001 & 8.7 & 9000 \end{bmatrix}$$

are examples of matrices in which the entries are from the field $\mathbf{R}$.

Consider now, for example, the $\mathbf{R}$-spaces $\mathbf{R} \oplus \mathbf{R}$ and $\mathbf{R} \oplus \mathbf{R} \oplus \mathbf{R}$, and let ϕ be a linear transformation from $\mathbf{R} \oplus \mathbf{R}$ to $\mathbf{R} \oplus \mathbf{R} \oplus \mathbf{R}$ such that

$$(1, 0)\phi = (2, 1, -5) \quad \text{and} \quad (0, 1)\phi = (4, -3, 18)$$

We can abbreviate this information by just writing

$$\phi = \begin{bmatrix} 2 & 1 & -5 \\ 4 & -3 & 18 \end{bmatrix}$$

By this notation we mean that the image of $(1, 0)$ under ϕ is the vector $(2, 1, -5)$ and the image of $(0, 1)$ under ϕ is $(4, -3, 18)$. Remember that because $\{(1, 0), (0, 1)\}$ is a basis for $\mathbf{R} \oplus \mathbf{R}$, ϕ is completely described once the images of these elements have been described.

If f is a linear transformation from F^n to F^m, the first row of the matrix for f has, as its entries, the components of $e_1 f$ and, in general, the ith row of the matrix has as its entries the components of $e_i f$, the image of e_i under f. Thus the matrix for f is a notational device which lists the images of each basis element $e_1, \ldots, e_n$ under the linear transformation f. Since the transformation is completely described once we know the image of each basis element, the matrix completely describes f.

As further examples, consider the following linear transformations.

(a) Let f be a linear transformation from $\mathbf{Q} \oplus \mathbf{Q} \oplus \mathbf{Q}$ to $\mathbf{Q} \oplus \mathbf{Q} \oplus \mathbf{Q}$ in which $(1, 0, 0)f = (4, \frac{2}{3}, 0.7)$, $(0, 1, 0)f = (-3, 2, \frac{9}{7})$, and $(0, 0, 1)f = (0, 1, -4)$. Then

$$f = \begin{bmatrix} 4 & \frac{2}{3} & 0.7 \\ -3 & 2 & \frac{9}{7} \\ 0 & 1 & -4 \end{bmatrix}$$

(b) Let g be a linear transformation from $\mathbf{C} \oplus \mathbf{C} \oplus \mathbf{C} \oplus \mathbf{C}$ to $\mathbf{C} \oplus \mathbf{C}$ in which $(1, 0, 0, 0)g = (4, 2)$, $(0, 1, 0, 0)g = (i, 1 + i)$, $(0, 0, 1, 0)g = (3, 4 - i)$, and $(0, 0, 0, 1)g = (0.7 + 2i, 0.9 - 0.8i)$. Then

$$g = \begin{bmatrix} 4 & 2 \\ i & 1+i \\ 3 & 4-i \\ 0.7+2i & 0.9-0.8i \end{bmatrix}$$

Remember that we are using a matrix as a notational device for describing linear transformations, much as we use polynomials as devices to describe certain functions. When we write

$$f = \begin{bmatrix} 2 & 3 \\ 4 & 7 \end{bmatrix}$$

we mean that f is the linear transformation from $\mathbf{R} \oplus \mathbf{R}$ to $\mathbf{R} \oplus \mathbf{R}$ described by the matrix

$$\begin{bmatrix} 2 & 3 \\ 4 & 7 \end{bmatrix}$$

If f and g are linear transformations from F^n to F^m, we know that $f + g$ is also a linear transformation from F^n to F^m. If $(1, 0, \ldots, 0)f = (\alpha_1, \ldots, \alpha_m)$ and $(1, 0, \ldots, 0)g = (\beta_1, \ldots, \beta_m)$, then

$$(1, 0, \ldots, 0)(f + g) = (1, 0, \ldots, 0)f + (1, 0, \ldots, 0)g$$
$$= (\alpha_1, \ldots, \alpha_m) + (\beta_1, \ldots, \beta_m)$$
$$= (\alpha_1 + \beta_1, \ldots, \alpha_m + \beta_m)$$

Thus, if

$$f = \begin{bmatrix} \alpha_1 & \alpha_2 & \cdots & \alpha_m \\ & \cdot & \cdot & \cdot \end{bmatrix} \quad \text{and} \quad g = \begin{bmatrix} \beta_1 & \beta_2 & \cdots & \beta_m \\ \cdot & & \cdot & \cdot \end{bmatrix}$$

$$f + g = \begin{bmatrix} \alpha_1 + \beta_1 & \alpha_2 + \beta_2 & \cdots & \alpha_m + \beta_m \\ \cdot & \cdot & \cdot & \cdot \end{bmatrix}$$

Thus, the first row of the matrix representation for $f + g$ can be obtained by adding, component-wise, the first rows of the matrices for f and g. We obtain the rest of the rows of the matrix for $f + g$ by a similar process; that is, the ith row of the matrix for $f + g$ is obtained by adding, component-wise, the ith rows of the matrices for f and g.

If f is a linear transformation from F^n to F^m, and β is an element of F, then βf is also a linear transformation. If $(1, 0, \ldots, 0)f = (\alpha_1, \ldots, \alpha_m)$, then

$$(1, 0, \ldots, 0)(\beta f) = \beta[(1, 0, \ldots, 0)f] = \beta(\alpha_1, \ldots, \alpha_m)$$
$$= (\beta\alpha_1, \ldots, \beta\alpha_m)$$

Thus

$$\beta f = \begin{bmatrix} \beta\alpha_1 & \cdots & \beta\alpha_j & \cdots & \beta\alpha_m \\ \cdot & & \cdot & & \cdot \end{bmatrix}$$

and the first row of the matrix for βf is obtained simply by multiplying each component of the first row of the matrix for f by the scalar β. A similar argument holds for the other rows, and the matrix for βf is obtained by multiplying each entry in the matrix for f by the scalar β.

For example, if

$$f = \begin{bmatrix} 4 & -3 & 7 \\ 2 & 9 & -8 \end{bmatrix}$$

then

$$4f = \begin{bmatrix} 16 & -12 & 28 \\ 8 & 36 & -32 \end{bmatrix}$$

and

$$0.1f = \begin{bmatrix} 0.4 & -0.3 & 0.7 \\ 0.2 & 0.9 & -0.8 \end{bmatrix}$$

Each vector in F^m has m components, and therefore a matrix for a linear transformation from F^n to F^m must have m columns. If f is a linear transformation from F^n to F^m, f is specified when we know the images of the basis $\{(1, 0, \ldots, 0), \ldots, (0, \ldots, 0, 1)\}$ for F^n, and therefore the matrix must have n rows. Thus, if f is a linear transformation from F^n to F^m, its matrix must have n rows and m columns. A matrix with n rows and m columns is said to be an n by m matrix. We usually denote the phrase "n by m" by "$n \times m$." Since addition of matrices is just addition of linear transformations, addition of matrices makes sense only when addition of the corresponding linear transformations makes sense. Thus, we could add the matrices

$$\begin{bmatrix} 1 & 3 & 5 \\ -2 & 4 & 7 \end{bmatrix} \text{ and } \begin{bmatrix} 4 & -1 & 3 \\ 8 & 1 & 9 \end{bmatrix}$$

because both represent linear transformations from $\mathbf{R} \oplus \mathbf{R}$ to $\mathbf{R} \oplus \mathbf{R} \oplus \mathbf{R}$. Addition of the matrices

$$\begin{bmatrix} 1 & 2 \\ 3 & 4 \\ 7 & 8 \end{bmatrix} \text{ and } \begin{bmatrix} 1 & 3 & 7 & 2 \\ 2 & 9 & 6 & 7 \end{bmatrix}$$

would have no meaning, because the first represents a linear transformation from $\mathbf{R} \oplus \mathbf{R} \oplus \mathbf{R}$ to $\mathbf{R} \oplus \mathbf{R}$ and the second represents a linear transformation from $\mathbf{R} \oplus \mathbf{R}$ to $\mathbf{R} \oplus \mathbf{R} \oplus \mathbf{R} \oplus \mathbf{R}$ and we have not defined a way of adding such transformations.

Let f be a linear transformation from F^2 to F^3 with the matrix of f given by

$$f = \begin{bmatrix} a & b & c \\ \alpha & \beta & \gamma \end{bmatrix}$$

If (x, y) is a vector in F^2, then

$$\begin{aligned} (x, y)f &= [(x, 0) + (0, y)]f \\ &= [x(1, 0) + y(0, 1)]f \\ &= x[(1, 0)f] + y[(0, 1)f] \qquad \text{because } f \text{ is linear} \\ &= x(a, b, c) + y(\alpha, \beta, \gamma) \\ &= (xa, xb, xc) + (y\alpha, y\beta, y\gamma) \\ &= (xa + y\alpha, xb + y\beta, xc + y\gamma) \end{aligned}$$

Thus,

$$(x, y)f = (xa + y\alpha, xb + y\beta, xc + y\gamma)$$

Using the matrix notation this equation becomes

$$(x, y) \begin{bmatrix} a & b & c \\ \alpha & \beta & \gamma \end{bmatrix} = (xa + y\alpha, xb + y\beta, xc + y\gamma)$$

Notice that the first component is obtained using (x, y) and the first column, the second using (x, y) and the second column, and the third using (x, y) and the third column. To obtain the third component, for example, we multiply the first component of the vector with the first entry in the third column and the second component with the second entry in the third column and then add the products. Thus, for example,

$$\begin{aligned} (2, 3) \begin{bmatrix} 4 & 7 & -2 \\ 3 & -1 & 0 \end{bmatrix} &= (2 \cdot 4 + 3 \cdot 3, 2 \cdot 7 + 3(-1), 2(-2) + 3 \cdot 0) \\ &= (17, \quad 11, \quad -4) \end{aligned}$$

This process generalizes, but we need to make some notational conventions before proceeding with the argument. If f is a linear transformation from F^n to F^m, we know that f is represented by an $n \times m$ matrix. We assign subscripts to matrix entries according to the row and column they are in. Thus, by the element α_{ij} in the matrix for f we mean the entry in the ith row and the jth column. We shall frequently denote the matrix by $[\alpha_{ij}]$. For example, α_{27} would be the entry in the second row and the seventh column. In writing down such a representation we would have

$$[\alpha_{ij}] = \begin{bmatrix} \alpha_{11} & \alpha_{12} & \alpha_{13} & \cdots & \alpha_{1m} \\ \alpha_{21} & \alpha_{22} & \alpha_{23} & \cdots & \alpha_{2m} \\ \vdots & \vdots & \vdots & & \vdots \\ \alpha_{n1} & \alpha_{n2} & \alpha_{n3} & \cdots & \alpha_{nm} \end{bmatrix}$$

In F^n, let e_i denote the vector that has 0's everywhere except for the ith component and a 1 for the ith component. Then, if $[\alpha_{ij}]$ is the matrix for the linear transformation f,

$$e_i f = (\alpha_{i1}, \ldots, \alpha_{im})$$

Now if $x = (x_1, \ldots, x_n)$ is a vector in F^n, and f is a linear transformation with matrix $[\alpha_{ij}]$,

$$\begin{aligned}
xf &= (x_1, \ldots, x_n)f \\
&= (x_1 e_1 + x_2 e_2 + \cdots + x_n e_n)f \\
&= x_1(e_1 f) + x_2(e_2 f) + \cdots + x_n(e_n f) \\
&= x_1(\alpha_{11}, \alpha_{12}, \ldots, \alpha_{1m}) + x_2(\alpha_{21}, \alpha_{22}, \ldots, \alpha_{2m}) \\
&\quad + \cdots + x_n(\alpha_{n1}, \alpha_{n2}, \ldots, \alpha_{nm}) \\
&= (x_1 \alpha_{11}, x_1 \alpha_{12}, \ldots, x_1 \alpha_{1m}) + (x_2 \alpha_{21}, x_2 \alpha_{22}, \ldots, x_2 \alpha_{2m}) \\
&\quad + \cdots + (x_n \alpha_{n1}, x_n \alpha_{n2}, \ldots, x_n \alpha_{nm}) \\
&= (x_1 \alpha_{11} + x_2 \alpha_{21} + \cdots + x_n \alpha_{n1}, x_1 \alpha_{12} + x_2 \alpha_{22} \\
&\quad + \cdots + x_n \alpha_{n2}, \ldots, x_1 \alpha_{1m} + x_2 \alpha_{2m} + \cdots + x_n \alpha_{nm})
\end{aligned}$$

Thus, the kth component of $(x_1, \ldots, x_n)f$ is

$$x_1 \alpha_{1k} + x_2 \alpha_{2k} + \cdots + x_n \alpha_{nk}$$

This scalar is obtained by forming the sum of the products obtained by multiplying x_1 by the first entry in the kth column, x_2 by the second entry in the kth column, and so on to multiplying x_n by the nth entry in the kth column. Using summation notation we have

$$(x_1, \ldots, x_n)[\alpha_{ij}] = \left(\sum_{i=1}^{n} x_i \alpha_{i1}, \sum_{i=1}^{n} x_i \alpha_{i2}, \ldots, \sum_{i=1}^{n} x_i \alpha_{im} \right)$$

For example,

$$(1, 7, 3) \begin{bmatrix} 1 & 3 & 8 & -4 & 9 \\ 4 & 9 & 4 & 6 & 4 \\ 2 & -3 & 7 & -2 & -1 \end{bmatrix}$$

$$\begin{aligned}
&= (1 \cdot 1 + 7 \cdot 4 + 3 \cdot 2, 1 \cdot 3 + 7 \cdot 9 + 3(-3), 1 \cdot 8 + 7 \cdot 4 \\
&\quad + 3 \cdot 7, 1(-4) + 7 \cdot 6 + 3(-2), 1 \cdot 9 + 7 \cdot 4 + 3(-1)) \\
&= (35, 57, 57, 32, 34)
\end{aligned}$$

Graphically this process is illustrated by a "turn, multiply, and then add" process:

$$(x_1, \ldots, x_n) \begin{bmatrix} \alpha_{11} & \alpha_{12} & \cdots & \alpha_{1m} \\ \alpha_{21} & \alpha_{22} & \cdots & \alpha_{2m} \\ \vdots & \vdots & & \vdots \\ \alpha_{n1} & \alpha_{n2} & \cdots & \alpha_{nm} \end{bmatrix} = \begin{bmatrix} x_1\alpha_{11} & x_1\alpha_{12} & \cdots & x_1\alpha_{1m} \\ + & + & & + \\ x_2\alpha_{21} & x_2\alpha_{22} & \cdots & x_2\alpha_{2m} \\ + & + & & + \\ x_3\alpha_{31} & x_3\alpha_{32} & \cdots & x_3\alpha_{3m} \\ + & + & & + \\ \vdots & \vdots & & \vdots \\ + & + & & + \\ x_n\alpha_{n1} & x_n\alpha_{n2} & \cdots & x_n\alpha_{nm} \end{bmatrix}$$

Thus, once we have a matrix representation for a linear transformation f from F^n to F^m, we can use this representation to compute the image of any vector under f by the computational process above.

Examples

(a) Let g be the linear transformation from $\mathbf{R}^3$ to $\mathbf{R}^2$ with matrix

$$\begin{bmatrix} 1 & 3 \\ \frac{1}{2} & 4 \\ -3 & 1 \end{bmatrix}$$

Then

$$(1, -3, 4)g = (1, -3, 4) \begin{bmatrix} 1 & 3 \\ \frac{1}{2} & 4 \\ -3 & 1 \end{bmatrix} = \left(-\tfrac{25}{2}, -5\right)$$

and

$$(2, 4, 1)g = (2, 4, 1) \begin{bmatrix} 1 & 3 \\ \frac{1}{2} & 4 \\ -3 & 1 \end{bmatrix} = (1, 23)$$

(b) Let T be the linear transformation from $\mathbf{Z}_5{}^3$ to $\mathbf{Z}_5{}^3$ with matrix

$$\begin{bmatrix} \bar{1} & \bar{3} & \bar{2} \\ \bar{2} & \bar{4} & \bar{2} \\ \bar{1} & \bar{3} & \bar{3} \end{bmatrix}$$

Then

$$(\bar{2}, \bar{3}, \bar{1})T = (\bar{2}, \bar{3}, \bar{1}) \begin{bmatrix} \bar{1} & \bar{3} & \bar{2} \\ \bar{2} & \bar{4} & \bar{2} \\ \bar{1} & \bar{3} & \bar{3} \end{bmatrix} = (\bar{4}, \bar{1}, \bar{3})$$

Suppose f is a linear transformation from F^2 to F^3 and g is a linear transformation from F^3 to F^4, and let the matrices for f and g be given by

$$f = \begin{bmatrix} a_{11} & a_{12} & a_{13} \\ a_{21} & a_{22} & a_{23} \end{bmatrix} \quad \text{and} \quad g = \begin{bmatrix} b_{11} & b_{12} & b_{13} & b_{14} \\ b_{21} & b_{22} & b_{23} & b_{24} \\ b_{31} & b_{32} & b_{33} & b_{34} \end{bmatrix}$$

We know that $f \circ g$ is a linear transformation from F^2 to F^4, and therefore that it will have a 2×4 matrix as its representation.

$$(1, 0)(f \circ g) = [(1, 0)f]g$$

$$= (a_{11}, a_{12}, a_{13})g$$

$$= (a_{11}, a_{12}, a_{13}) \begin{bmatrix} b_{11} & b_{12} & b_{13} & b_{14} \\ b_{21} & b_{22} & b_{23} & b_{24} \\ b_{31} & b_{32} & b_{33} & b_{34} \end{bmatrix}$$

$$= (a_{11}b_{11} + a_{12}b_{21} + a_{13}b_{31}, a_{11}b_{12} + a_{12}b_{22} + a_{13}b_{32},$$
$$a_{11}b_{13} + a_{12}b_{23} + a_{13}b_{33}, a_{11}b_{14} + a_{12}b_{24} + a_{13}b_{34})$$

Thus, $(1, 0)(f \circ g)$ is just the image of (a_{11}, a_{12}, a_{13}) under g, and this is obtained in the "turn, multiply, and add" process described above. A similar computation holds for $(0, 1)(f \circ g)$ and the matrix for $f \circ g$ is then

$$\begin{bmatrix} a_{11}b_{11}+a_{12}b_{21}+a_{13}b_{31} & a_{11}b_{12}+a_{12}b_{22}+a_{13}b_{32} & a_{11}b_{13}+a_{12}b_{23}+a_{13}b_{33} & a_{11}b_{14}+a_{12}b_{24}+a_{13}b_{34} \\ a_{21}b_{11}+a_{22}b_{21}+a_{23}b_{31} & a_{21}b_{12}+a_{22}b_{22}+a_{23}b_{32} & a_{21}b_{13}+a_{22}b_{23}+a_{23}b_{33} & a_{21}b_{14}+a_{22}b_{24}+a_{23}b_{34} \end{bmatrix}$$

Notice that in each product computed to obtain this matrix, the second subscript of a_{ij} is the first subscript of b_{pq}.

In general, if f is a linear transformation from F^n to F^m with $n \times m$ matrix $[\alpha_{ij}]$, and g is a linear transformation from F^m to F^t with $m \times t$ matrix $[\beta_{ij}]$, then the matrix for $f \circ g$ will be an $n \times t$ matrix $[\gamma_{ij}]$. Let us determine the form of this matrix. We know that $(e_i)(f \circ g) = (\gamma_{i1}, \gamma_{i2}, \ldots, \gamma_{it})$, the ith row of the matrix. Now,

$$(e_i)(f \circ g) = (e_i f)g$$

$$= (\alpha_{i1}, \ldots, \alpha_{im})g$$

$$= (\alpha_{i1}, \ldots, \alpha_{im})[\beta_{pq}]$$

$$= (\alpha_{i1}\beta_{11} + a_{i2}\beta_{21} + \alpha_{i3}\beta_{31} + \cdots + \alpha_{im}\beta_{m1}, \ldots,$$
$$\alpha_{i1}\beta_{it} + \alpha_{i2}\beta_{2t} + \alpha_{i3}\beta_{3t} + \cdots + \alpha_{im}\beta_{mt})$$

$$= \left(\sum_{k=1}^{m} \alpha_{ik}\beta_{k1}, \sum_{k=1}^{m} \alpha_{ik}\beta_{k2}, \ldots, \sum_{k=1}^{m} \alpha_{ik}\beta_{kt} \right)$$

Thus, the ith row of the matrix for $f \circ g$ is

$$\left(\sum_{k=1}^{m} \alpha_{ik}\beta_{k1}, \sum_{k=1}^{m} \alpha_{ik}\beta_{k2}, \sum_{k=1}^{m} \alpha_{ik}\beta_{k3}, \ldots, \sum_{k=1}^{m} \alpha_{ik}\beta_{kt} \right)$$

and therefore, for each i, j with $1 \leq i \leq n, 1 \leq j \leq t$,

$$\gamma_{ij} = \sum_{k=1}^{m} \alpha_{ik} \beta_{kj}$$

that is,

$$\gamma_{ij} = \alpha_{i1} \beta_{1j} + \alpha_{i2} \beta_{2j} + \cdots + \alpha_{im} \beta_{mj}$$

Hence, γ_{ij} is obtained by the "turn, multiply, and add" process using the ith row of $[\alpha_{ij}]$ and the jth column of $[\beta_{ij}]$.

In general,

$$[\alpha_{ij}][\beta_{ij}] = \left[\sum_{k=1}^{m} \alpha_{ik} \beta_{kj} \right]$$

Examples

$$\begin{bmatrix} 2 & 3 \\ 4 & -1 \end{bmatrix} \begin{bmatrix} 1 & 7 & 3 \\ 2 & 9 & 1 \end{bmatrix} = \begin{bmatrix} 2 \cdot 1 + 3 \cdot 2 & 2 \cdot 7 + 3 \cdot 9 & 2 \cdot 3 + 3 \cdot 1 \\ 4 \cdot 1 + (-1)2 & 4 \cdot 7 + (-1)9 & 4 \cdot 3 + (-1)1 \end{bmatrix}$$

$$= \begin{bmatrix} 8 & 41 & 9 \\ 2 & 19 & 11 \end{bmatrix}$$

$$\begin{bmatrix} 1 & 7 \\ -3 & 4 \\ 7 & 9 \end{bmatrix} \begin{bmatrix} 9 & 4 & -3 \\ 7 & 1 & 5 \end{bmatrix} = \begin{bmatrix} 1 \cdot 9 + 7 \cdot 7 & 1 \cdot 4 + 7 \cdot 1 & 1(-3) + 7 \cdot 5 \\ (-3)9 + 4 \cdot 7 & (-3)4 + 4 \cdot 1 & (-3)(-3) + 4 \cdot 5 \\ 7 \cdot 9 + 9 \cdot 7 & 7 \cdot 4 + 9 \cdot 1 & 7(-3) + 9 \cdot 5 \end{bmatrix}$$

$$= \begin{bmatrix} 58 & 11 & 32 \\ 1 & -8 & 29 \\ 126 & 37 & 24 \end{bmatrix}$$

The arithmetic of matrices is completely determined by the linear transformations they represent. Thus, since addition of linear transformations is associative and commutative, the addition of the corresponding matrices must be also; that is, if $f = [\alpha_{ij}]$, $g = [\beta_{ij}]$ and $h = [\gamma_{ij}]$, since

$$(f + g) + h = f + (g + h)$$

it follows that

$$([\alpha_{ij}] + [\beta_{ij}]) + [\gamma_{ij}] = [\alpha_{ij}] + ([\beta_{ij}] + [\gamma_{ij}])$$

and since $f + g = g + f$,

$$[\alpha_{ij}] + [\beta_{ij}] = [\beta_{ij}] + [\alpha_{ij}]$$

Since composition of functions is associative, the multiplication of their corresponding matrices must also be associative.

$$[\alpha_{ij}]([\beta_{ij}][\gamma_{ij}]) = ([\alpha_{ij}][\beta_{ij}])[\gamma_{ij}]$$

From previous work we know that if f, g, and h are linear transformations such that $(f + g)h$ or $h(f + g)$ is defined, then $(f + g)h = fh + gh$ or $h(f + g) = hf + hg$. Thus, if $[\alpha_{ij}]$, $[\beta_{ij}]$, and $[\gamma_{ij}]$ are matrices such that $([\alpha_{ij}] + [\beta_{ij}])[\gamma_{ij}]$ or $[\gamma_{ij}]([\alpha_{ij}] + [\beta_{ij}])$ is defined, then

$$([\alpha_{ij}] + [\beta_{ij}])[\gamma_{ij}] = [\alpha_{ij}][\gamma_{ij}] + [\beta_{ij}][\gamma_{ij}]$$

or

$$[\gamma_{ij}]([\alpha_{ij}] + [\beta_{ij}]) = [\gamma_{ij}][\alpha_{ij}] + [\gamma_{ij}][\beta_{ij}]$$

and the Distributive laws hold when they are applicable, that is, when the product and sums involved make sense.

One can start with purely formal definitions of square matrices and matrix operations and prove that the structure is, in fact, a ring and is isomorphic to the ring of linear transformations from F^n to itself. Our viewpoint is that matrices represent linear transformations and, since the linear transformations from F^n to F^n form a ring, the set of $n \times n$ matrices with the induced operations is just a convenient tool for describing the elements of this ring and for computing with these elements. Thus, when we speak of the ring of $n \times n$ matrices over F we actually mean the ring $\mathcal{L}(F^n)$ represented by $n \times n$ matrices. Similarly, the set of $n \times m$ matrices can formally be shown to be an F-space, but our viewpoint is that the $n \times m$ matrices are just a tool for describing the F-space $\mathcal{L}(F^n, F^m)$.

EXERCISE SET A

1. Compute the matrix products below:

$$\begin{bmatrix} 1 & 2 & 7 \\ -2 & 3 & 4 \end{bmatrix} \begin{bmatrix} 4 & 9 & 3 \\ 2 & -1 & 1 \\ 0 & 4 & 1 \end{bmatrix} =$$

$$\begin{bmatrix} 1 & 0 & 4 \\ 4 & 0 & 1 \end{bmatrix} \begin{bmatrix} 1 & 1 \\ 2 & 2 \\ 3 & 3 \end{bmatrix} =$$

$$\begin{bmatrix} 2 & 1 \\ 3 & 2 \end{bmatrix} \begin{bmatrix} 2 & -1 \\ -3 & 2 \end{bmatrix} =$$

$$\begin{bmatrix} 0 & 1 & 0 \\ 0 & 0 & 1 \\ 1 & 0 & 0 \end{bmatrix} \begin{bmatrix} 0 & 1 & 0 \\ 0 & 0 & 1 \\ 1 & 0 & 0 \end{bmatrix} =$$

2. If I denotes the identity function from F^4 to F^4, what is the matrix for I?

3. Let α be an element of the field F. If T is the function from F^n to F^n defined by $xT = \alpha x$, what is the matrix for T?

4. (a) Let P be the linear transformation from F^4 to F^4 defined by $e_1 P = e_2$, $e_2 P = e_3$, $e_3 P = e_4$, and $e_4 P = e_1$. What is the matrix for P?

   (b) Consider now the permutation (1 2 3 4) in S_4. Define a linear transformation T from F^4 to F^4 by defining $e_i T = e_{i(1\ 2\ 3\ 4)}$. For example, $e_2 T = e_{2(1\ 2\ 3\ 4)} = e_3$. What is the matrix for T?

   (c) If $(a\ b)$ is a transposition in S_4 and we define a linear transformation Y from F^4 to F^4 by $e_i Y = e_{i(a\ b)}$, what will be the form of matrix for Y?

   (d) If p is a permutation in S_n and we define a linear transformation T from F^n to F^n by $e_i T = e_{ip}$, what will be the general form of the matrix for T?

5. Let $\beta_1, \ldots, \beta_n$ be elements of F and define a linear transformation A from F^n to F^n by $e_i A = \beta_i e_i$. What is the form of the matrix for A?

**6. If we require $1 \le p \le 3$, $1 \le q \le 3$, $p \ne q$ and we define $E_{pq}{}^1$ from F^3 to F^3 by

$$e_i E_{pq}{}^1 = e_i \quad \text{if } i \ne p$$
$$e_p E_{pq}{}^1 = e_p + e_q$$

   what is the matrix for $E_{23}{}^1$? For $E_{32}{}^1$? For $E_{pq}{}^1$ in general?

7. If f is the linear transformation from $\mathbf{R}^2$ to $\mathbf{R}^2$ having as its matrix

$$\begin{bmatrix} 1 & 2 \\ 1 & 2 \end{bmatrix}$$

   what is the kernel of f? Can you describe it geometrically?

8. Prove: If $[\alpha_{ij}]$ is a linear transformation from F^n to F^m, the image of $[\alpha_{ij}]$ (or image of F^n under $[\alpha_{ij}]$) is the subspace of F^m spanned by $\{(\alpha_{11}, \alpha_{12}, \ldots, \alpha_{1m}), (\alpha_{21}, \alpha_{22}, \ldots, \alpha_{2m}), \ldots, (\alpha_{n1}, \ldots, \alpha_{nm})\}$. Show that $[\alpha_{ij}]$ is one-to-one iff this set is linearly independent.

9. Prove: If $[\alpha_{ij}]$ is an $n \times m$ matrix and $n \ne m$, then the linear transformation represented by $[\alpha_{ij}]$ is either not one-to-one or is not onto.

10. If g is a linear transformation from $\mathbf{R}^2$ to $\mathbf{R}^2$ having as its matrix

$$\begin{bmatrix} 2 & 1 \\ 3 & 2 \end{bmatrix}$$

   what is the kernel of g?

11. If T is the linear transformation from $\mathbf{R}^2$ to $\mathbf{R}^2$ having as its matrix

$$\begin{bmatrix} a & b \\ c & d \end{bmatrix}$$

   what is the kernel of T? Under what conditions on a, b, c, d is the kernel of $T = \{(0, 0)\}$?

12. Use systems of equations in two variables to
    (a) find a matrix

$$\begin{bmatrix} x & y \\ z & w \end{bmatrix} \quad \text{such that} \quad \begin{bmatrix} 2 & 3 \\ 1 & 2 \end{bmatrix}\begin{bmatrix} x & y \\ z & w \end{bmatrix} = \begin{bmatrix} 1 & 0 \\ 0 & 1 \end{bmatrix}$$

   (if such a matrix exists)
    (b) find a matrix

$$\begin{bmatrix} x & y \\ z & w \end{bmatrix} \quad \text{such that} \quad \begin{bmatrix} 2 & 1 \\ 2 & 1 \end{bmatrix}\begin{bmatrix} x & y \\ z & w \end{bmatrix} = \begin{bmatrix} 1 & 0 \\ 0 & 1 \end{bmatrix}$$

   (if such a matrix exists)
    (c) find a matrix

$$\begin{bmatrix} x & y \\ z & w \end{bmatrix} \quad \text{such that} \quad \begin{bmatrix} 5 & 2 \\ 3 & 4 \end{bmatrix}\begin{bmatrix} x & y \\ z & w \end{bmatrix} = \begin{bmatrix} 1 & 0 \\ 0 & 1 \end{bmatrix}$$

   (if such a matrix exists)
    (d) find a matrix

$$\begin{bmatrix} x & y \\ z & w \end{bmatrix} \quad \text{such that} \quad \begin{bmatrix} a & b \\ c & d \end{bmatrix}\begin{bmatrix} x & y \\ z & w \end{bmatrix} = \begin{bmatrix} 1 & 0 \\ 0 & 1 \end{bmatrix}$$

   (if such a matrix exists)

§2. Inverses of Matrices

If A is an automorphism of F^n, we know that A^{-1} is also an automorphism. A linear transformation or matrix that has an inverse is said to be *invertible*. If we know that $[\alpha_{ij}]$ is invertible, how can we determine $[\alpha_{ij}]^{-1}$? There are several techniques available; the ones we shall describe here can, in the case of matrices over $\mathbf{R}$, be adapted rather easily for use with computers.

The problem can be viewed as the problem of solving the matrix equation

$$XA = I \quad \text{or} \quad [x_{ij}][\alpha_{ij}] = I$$

If we multiply on the right by A^{-1} or $[\alpha_{ij}]^{-1}$ we get $XAA^{-1} = IA^{-1}$ or

$$[x_{ij}][\alpha_{ij}][\alpha_{ij}]^{-1} = I[\alpha_{ij}]^{-1}$$

that is, $X = A^{-1}$ or $[x_{ij}] = [\alpha_{ij}]^{-1}$.

If we multiply both sides of the matrix equation on the right by an invertible matrix $[\beta_{ij}]$, the resulting matrix equation

$$[x_{ij}][\alpha_{ij}][\beta_{ij}] = I[\beta_{ij}]$$

is equivalent to the original, since we can obtain the original again just by multiplying on the right by $[\beta_{ij}]^{-1}$. The technique we develop for computing inverses is based solely on this idea but using simple invertible matrices. If we could find invertible matrices

$$[\gamma_{ij}],[\epsilon_{ij}],\cdots,[\mu_{ij}]$$

such that

$$[\alpha_{ij}][\gamma_{ij}][\epsilon_{ij}]\cdots[\mu_{ij}] = I$$

then

$$[\gamma_{ij}][\epsilon_{ij}]\cdots[\mu_{ij}] = [\alpha_{ij}]^{-1}$$

Before proceeding to the general technique, let us look at some examples. The problem of finding the inverse of the matrix

$$\begin{bmatrix} 2 & 1 \\ -1 & 1 \end{bmatrix}$$

is equivalent to solving the equation

$$\begin{bmatrix} x_{11} & x_{12} \\ x_{21} & x_{22} \end{bmatrix}\begin{bmatrix} 2 & 1 \\ -1 & 1 \end{bmatrix} = \begin{bmatrix} 1 & 0 \\ 0 & 1 \end{bmatrix}$$

If we multiply both sides on the right by

$$\begin{bmatrix} 1 & 0 \\ 1 & 1 \end{bmatrix}$$

we obtain

$$\begin{bmatrix} x_{11} & x_{12} \\ x_{21} & x_{22} \end{bmatrix}\begin{bmatrix} 2 & 1 \\ -1 & 1 \end{bmatrix}\begin{bmatrix} 1 & 0 \\ 1 & 1 \end{bmatrix} = \begin{bmatrix} 1 & 0 \\ 0 & 1 \end{bmatrix}\begin{bmatrix} 1 & 0 \\ 1 & 1 \end{bmatrix}$$

or

$$\begin{bmatrix} x_{11} & x_{12} \\ x_{21} & x_{22} \end{bmatrix}\begin{bmatrix} 3 & 1 \\ 0 & 1 \end{bmatrix} = \begin{bmatrix} 1 & 0 \\ 1 & 1 \end{bmatrix}$$

If we multiply both sides on the right by

$$\begin{bmatrix} \frac{1}{3} & -\frac{1}{3} \\ 0 & 1 \end{bmatrix}$$

we obtain

$$\begin{bmatrix} x_{11} & x_{12} \\ x_{21} & x_{22} \end{bmatrix}\begin{bmatrix} 3 & 1 \\ 0 & 1 \end{bmatrix}\begin{bmatrix} \frac{1}{3} & -\frac{1}{3} \\ 0 & 1 \end{bmatrix} = \begin{bmatrix} 1 & 0 \\ 1 & 1 \end{bmatrix}\begin{bmatrix} \frac{1}{3} & -\frac{1}{3} \\ 0 & 1 \end{bmatrix}$$

which yields

$$\begin{bmatrix} x_{11} & x_{12} \\ x_{21} & x_{22} \end{bmatrix} \begin{bmatrix} 1 & 0 \\ 0 & 1 \end{bmatrix} = \begin{bmatrix} \frac{1}{3} & -\frac{1}{3} \\ \frac{1}{3} & \frac{2}{3} \end{bmatrix}$$

or

$$\begin{bmatrix} x_{11} & x_{12} \\ x_{21} & x_{22} \end{bmatrix} = \begin{bmatrix} \frac{1}{3} & -\frac{1}{3} \\ \frac{1}{3} & \frac{2}{3} \end{bmatrix}$$

and

$$\begin{bmatrix} 2 & 1 \\ -1 & 1 \end{bmatrix}^{-1} = \begin{bmatrix} \frac{1}{3} & -\frac{1}{3} \\ \frac{1}{3} & \frac{2}{3} \end{bmatrix}$$

In the above example we used the two matrices

$$\begin{bmatrix} 1 & 0 \\ 1 & 1 \end{bmatrix} \text{ and } \begin{bmatrix} \frac{1}{3} & -\frac{1}{3} \\ 0 & 1 \end{bmatrix}$$

which can easily be shown to have trivial kernels and, therefore, are invertible. Notice that multiplying on the right by

$$\begin{bmatrix} 1 & 0 \\ 1 & 1 \end{bmatrix}$$

has the effect of adding the second column to the first and fixing the second column. The result of multiplying on the right by the matrix

$$\begin{bmatrix} \frac{1}{3} & -\frac{1}{3} \\ 0 & 1 \end{bmatrix}$$

is equivalent to multiplying the first column by $\frac{1}{3}$ and to multiplying the first column by $-\frac{1}{3}$ and adding it to the second column.

As a further illustration, consider the matrix

$$\begin{bmatrix} \alpha_{11} & \alpha_{12} & \alpha_{13} \\ \alpha_{21} & \alpha_{22} & \alpha_{23} \\ \alpha_{31} & \alpha_{32} & \alpha_{33} \end{bmatrix}$$

and form the product with

$$\begin{bmatrix} 1 & 0 & 0 \\ 0 & 1 & 0 \\ b & 0 & 1 \end{bmatrix}$$

$$\begin{bmatrix} \alpha_{11} & \alpha_{12} & \alpha_{13} \\ \alpha_{21} & \alpha_{22} & \alpha_{23} \\ \alpha_{31} & \alpha_{32} & \alpha_{33} \end{bmatrix} \begin{bmatrix} 1 & 0 & 0 \\ 0 & 1 & 0 \\ b & 0 & 1 \end{bmatrix} = \begin{bmatrix} \alpha_{11} + b\alpha_{13} & \alpha_{12} & \alpha_{13} \\ \alpha_{21} + b\alpha_{23} & \alpha_{22} & \alpha_{23} \\ \alpha_{31} + b\alpha_{33} & \alpha_{32} & \alpha_{33} \end{bmatrix}$$

Thus, the result of multiplying these matrices together is to multiply the *third column* of $[\alpha_{ij}]$ by b and add it to the *first column*. Notice that b is in the *third row* in the *first column*. What happens with b in the *third row* and *second column*?

$$\begin{bmatrix} \alpha_{11} & \alpha_{12} & \alpha_{13} \\ \alpha_{21} & \alpha_{22} & \alpha_{23} \\ \alpha_{31} & \alpha_{32} & \alpha_{33} \end{bmatrix} \begin{bmatrix} 1 & 0 & 0 \\ 0 & 1 & 0 \\ 0 & b & 1 \end{bmatrix} = \begin{bmatrix} \alpha_{11} & \alpha_{12} + b\alpha_{13} & \alpha_{13} \\ \alpha_{21} & \alpha_{22} + b\alpha_{23} & \alpha_{23} \\ \alpha_{31} & \alpha_{32} + b\alpha_{33} & \alpha_{33} \end{bmatrix}$$

This multiplies the *third column* by b and adds it to the *second column*. Suppose we put b in the *first row* and *second column*.

$$\begin{bmatrix} \alpha_{11} & \alpha_{12} & \alpha_{13} \\ \alpha_{21} & \alpha_{22} & \alpha_{23} \\ \alpha_{31} & \alpha_{32} & \alpha_{33} \end{bmatrix} \begin{bmatrix} 1 & b & 0 \\ 0 & 1 & 0 \\ 0 & 0 & 1 \end{bmatrix} = \begin{bmatrix} \alpha_{11} & b\alpha_{11} + \alpha_{12} & \alpha_{13} \\ \alpha_{21} & b\alpha_{21} + \alpha_{22} & \alpha_{23} \\ \alpha_{31} & b\alpha_{31} + \alpha_{32} & \alpha_{33} \end{bmatrix}$$

In this case the result is to multiply the *first column* by b and add it to the *second column*.

If we put b on the diagonal, say in the first row, we obtain

$$\begin{bmatrix} \alpha_{11} & \alpha_{12} & \alpha_{13} \\ \alpha_{21} & \alpha_{22} & \alpha_{23} \\ \alpha_{31} & \alpha_{32} & \alpha_{33} \end{bmatrix} \begin{bmatrix} b & 0 & 0 \\ 0 & 1 & 0 \\ 0 & 0 & 1 \end{bmatrix} = \begin{bmatrix} b\alpha_{11} & \alpha_{12} & \alpha_{13} \\ b\alpha_{21} & \alpha_{22} & \alpha_{23} \\ b\alpha_{31} & \alpha_{32} & \alpha_{33} \end{bmatrix}$$

which is equivalent to multiplying the first column by b.

Now let us define two special types of linear transformations that have special types of matrices. The Kronecker δ is a convenient bit of notation and is defined by

$$\delta_{ij} = 1 \quad \text{if } i = j \quad \text{and} \quad \delta_{ij} = 0 \quad \text{if } i \neq j$$

Using this notation, the matrix for the identity transformation is just $[\delta_{ij}]$. Now for each pair of positive integers p and q with $1 \leq p \leq n$, $1 \leq q \leq n$, $p \neq q$ define $E_{pq}{}^1$ to be the matrix $[\epsilon_{ij}]$ defined by $\epsilon_{ij} = \delta_{ij}$ except for $\epsilon_{pq} = 1$. Thus, $E_{pq}{}^1$ has 1's on the diagonal, a 1 in the pth row in the qth column, and zeros elsewhere.

If we multiply $[\alpha_{ij}]E_{pq}{}^1$ we get

$$[\alpha_{ij}][\epsilon_{ij}] = [\gamma_{ij}] \quad \text{where} \quad \gamma_{ij} = \sum_{k=1}^{n} \alpha_{ik}\epsilon_{kj}$$

For $j \neq q$,

$$\gamma_{ij} -= \sum_{k=1}^{n} \alpha_{ik} \delta_{kj} = \alpha_{ij}$$

and

$$\gamma_{iq} = \sum_{k=1}^{n} \alpha_{ik} \epsilon_{kq}$$

$$= \alpha_{i1}\epsilon_{iq} + \alpha_{i2}\epsilon_{2q} + \cdots + \alpha_{in}\epsilon_{nq}$$

$$= \alpha_{iq}\delta_{qq} + \alpha_{ip}\epsilon_{pq}$$

since

$$\epsilon_{kq} = 0 \text{ for } k \neq q, k \neq p$$

Thus

$$\gamma_{iq} = \alpha_{iq} + \alpha_{ip}$$

Thus, the columns of $[\alpha_{ij}][\epsilon_{ij}] = [\gamma_{ij}]$ are the same as the columns of $[\alpha_{ij}]$ except for the qth column. The qth column of $[\gamma_{ij}]$ is obtained by adding the pth column of $[\alpha_{ij}]$ to the qth column of $[\alpha_{ij}]$.

If we wish to multiply a column of $[\alpha_{ij}]$ by a scalar b, this can be accomplished by using the matrix that has 1's on the diagonal except for one entry that is b. We can make this precise by defining for each scalar b a matrix $(b)_p$ by $(b)_p = [b_{ij}]$, where $b_{ij} = \delta_{ij}$ except for $b_{pp} = b$.

We leave it to the reader to show that $[\alpha_{ij}](b)_p = [\gamma_{ij}]$, where

$$\gamma_{ij} = \alpha_{ij} \qquad \text{for } j \neq p$$

and

$$\gamma_{ip} = b\alpha_{ip}$$

It is easy to show that for $b \neq 0$, $(b)_p^{-1} = (b^{-1})_p$. Thus, the two types of transformations E_{pq}^{1} and $(b)_p$ enable us to multiply columns of a matrix by scalars and to add one column to another.

The result of multiplying $[\alpha_{ij}]$ first by $(b)_p$, then by E_{pq}^{1}, and finally by $(b)_p^{-1}$ is equivalent to multiplying the pth column by b and adding it to the qth column. Using the definitions, we can verify that the matrix for $(b)_p E_{pq}^{1}(b)_p^{-1}$ is the matrix with 1's on the diagonal and zeros elsewhere, except for a b in the pth row in the qth column. We shall denote this transformation by E_{pq}^{b}, that is, $E_{pq}^{b} = (b)_p E_{pq}(b)_p^{-1}$.

Consider again the general matrix equation

$$\begin{bmatrix} x_{11} & x_{12} & \cdots & x_{1n} \\ \vdots & \vdots & & \vdots \\ x_{n1} & x_{n-2} & \cdots & x_{nn} \end{bmatrix} \begin{bmatrix} \alpha_{11} & \alpha_{12} & \cdots & \alpha_{1n} \\ \vdots & \vdots & & \vdots \\ \alpha_{n1} & \alpha_{n2} & \cdots & \alpha_{nn} \end{bmatrix} = \begin{bmatrix} 1 & 0 & 0 & \cdots & 0 \\ 0 & 1 & 0 & \cdots & 0 \\ \vdots & \vdots & \vdots & & \vdots \\ 0 & & \cdots & & 1 \end{bmatrix}$$

Assume for now that $\alpha_{11} \neq 0$.

If we first multiply by $(\alpha_{11}^{-1})_1$, this multiplies the first column by α_{11}^{-1} and the resulting matrix has a 1 for the first entry in the first row. If we now multiply on the right by $E_{12}^{-\alpha_{12}}$ the resultant matrix has a zero in the first row, second column since the second column in the product has as its first entry $(-\alpha_{12}) + \alpha_{12}$. Furthermore, if we multiply this resultant matrix by $E_{13}^{-\alpha_{13}}$, this product will have $0 = (-\alpha_{13}) + \alpha_{13}$ in the first row, third column.

For example,

$$
\begin{bmatrix} 2 & 3 & 1 \\ 1 & 2 & 4 \\ 1 & 4 & 7 \end{bmatrix} (\tfrac{1}{2})_1 E_{12}^{-3} E_{13}^{-1} = \begin{bmatrix} 2 & 3 & 1 \\ 1 & 2 & 4 \\ 1 & 4 & 7 \end{bmatrix} \begin{bmatrix} \tfrac{1}{2} & 0 & 0 \\ 0 & 1 & 0 \\ 0 & 0 & 1 \end{bmatrix} E_{12}^{-3} E_{13}^{-1}
$$

$$
= \begin{bmatrix} 1 & 3 & 1 \\ \tfrac{1}{2} & 2 & 4 \\ \tfrac{1}{2} & 4 & 7 \end{bmatrix} E_{12}^{-3} E_{13}^{-1}
$$

$$
= \begin{bmatrix} 1 & 3 & 1 \\ \tfrac{1}{2} & 2 & 4 \\ \tfrac{1}{2} & 4 & 7 \end{bmatrix} \begin{bmatrix} 1 & -3 & 0 \\ 0 & 1 & 0 \\ 0 & 0 & 1 \end{bmatrix} E_{13}^{-1}
$$

$$
= \begin{bmatrix} 1 & 0 & 1 \\ \tfrac{1}{2} & \tfrac{1}{2} & 4 \\ \tfrac{1}{2} & \tfrac{5}{2} & 7 \end{bmatrix} E_{13}^{-1}
$$

$$
= \begin{bmatrix} 1 & 0 & 1 \\ \tfrac{1}{2} & \tfrac{1}{2} & 4 \\ \tfrac{1}{2} & \tfrac{5}{2} & 7 \end{bmatrix} \begin{bmatrix} 1 & 0 & -1 \\ 0 & 1 & 0 \\ 0 & 0 & 1 \end{bmatrix}
$$

$$
= \begin{bmatrix} 1 & 0 & 0 \\ \tfrac{1}{2} & \tfrac{1}{2} & \tfrac{7}{2} \\ \tfrac{1}{2} & \tfrac{5}{2} & \tfrac{13}{2} \end{bmatrix}
$$

The equation

$$
\begin{bmatrix} x_{11} & x_{12} & x_{13} \\ x_{21} & x_{22} & x_{23} \\ x_{31} & x_{32} & x_{33} \end{bmatrix} \begin{bmatrix} 2 & 3 & 1 \\ 1 & 2 & 4 \\ 1 & 4 & 7 \end{bmatrix} = \begin{bmatrix} 1 & 0 & 0 \\ 0 & 1 & 0 \\ 0 & 0 & 1 \end{bmatrix}
$$

becomes

$$\begin{bmatrix} x_{11} & x_{12} & x_{13} \\ x_{21} & x_{22} & x_{23} \\ x_{31} & x_{32} & x_{33} \end{bmatrix} \begin{bmatrix} 2 & 3 & 1 \\ 1 & 2 & 4 \\ 1 & 4 & 7 \end{bmatrix} (\tfrac{1}{2})_1 E_{12}^{-3} E_{13}^{-1} = \begin{bmatrix} 1 & 0 & 0 \\ 0 & 1 & 0 \\ 0 & 0 & 1 \end{bmatrix} (\tfrac{1}{2})_1 E_{12}^{-3} E_{13}^{-1}$$

Multiplying out yields

$$\begin{bmatrix} x_{11} & x_{12} & x_{13} \\ x_{21} & x_{22} & x_{23} \\ x_{31} & x_{32} & x_{33} \end{bmatrix} \begin{bmatrix} 1 & 3 & 1 \\ \tfrac{1}{2} & 2 & 4 \\ \tfrac{1}{2} & 4 & 7 \end{bmatrix} E_{12}^{-3} E_{13}^{-1} = \begin{bmatrix} \tfrac{1}{2} & 0 & 0 \\ 0 & 1 & 0 \\ 0 & 0 & 1 \end{bmatrix} E_{12}^{-3} E_{13}^{-1}$$

$$\begin{bmatrix} x_{11} & x_{12} & x_{13} \\ x_{21} & x_{22} & x_{23} \\ x_{31} & x_{32} & x_{33} \end{bmatrix} \begin{bmatrix} 1 & 0 & 1 \\ \tfrac{1}{2} & \tfrac{1}{2} & 4 \\ \tfrac{1}{2} & \tfrac{5}{2} & 7 \end{bmatrix} E_{13}^{-1} = \begin{bmatrix} \tfrac{1}{2} & -\tfrac{3}{2} & 0 \\ 0 & 1 & 0 \\ 0 & 0 & 1 \end{bmatrix} E_{13}^{-1}$$

$$\begin{bmatrix} x_{11} & x_{12} & x_{13} \\ x_{21} & x_{22} & x_{23} \\ x_{31} & x_{32} & x_{33} \end{bmatrix} \begin{bmatrix} 1 & 0 & 0 \\ \tfrac{1}{2} & \tfrac{1}{2} & \tfrac{7}{2} \\ \tfrac{1}{2} & \tfrac{5}{2} & \tfrac{13}{2} \end{bmatrix} = \begin{bmatrix} \tfrac{1}{2} & -\tfrac{3}{2} & -\tfrac{1}{2} \\ 0 & 1 & 0 \\ 0 & 0 & 1 \end{bmatrix}$$

In general, if $\alpha_{11} \neq 0$,

$[\alpha_{ij}](\alpha_{11}^{-1})_1 E_{12}^{-\alpha_{12}} E_{13}^{-\alpha_{13}} \cdots E_{1n}^{-\alpha_{1n}}$ will be a matrix with first row $(1, 0, 0, \ldots, 0)$, since the first entry in the jth column is always $0 = (-\alpha_{1j}) + \alpha_{1j}$.

Thus, the original equation becomes

$$[x_{ij}][\alpha_{ij}](\alpha_{11}^{-1})_1 E_{12}^{-\alpha_{12}} \cdots E_{1n}^{-\alpha_{1n}} = [\delta_{ij}](\alpha_{11}^{-1})_1 E_{12}^{-\alpha_{12}} \cdots E_{1n}^{-\alpha_{1n}}$$

which now has the form

$$\begin{bmatrix} x_{11} & \cdots & x_{1n} \\ \vdots & & \vdots \\ x_{n1} & \cdots & x_{nn} \end{bmatrix} \begin{bmatrix} 1 & 0 & 0 & \cdots & 0 \\ \beta_{21} & \beta_{22} & & \cdots & \beta_{2n} \\ \vdots & \vdots & & & \vdots \\ \beta_{n1} & \beta_{n2} & & & \beta_{nn} \end{bmatrix} = \begin{bmatrix} \gamma_{11} & \gamma_{12} & \cdots & \gamma_{1n} \\ \vdots & & & \vdots \\ \gamma_{n1} & \gamma_{n2} & \cdots & \gamma_{nn} \end{bmatrix}$$

where $[\gamma_{ij}] = [\delta_{ij}](\alpha_{11}^{-1})_1 E_{12}^{-\alpha_{12}} \cdots E_{1n}^{-\alpha_{1n}}$

Assume for now that $\beta_{22} \neq 0$. If we multiply on the right by $(\beta_{22}^{-1})_2$, the second entry in the second column is 1, and if we then multiply on the right by $E_{21}^{-\beta_{21}} E_{23}^{-\beta_{23}} E_{24}^{-\beta_{24}} \cdots E_{2n}^{-\beta_{2n}}$, the resulting equation will have the form

$$\begin{bmatrix} x_{11} & \cdots & x_{1n} \\ & & \\ \vdots & & \vdots \\ & & \\ x_{n1} & \cdots & x_{nn} \end{bmatrix} \begin{bmatrix} 1 & 0 & 0 & \cdots & 0 \\ 0 & 1 & 0 & \cdots & 0 \\ \mu_{31} & \mu_{32} & \mu_{33} & \cdots & \mu_{3n} \\ & & & \vdots & \\ \mu_{n1} & & \cdots & & \mu_{nn} \end{bmatrix} = \begin{bmatrix} \rho_{11} & \cdots & \rho_{1n} \\ & & \\ \vdots & & \vdots \\ & & \\ \rho_{n1} & \cdots & \rho_{nn} \end{bmatrix}$$

Look again at our example:

$$[x_{ij}] \begin{bmatrix} 1 & 0 & 0 \\ \frac{1}{2} & \frac{1}{2} & \frac{7}{2} \\ \frac{1}{2} & \frac{5}{2} & \frac{13}{2} \end{bmatrix} (2)_2 E_{21}{}^{-1/2} E_{23}{}^{-7/2} = \begin{bmatrix} \frac{1}{2} & -\frac{3}{2} & -\frac{1}{2} \\ 0 & 1 & 0 \\ 0 & 0 & 1 \end{bmatrix} (2)_2 E_{21}{}^{-1/2} E_{23}{}^{-7/2}$$

$$[x_{ij}] \begin{bmatrix} 1 & 0 & 0 \\ \frac{1}{2} & 1 & \frac{7}{2} \\ \frac{1}{2} & 5 & \frac{13}{2} \end{bmatrix} E_{21}{}^{-1/2} E_{23}{}^{-7/2} = \begin{bmatrix} \frac{1}{2} & -3 & -\frac{1}{2} \\ 0 & 2 & 0 \\ 0 & 0 & 1 \end{bmatrix} E_{21}{}^{-1/2} E_{23}{}^{-7/2}$$

$$[x_{ij}] \begin{bmatrix} 1 & 0 & 0 \\ 0 & 1 & \frac{7}{2} \\ -2 & 5 & \frac{13}{2} \end{bmatrix} E_{23}{}^{-7/2} = \begin{bmatrix} 2 & -3 & -\frac{1}{2} \\ -1 & 2 & 0 \\ 0 & 0 & 1 \end{bmatrix} E_{23}{}^{-7/2}$$

$$[x_{ij}] \begin{bmatrix} 1 & 0 & 0 \\ 0 & 1 & 0 \\ -2 & 5 & -11 \end{bmatrix} = \begin{bmatrix} 2 & -3 & 10 \\ -1 & 2 & -7 \\ 0 & 0 & 1 \end{bmatrix}$$

Now if we multiply both sides on the right by $(-\frac{1}{11})_3 E_{31}{}^2 E_{32}{}^{-5}$ we obtain the desired equation.

$$[x_{ij}] \begin{bmatrix} 1 & 0 & 0 \\ 0 & 1 & 0 \\ -2 & 5 & -11 \end{bmatrix} (-\tfrac{1}{11})_3 E_{31}{}^2 E_{32}{}^{-5} = \begin{bmatrix} 2 & -3 & 10 \\ -1 & 2 & -7 \\ 0 & 0 & 1 \end{bmatrix} (-\tfrac{1}{11})_3 E_{31}{}^2 E_{32}{}^{-5}$$

$$[x_{ij}] \begin{bmatrix} 1 & 0 & 0 \\ 0 & 1 & 0 \\ -2 & 5 & 1 \end{bmatrix} E_{31}{}^2 E_{32}{}^{-5} = \begin{bmatrix} 2 & -3 & -\frac{10}{11} \\ -1 & 2 & \frac{7}{11} \\ 0 & 0 & -\frac{1}{11} \end{bmatrix} E_{31}{}^2 E_{32}{}^{-5}$$

$$[x_{ij}] \begin{bmatrix} 1 & 0 & 0 \\ 0 & 1 & 0 \\ 0 & 5 & 1 \end{bmatrix} E_{32}{}^{-5} = \begin{bmatrix} \frac{2}{11} & -1 & -\frac{10}{11} \\ \frac{3}{11} & 2 & \frac{7}{11} \\ -\frac{2}{11} & 0 & -\frac{1}{11} \end{bmatrix} E_{32}{}^{-5}$$

$$[x_{ij}] \begin{bmatrix} 1 & 0 & 0 \\ 0 & 1 & 0 \\ 0 & 0 & 1 \end{bmatrix} = \begin{bmatrix} \frac{2}{11} & \frac{17}{11} & -\frac{10}{11} \\ \frac{3}{11} & -\frac{13}{11} & \frac{7}{11} \\ -\frac{2}{11} & \frac{5}{11} & -\frac{1}{11} \end{bmatrix}$$

Therefore,

$$\begin{bmatrix} 2 & 3 & 1 \\ 1 & 2 & 4 \\ 1 & 4 & 7 \end{bmatrix}^{-1} = \begin{bmatrix} \frac{2}{11} & \frac{17}{11} & -\frac{10}{11} \\ \frac{3}{11} & -\frac{13}{11} & \frac{7}{11} \\ -\frac{2}{11} & \frac{5}{11} & -\frac{1}{11} \end{bmatrix}$$

In the general situation we obtained the general matrix equation

$$\begin{bmatrix} x_{11} & x_{12} & \cdots & x_{1n} \\ & & & \vdots \\ \vdots & & & \\ x_{n1} & \cdots & & x_{nn} \end{bmatrix} \begin{bmatrix} 1 & 0 & 0 & \cdots & 0 \\ 0 & 1 & 0 & \cdots & 0 \\ \mu_{31} & \mu_{32} & \mu_{33} & \cdots & \mu_{3n} \\ & & \vdots & & \\ \mu_{n1} & \cdots & & & \mu_{nn} \end{bmatrix} = \begin{bmatrix} \rho_{11} & \cdots & \rho_{1n} \\ & & \\ \vdots & & \\ \rho_{n1} & \cdots & \rho_{nn} \end{bmatrix}$$

At this point, if $\mu_{33} \neq 0$, we can multiply on the right by

$$(\mu_{33}^{-1})_3 E_{31}^{-\mu_{31}} E_{32}^{-\mu_{32}} \cdots E_{3n}^{-\mu_{3n}}$$

and obtain a matrix of the form

$$\begin{bmatrix} 1 & 0 & 0 & 0 & \cdots & 0 \\ 0 & 1 & 0 & 0 & \cdots & 0 \\ 0 & 0 & 1 & 0 & \cdots & 0 \\ y_{41} & y_{42} & y_{43} & y_{44} & \cdots & y_{4n} \\ & \vdots & & & & \\ y_{n1} & & \cdots & & & y_{nn} \end{bmatrix}$$

If, at each step in the process, we have a suitable nonzero entry we can continue and each time get a matrix in which the first rows are as in

$$\begin{bmatrix} 1 & 0 & 0 & \cdots & 0 \\ 0 & 1 & 0 & \cdots & 0 \\ 0 & 0 & 1 & 0 & \cdots & 0 \\ & & \cdots & & \end{bmatrix}$$

that is, we have 1's on the diagonal and 0's elsewhere in the first t rows.

If we have the suitable nonzero entries then, after going through the process n times, the resultant matrix on the left is just the identity matrix and the matrix on the right is the desired inverse. Suppose we do not have the suitable nonzero entries, then what do we do? If, at some point, we obtain a matrix in which some row, say the mth row, is all zeros, then $e_m = (0, 0, 0, \ldots, 1, 0, \ldots, 0)$ is in the kernel of this matrix. But since each of our matrices $(b)_p$ and $E_{pq}{}^b$ is invertible, e_m must be in the kernel of $[\alpha_{ij}]$ and therefore $[\alpha_{ij}]$ has no inverse. Similarly, if we obtain a matrix $[\beta_{ij}]$ that has two rows, say the mth and pth rows that have zeros everywhere but in the qth column, and $\beta_{mq} = 1$ and $\beta_{pq} = a$, then $ae_m - e_p$ is in the kernel of $[\beta_{ij}]$ and therefore in the kernel of $[\alpha_{ij}]$, and again $[\alpha_{ij}]$ is not invertible. Thus, if $[\alpha_{ij}]^{-1}$ exists, at every step in the process there is a nonzero entry in the next row and it is not under one of the 1's thus far obtained. To transform this general situation into the situation described above, we need one more type of elementary linear transformation.

If $\rho \in S_n$, then ρ determines an automorphism P_ρ on F^n just by permuting the components; that is

$$e_i P_\rho = e_{i\rho}$$

where $e_i = (0, 0, \ldots, 1, \ldots, 0)$ as before. For example, in $\mathbf{R}^3$, $(1, 0, 0)P_{(1\ 2\ 3)} = (0, 1, 0)$ because $e_1 P_{(1\ 2\ 3)} = e_{1(1\ 2\ 3)} = e_2$.

The matrix for P_ρ is just the matrix for I with the columns permuted according to ρ. For example,

$$
\begin{aligned}
(x, y, z)P_{(1\ 2\ 3)} &= (xe_1 + ye_2 + ze_3)P_{(1\ 2\ 3)} \\
&= xe_{1(1\ 2\ 3)} + ye_{2(1\ 2\ 3)} + ze_{3(1\ 2\ 3)} \\
&= xe_2 + ye_3 + ze_1 \\
&= (z, x, y)
\end{aligned}
$$

and the matrix for $P_{(1\ 2\ 3)}$ is

$$
\begin{bmatrix}
0 & 1 & 0 \\
0 & 0 & 1 \\
1 & 0 & 0
\end{bmatrix}
$$

which is obtained from $[\delta_{ij}]$ by putting the first column where the second was, the second where the third was, and the third where the first was: 1 to 2, 2 to 3, 3 to 1 according to $(1\ 2\ 3)$.

To exchange the pth and qth columns of a matrix we need only multiply on the right by $P_{(p\ q)}$. If $[\rho_{ij}] = P_{(p\ q)}$, then $\rho_{pq} = 1 = \rho_{qp}$, $\rho_{pp} = 0 = \rho_{qq}$ and otherwise $\rho_{ij} = \delta_{ij}$, because the matrix for $P_{(p\ q)}$ exchanges the pth and qth columns of $[\delta_{ij}]$. For example, in 6×6 matrices, the matrix for $P_{(2\ 4)}$ is just

$$[\rho_{ij}] = \begin{bmatrix} 1 & 0 & 0 & 0 & 0 & 0 \\ 0 & 0 & 0 & 1 & 0 & 0 \\ 0 & 0 & 1 & 0 & 0 & 0 \\ 0 & 1 & 0 & 0 & 0 & 0 \\ 0 & 0 & 0 & 0 & 1 & 0 \\ 0 & 0 & 0 & 0 & 0 & 1 \end{bmatrix} \quad \text{and} \quad \rho_{24} = 1 = \rho_{42}$$

If we multiply $[\alpha_{ij}]$ by $P_{(p\ q)}$ we get $[\alpha_{ij}]$ with the pth and qth columns exchanged, because if $[\alpha_{ij}]P_{(p\ q)} = [\alpha_{ij}][\rho_{ij}] = [\gamma_{ij}]$, then each $\gamma_{ij} = \sum_{k=1}^{n} \alpha_{ik}\rho_{kj}$. If $j \neq p$ or $j \neq q$, then $\rho_{kj} = \delta_{kj}$ and $\gamma_{ij} = \alpha_{ij}$. But $\gamma_{ip} = \sum_{k=1}^{n} \alpha_{ik}\rho_{kp} = \alpha_{iq}$, because $\rho_{qp} = 1$ and $\rho_{kp} = 0$ if $k \neq q$. Also

$$\gamma_{iq} = \sum_{k=1}^{n} \alpha_{ik}\rho_{kq} = \alpha_{ip}$$

for similar reasons. For example,

$$\begin{bmatrix} \alpha_{11} & \alpha_{12} & \cdots & \alpha_{1n} \\ \alpha_{21} & \alpha_{22} & \cdots & \alpha_{2n} \\ \vdots & & & \\ \alpha_{n1} & \alpha_{n2} & \cdots & \alpha_{nn} \end{bmatrix} P_{(1\ 2)} = \begin{bmatrix} \alpha_{12} & \alpha_{11} & \alpha_{13} & \cdots & \alpha_{1n} \\ \alpha_{22} & \alpha_{21} & \alpha_{23} & \cdots & \alpha_{2n} \\ \vdots & & & \\ \alpha_{n2} & \alpha_{n1} & \alpha_{n3} & \cdots & \alpha_{nn} \end{bmatrix}$$

In general, if $\rho \in S_n$ and $[\alpha_{ij}]$ is an $n \times n$ matrix then $[\alpha_{ij}]P_\rho = [\alpha_{i_{j\rho^{-1}}}]$

If we apply the previous technique in the following example, we do not immediately obtain the wanted nonzero entry in the second row.

$$\begin{bmatrix} 1 & 1 & 3 \\ 2 & 2 & 2 \\ 4 & 5 & 1 \end{bmatrix} E_{12}^{-1} E_{13}^{-3} = \begin{bmatrix} 1 & 0 & 0 \\ 2 & 0 & -4 \\ 4 & 1 & -11 \end{bmatrix}$$

The second entry in the second row is 0 and the method used before stops. The situation is, however, not a hindrance but actually a simplification, since just by permuting the second and third columns we get

$$\begin{bmatrix} 1 & 1 & 3 \\ 2 & 2 & 2 \\ 4 & 5 & 1 \end{bmatrix} E_{12}^{-1} E_{13}^{-3} P_{(2\ 3)} = \begin{bmatrix} 1 & 0 & 0 \\ 2 & -4 & 0 \\ 4 & -11 & 1 \end{bmatrix}$$

If we now proceed by multiplying by $E_{31}^{-4} E_{32}^{11}$ we shall have

$$\begin{bmatrix} 1 & 1 & 3 \\ 2 & 2 & 2 \\ 4 & 5 & 1 \end{bmatrix} E_{12}^{-1} E_{13}^{-3} P_{(2\ 3)} E_{31}^{-4} E_{32}^{11} = \begin{bmatrix} 1 & 0 & 0 \\ 2 & -4 & 0 \\ 0 & 0 & 1 \end{bmatrix}$$

and multiplying now by $(-\frac{1}{4})_2 E_{21}^{-2}$ we obtain the identity matrix. All that we needed in the above example was a nonzero entry in the second row outside of the first column. By using a suitable permutation we exchanged the necessary columns and obtained a matrix with a nonzero entry in the second row, second column.

In the general equation $[x_{ij}][\alpha_{ij}] = [\delta_{ij}]$, if $[\alpha_{ij}]$ is invertible, each row contains nonzero entries and, therefore, if $\alpha_{11} = 0$ we know $\alpha_{1q} \neq 0$ for some q, and therefore $[\alpha_{ij}]P_{(1\ q)}$ has a nonzero entry in the first row, first column and we can apply our method to this matrix. If at some point in the process we obtain a matrix such as

$$\begin{bmatrix} 1 & 0 & 0 & \cdots & 0 & 0 & \cdots & 0 \\ 0 & 1 & 0 & \cdots & 0 & 0 & \cdots & 0 \\ 0 & 0 & 1 & \cdots & 0 & 0 & \cdots & 0 \\ \vdots & & \ddots & & \vdots & \vdots & & \\ 0 & & & 1 & 0 & 0 & \cdots & \\ \beta_{k1} & \cdots & & \beta_{kk-1} & 0 & \beta_{kk+1} & \cdots & \beta_{kn} \\ \vdots & & & & & & & \\ \beta_{n1} & & & \cdots & & & & \beta_{nn} \end{bmatrix}$$

from previous arguments we know that one of $\beta_{kk+1}, \ldots, \beta_{kn}$ must be nonzero if $[\alpha_{ij}]$ is invertible. (Not all β_{kj} are 0 and if β_{ki} is the only nonzero entry for some $i < k$, then $\beta_{ki} e_i - e_k$ is in the kernel of $[\beta_{ij}]$.) If $\beta_{kq} \neq 0$, then $[\beta_{ij}]P_{(k\ q)}$ is a matrix with a nonzero entry in the kkth component and zeros above this component. We can now continue our process using the transformations of the types $(b)_p$, $E_{pq}^{\ b}$, and P_p until we obtain the identity matrix. Our technique of working on both sides of the equation $[x_{ij}][\alpha_{ij}] = [\delta_{ij}]$ then yields the inverse of $[\alpha_{ij}]$ on the right. In practice we usually do not work with the equation and the matrices for $(b)_p$, $E_{pq}^{\ b}$, and P_p but rather just "do the same thing" to $[\alpha_{ij}]$ and $[\delta_{ij}]$ and thus obtain the inverse for $[\alpha_{ij}]$ by "doing to $[\delta_{ij}]$ what it takes to make $[\alpha_{ij}]$ into $[\delta_{ij}]$."

Example 1. Find the inverse of

$$\begin{bmatrix} 4 & 2 \\ 3 & 1 \end{bmatrix}$$

We write down the given matrix and the identity matrix

$$\begin{bmatrix} 4 & 2 \\ 3 & 1 \end{bmatrix} \quad \begin{bmatrix} 1 & 0 \\ 0 & 1 \end{bmatrix}$$

Multiply the first column by $\frac{1}{4}$.

$$\begin{bmatrix} 1 & 2 \\ \frac{3}{4} & 1 \end{bmatrix} \quad \begin{bmatrix} \frac{1}{4} & 0 \\ 0 & 1 \end{bmatrix}$$

Multiply the first column by -2 and add to the second column.

$$\begin{bmatrix} 1 & 0 \\ \frac{3}{4} & -\frac{1}{2} \end{bmatrix} \quad \begin{bmatrix} \frac{1}{4} & -\frac{1}{2} \\ 0 & 1 \end{bmatrix}$$

Multiply the second column by -2.

$$\begin{bmatrix} 1 & 0 \\ \frac{3}{4} & 1 \end{bmatrix} \quad \begin{bmatrix} \frac{1}{4} & 1 \\ 0 & -2 \end{bmatrix}$$

Multiply the second column by $-\frac{3}{4}$ and add to the first column.

$$\begin{bmatrix} 1 & 0 \\ 0 & 1 \end{bmatrix} \quad \begin{bmatrix} -\frac{1}{2} & 1 \\ \frac{3}{2} & -2 \end{bmatrix}$$

Therefore

$$\begin{bmatrix} 4 & 2 \\ 3 & 1 \end{bmatrix}^{-1} = \begin{bmatrix} -\frac{1}{2} & 1 \\ \frac{3}{2} & -2 \end{bmatrix}$$

Example 2. To find the inverse of

$$\begin{bmatrix} 2 & 2 & 5 \\ 4 & 4 & 2 \\ 4 & 5 & 1 \end{bmatrix}$$

we write down the matrix and the identity matrix.

$$\text{(a)} \begin{bmatrix} 2 & 2 & 5 \\ 4 & 4 & 2 \\ 4 & 5 & 1 \end{bmatrix} \quad \begin{bmatrix} 1 & 0 & 0 \\ 0 & 1 & 0 \\ 0 & 0 & 1 \end{bmatrix}$$

Multiply the first column by $\frac{1}{2}$.

$$(b) \quad \begin{bmatrix} 1 & 2 & 5 \\ 2 & 4 & 2 \\ 2 & 5 & 1 \end{bmatrix} \quad \begin{bmatrix} \frac{1}{2} & 0 & 0 \\ 0 & 1 & 0 \\ 0 & 0 & 1 \end{bmatrix}$$

Multiply the first column by (-2) and add to the second column.

$$(c) \quad \begin{bmatrix} 1 & 0 & 5 \\ 2 & 0 & 2 \\ 2 & 1 & 1 \end{bmatrix} \quad \begin{bmatrix} \frac{1}{2} & -1 & 0 \\ 0 & 1 & 0 \\ 0 & 0 & 1 \end{bmatrix}$$

Multiply the first column by (-5) and add to the third column.

$$(d) \quad \begin{bmatrix} 1 & 0 & 0 \\ 2 & 0 & -8 \\ 2 & 1 & -9 \end{bmatrix} \quad \begin{bmatrix} \frac{1}{2} & -1 & -\frac{5}{2} \\ 0 & 1 & 0 \\ 0 & 0 & 1 \end{bmatrix}$$

Permute second and third columns.

$$(e) \quad \begin{bmatrix} 1 & 0 & 0 \\ 2 & -8 & 0 \\ 2 & -9 & 1 \end{bmatrix} \quad \begin{bmatrix} \frac{1}{2} & -\frac{5}{2} & -1 \\ 0 & 0 & 1 \\ 0 & 1 & 0 \end{bmatrix}$$

Multiply the third column by (-2) and add to the first.

$$(f) \quad \begin{bmatrix} 1 & 0 & 0 \\ 2 & -8 & 0 \\ 0 & -9 & 1 \end{bmatrix} \quad \begin{bmatrix} \frac{5}{2} & -\frac{5}{2} & -1 \\ -2 & 0 & 1 \\ 0 & 1 & 0 \end{bmatrix}$$

Multiply the third column by 9 and add to the second.

$$(g) \quad \begin{bmatrix} 1 & 0 & 0 \\ 2 & -8 & 0 \\ 0 & 0 & 1 \end{bmatrix} \quad \begin{bmatrix} \frac{5}{2} & -\frac{23}{2} & -1 \\ -2 & 9 & 1 \\ 0 & 1 & 0 \end{bmatrix}$$

Multiply the second column by $-\frac{1}{8}$.

$$(h) \quad \begin{bmatrix} 1 & 0 & 0 \\ 2 & 1 & 0 \\ 0 & 0 & 1 \end{bmatrix} \quad \begin{bmatrix} \frac{5}{2} & \frac{23}{16} & -1 \\ -2 & -\frac{9}{8} & 1 \\ 0 & -\frac{1}{8} & 0 \end{bmatrix}$$

Multiply the second column by (-2) and add to the first.

$$\text{(i)} \quad \begin{bmatrix} 1 & 0 & 0 \\ 0 & 1 & 0 \\ 0 & 0 & 1 \end{bmatrix} \quad \begin{bmatrix} -\frac{3}{8} & \frac{23}{16} & -1 \\ \frac{1}{4} & -\frac{9}{8} & 1 \\ \frac{1}{4} & -\frac{1}{8} & 0 \end{bmatrix}$$

Therefore,

$$\begin{bmatrix} 2 & 2 & 5 \\ 4 & 4 & 2 \\ 4 & 5 & 1 \end{bmatrix}^{-1} = \begin{bmatrix} -\frac{3}{8} & \frac{23}{16} & -1 \\ \frac{1}{4} & -\frac{9}{8} & 1 \\ \frac{1}{4} & -\frac{1}{8} & 0 \end{bmatrix}$$

In practice we frequently combine some of these steps. For example, steps (c) and (d) in Example 2 could easily be combined. In the exercises that follow, the student is asked to show that everything done in the text with columns can be done with rows if he multiplies on the left of the equation

$$[\alpha_{ij}][x_{ij}] = [\delta_{ij}]$$

and the result will again be the matrix for $[\alpha_{ij}]^{-1}$.

We now state the results of this section in the following theorem. The proof is contained in the preceding discussion.

Theorem 1

If $[\alpha_{ij}]$ is an invertible matrix, then $[\alpha_{ij}]^{-1}$ is expressible as the product of matrices of the forms $(b)_p$, $E_{pq}{}^b$, and P_p.

EXERCISE SET B

1. Compute the inverses of the following matrices by working with the columns and the technique exhibited in the examples.

$$\text{(a)} \begin{bmatrix} 1 & 2 \\ 3 & 4 \end{bmatrix} \qquad \text{(b)} \begin{bmatrix} 4 & 7 \\ -3 & 1 \end{bmatrix} \qquad \text{(c)} \begin{bmatrix} \frac{1}{2} & \frac{1}{3} \\ 4 & \frac{2}{3} \end{bmatrix}$$

$$\text{(d)} \begin{bmatrix} 1 & 2 & 3 \\ 1 & -2 & 4 \\ 0 & 5 & 1 \end{bmatrix} \qquad \text{(e)} \begin{bmatrix} 4 & 4 & 3 \\ 6 & 6 & 8 \\ 1 & 2 & 3 \end{bmatrix} \qquad \text{(f)} \begin{bmatrix} 1 & 2 & 5 \\ 4 & 9 & -1 \\ 7 & 3 & -2 \end{bmatrix}$$

2. Use the "inverse computing technique" to solve the equation

$$(x_1, x_2, x_3) \begin{bmatrix} 1 & 2 & 0 \\ 1 & 3 & 4 \\ 0 & 1 & 2 \end{bmatrix} = (4, 5, 6)$$

(Multiply both sides on the right by suitable matrices.)

3. Try to apply the "inverse computing technique" to the matrices

$$\begin{bmatrix} 1 & 2 & 2 \\ 2 & 5 & 4 \\ 3 & 1 & 6 \end{bmatrix} \text{ and } \begin{bmatrix} 2 & 1 & 1 \\ -3 & 1 & 4 \\ 5 & 3 & -2 \end{bmatrix}$$

What goes wrong? Why doesn't it work?

4. Prove that if $\rho \in S_n$ and $[\alpha_{ij}]$ is an $n \times n$ matrix, then $P_\rho[\alpha_{ij}] = [\alpha_{i\rho\ j}]$.

**5. Prove that the result of multiplying $E_{pq}{}^b[\alpha_{ij}]$ is a matrix with the same rows as $[\alpha_{ij}]$ except for the pth row, which is the pth row of $[\alpha_{ij}]$ plus b times the qth row.

6. Prove that multiplying $(b)_\rho[\alpha_{ij}]$ yields a matrix with the same rows as $[\alpha_{ij}]$ except for the pth row, which has been multiplied by b.

7. Do Exercise 1 working with rows instead of columns.

**8. We denote the *transpose* of $[\alpha_{ij}]$ by $[\alpha_{ij}]^t$ and define it by $[\alpha_{ij}]^t = [\alpha_{ji}]$; that is, the rows of $[\alpha_{ij}]^t$ are the columns of $[\alpha_{ij}]$. Prove that if $[\alpha_{ij}]$ and $[\beta_{ij}]$ are matrices, then $([\alpha_{ij}][\beta_{ij}])^t = [\beta_{ij}]^t[\alpha_{ij}]^t$.

**9. What is $(E_{pq}{}^b)^t$? What is $(P_\rho)^t$ if $\rho \in S_n$?

**10. Use Exercises 8 and 9 to do Exercise 5.

§3. Determinants

In studying elementary algebra we encounter the idea of a determinant in connection with solving systems of linear equations. It can be shown (Exercise 20, Set E, Chapter 5) that the system of equations

$$a_{11}x_1 + a_{21}x_2 = c_1$$
$$a_{12}x_1 + a_{22}x_2 = c_2$$

with $a_{11}, a_{21}, a_{12}, a_{22}, c_1$, and c_2 elements of some field, has a unique solution iff $a_{11}a_{22} - a_{21}a_{12} \neq 0$ and otherwise has either no solution or as many solutions as there are elements in the field. The system can be viewed as

$$(x_1, x_2) \begin{bmatrix} a_{11} & a_{12} \\ a_{21} & a_{22} \end{bmatrix} = (c_1, c_2)$$

and $a_{11}a_{22} - a_{21}a_{12}$ is called the *determinant* of the matrix

$$\begin{bmatrix} a_{11} & a_{12} \\ a_{21} & a_{22} \end{bmatrix}$$

Similarly, if we have a system of three equations

$$a_{11}x_1 + a_{21}x_2 + a_{31}x_3 = c_1$$
$$a_{12}x_1 + a_{22}x_2 + a_{32}x_3 = c_2$$
$$a_{13}x_1 + a_{23}x_2 + a_{33}x_3 = c_3$$

the system can be shown to have a unique solution if

$$a_{11}a_{22}a_{33} - a_{11}a_{23}a_{32} + a_{12}a_{23}a_{31} - a_{12}a_{21}a_{33}$$
$$+ a_{13}a_{21}a_{32} - a_{13}a_{22}a_{31} \neq 0$$

This field element is called the determinant of the matrix

$$\begin{bmatrix} a_{11} & a_{12} & a_{13} \\ a_{21} & a_{22} & a_{23} \\ a_{31} & a_{32} & a_{33} \end{bmatrix}$$

of coefficients of the system of equations.

If we look at the summands in the expression

$$a_{11}a_{22}a_{33} - a_{11}a_{23}a_{32} + a_{12}a_{23}a_{31} - a_{12}a_{21}a_{33} + a_{13}a_{21}a_{32} - a_{13}a_{22}a_{31}$$

we see that each summand is of the form

$$a_{1p}a_{2q}a_{3t}$$

and p, q, and t are each one of the numbers 1, 2, and 3; that is,

$$a_{1p}a_{2q}a_{3t}$$

can be obtained from

$$a_{11}a_{22}a_{33}$$

by permuting the second subscript. Furthermore, there are six summands and each possible permutation of the second subscripts is used. Thus, the summands are precisely all the expressions we can obtain from $a_{11}a_{22}a_{33}$ by permuting the second subscripts. Using the permutation notation we obtain, then, the following expression for the determinant

$$a_{11}a_{22}a_{33} - a_{1,1(2\ 3)}a_{2,2(2\ 3)}a_{3,3(2\ 3)} + a_{1,1(1\ 2\ 3)}a_{2,2(1\ 2\ 3)}a_{3,3(1\ 2\ 3)}$$
$$- a_{1,1(1\ 2)}a_{2,2(1\ 2)}a_{3,3(1\ 2)} + a_{1,1(1\ 3\ 2)}a_{2,2(1\ 3\ 2)}a_{3,3(1\ 3\ 2)}$$
$$- a_{1,1(1\ 3)}a_{2,2(1\ 3)}a_{3,3(1\ 3)}$$

Notice that the terms preceded by a minus sign are those in which the permutation applied is odd, those with an even permutation are preceded by a plus sign.

We shall use these notions to define determinants of all $n \times n$ matrices. To insert the minus signs more easily in the proper places we define the sign or signum of a permutation p in S_n to be 1 if p is even and -1 if p is odd and denote it by sgn p. Thus, sgn $p = 1$ if p is even and sgn $p = -1$ if p is odd. Notice that, in the notation of Chapter 3, sgn $p = \Delta p/\Delta$ and that sgn $(p\sigma) = (\text{sgn } p)(\text{sgn } \sigma)$ for any permutations p and σ in S_n. The expression for the determinant of a 3×3 matrix then becomes

$$\sum_{p \in S_3} (\text{sgn } p) a_{1,1p} a_{2,2p} a_{3,3p}$$

If $[\alpha_{ij}]$ is an $n \times n$ matrix with entries from the field F, then the *determinant* of $[\alpha_{ij}]$ is denoted by $|[\alpha_{ij}]|$, $|\alpha_{ij}|$, or by det $[\alpha_{ij}]$ and defined by

$$|[\alpha_{ij}]| = \sum_{p \in S_n} (\text{sgn } p) \alpha_{1,1p} \alpha_{2,2p} \cdots \alpha_{n,np}$$

Notice that determinants are defined only for square matrices.

The determinant of an $n \times n$ matrix can, of course, be determined by straightforward computations, but since $|S_n| = n!$ we would have to compute $n!$ products each with n factors and then sum them. If, for example, $[\alpha_{ij}]$ is a 5×5 matrix we must compute 120 products each having 5 factors and then sum. If it takes 10 seconds to compute each product, it will take over 20 minutes to compute the determinant.

The fact of the matter is that, in practice, when computing determinants of matrices larger than 3×3 or 4×4, we make some modifications to obtain a matrix in which the computation is simpler. The following sequence of assertions will provide such techniques.

Proposition 2

If $[\alpha_{ij}]$ is an $n \times n$ matrix, then

$$|[\alpha_{ij}]| = \sum_{p \in S_n} (sgn\ p)(\alpha_{1p,1} \cdots \alpha_{np,n})$$

Proof

$$|[\alpha_{ij}]| = \sum_{p \in S_n} (\text{sgn } p)(\alpha_{1,1p} \cdots \alpha_{n,np})$$

If we rearrange the factors in a summand according to the order of the second subscript, this is equivalent to applying p^{-1} to both subscripts of each factor but the products are just the same. Therefore,

$$|[\alpha_{ij}]| = \sum_{p \in S_n} (\text{sgn } p)(\alpha_{1p^{-1},1} \cdots \alpha_{np^{-1}n})$$

Since sgn p = sgn p^{-1} we have

$$|[\alpha_{ij}]| = \sum_{p \in S_n} (\text{sgn } p^{-1})(\alpha_{1p^{-1},1} \cdots \alpha_{np^{-1},n})$$

But as p ranges over S_n, so does p^{-1} and we have

$$|[\alpha_{ij}]| = \sum_{p \in S_n} (\text{sgn } p)(\alpha_{1p,1} \cdots \alpha_{np,n}) \quad \square$$

Corollary 3

If $[\alpha_{ij}]$ *is an* $n \times n$ *matrix, then* $|[\alpha_{ij}]| = |[\alpha_{ij}]^t|$.

Proof: Exercise. ($[\alpha_{ij}]^t$ was defined in Exercise 8, Set B.)

Proposition 4

If $[\alpha_{ij}]$ *is an* $n \times n$ *matrix and one of the columns or rows of* $[\alpha_{ij}]$ *is all zeros, then* $|[\alpha_{ij}]| = 0$.

Proof: If a row is all zeros, say the qth row, then since

$$|[\alpha_{ij}]| = \sum_{p \in S_n} (\text{sgn } p)(\alpha_{1,1p} \cdots \alpha_{n,np})$$

each summand has a factor from the qth row and therefore is 0. The argument is analogous for a column of zeros. $\square$

Notice that

$$\begin{vmatrix} 2 & 0 \\ 1 & 3 \end{vmatrix} = 6 \quad \text{and} \quad \begin{vmatrix} 2 & 1 & 3 \\ 0 & 4 & 2 \\ 0 & 0 & 5 \end{vmatrix} = 40$$

A matrix $[\alpha_{ij}]$ is said to be *upper triangular* if $\alpha_{ij} = 0$ for all $i > j$ and is said to be *lower triangular* if $\alpha_{ij} = 0$ for all $i < j$. Thus, a matrix is upper triangular if all entries below the diagonal are zero and lower triangular if all entries above the diagonal are zero. A matrix is said to be *triangular* if it is either lower or upper triangular. If all the entries above and below the diagonal are zero, the matrix is said to be *diagonal*. The two matrices above are triangular, the first is lower and the second upper. Notice that in these cases the determinant is the product of the elements on the diagonal. We now show that this is the case for all matrices.

Proposition 5

If $[\alpha_{ij}]$ *is a triangular* $n \times n$ *matrix, then* $|[\alpha_{ij}]| = \alpha_{11}\alpha_{22}\alpha_{33} \cdots \alpha_{nn}$.

Proof: Assume $[\alpha_{ij}]$ is lower triangular. By definition,

$$|[\alpha_{ij}]| = \sum_{p \in S_n} (\text{sgn } p)\alpha_{1,1p}\alpha_{2,2p} \cdots \alpha_{n,np}$$

If p is not the identity permutation, then for some j, $1 \leq j \leq n$, $jp > j$ for otherwise $\{1, \ldots, n\}p$ would be a proper subset of $\{1, \ldots, n\}$. Thus, for p different from the identity, each product contains some factor $\alpha_{j,jp}$ in which $jp > j$. Since $[\alpha_{ij}]$ is lower triangular, $\alpha_{j,jp} = 0$ and the product is zero. Therefore,

$$|[\alpha_{ij}]| = \sum_{p \in S_n} (\text{sgn } p)\alpha_{1,1p}\alpha_{2,2p} \cdots \alpha_{n,np}$$

$$= (1)\alpha_{11}\alpha_{22} \cdots \alpha_{nn} + 0$$

$$= \alpha_{11}\alpha_{22} \cdots \alpha_{nn}$$

The proof for upper triangular matrices is similar to the above argument.
$\square$

Corollary 6

The determinant of a matrix of the form E_{pq}^{1} is 1.

Proof: E_{pq}^{1} is either upper or lower triangular and the diagonal elements are all 1's. $\square$

Corollary 7

The determinant of a matrix of the form $(b)_p$ is b.

Proof: The matrix $(b)_p$ is triangular with all diagonal elements 1 except one, which is b. $\square$

Proposition 8

If $\sigma \in S_n$, the determinant of the matrix for P_σ is sgn σ.

Proof: The matrix for P_σ is just $[\delta_{i,j\sigma^{-1}}]$ and

$$|[\delta_{i,j\sigma^{-1}}]| = \sum_{p \in S_n} (\text{sgn } p)\delta_{1,(1\sigma^{-1})p}\delta_{2,(2\sigma^{-1})p} \cdots \delta_{n,(n\sigma^{-1})p}$$

For each $p \neq \sigma$, the product will have 0 as a factor and therefore will be 0. Therefore

$$|[\delta_{i,j\sigma^{-1}}]| = (\text{sgn } \sigma) \cdot 1 = \text{sgn } \sigma \quad \square$$

We now know how to compute easily the determinants of the elementary matrices of the form $(b)_p$, E_{pq}^{1}, and P_σ and we know that certain matrices, namely the invertible ones, can be expressed as products of such matrices. We now proceed to show that this knowledge enables us to compute all determinants.

Proposition 9

If $[\alpha_{ij}]$ is an $n \times n$ matrix over the field F and $(b)_p$ is an $n \times n$ matrix over F as defined in §2, then

$$|[\alpha_{ij}](b)_p| = |[\alpha_{ij}]| b = |[\alpha_{ij}]| \, |(b)_p|$$

Proof: Let $[\beta_{ij}] = [\alpha_{ij}](b)_p$.

Then $\beta_{ij} = \alpha_{ij}$ for $j \neq p$ and $\beta_{ip} = b\alpha_{ip}$.

Thus $\beta_{1,1\sigma}\beta_{2,2\sigma} \cdots \beta_{n,n\sigma} = b(\alpha_{1,1\sigma} \cdots \alpha_{n,n\sigma})$

since $\beta_{i,i\sigma} = \alpha_{i,i\sigma}$, except when $i\sigma = p$

and then $\beta_{i,p} = b\alpha_{i,p}$.

Therefore,

$$|[\beta_{ij}]| = \sum_{\sigma \in S_n} (\text{sgn } \sigma)\beta_{1,1\sigma} \cdots \beta_{n,n\sigma}$$

$$= \sum_{\sigma \in S_n} (\text{sgn } \sigma)b(\alpha_{1,1\sigma} \cdots \alpha_{n,n\sigma})$$

$$= b \sum_{\sigma \in S_n} (\text{sgn } \sigma)(\alpha_{1,1\sigma} \cdots \alpha_{n,n\sigma})$$

$$= b|[\alpha_{ij}]|$$

$$= |[\alpha_{ij}]|b = |[\alpha_{ij}]| \, |(b)_p| \quad \square$$

Proposition 10

If $[\alpha_{ij}]$ is an $n \times n$ matrix over the field F, ρ is in S_n, and P_ρ denotes the $n \times n$ matrix associated with ρ as defined in §2, then

$$|[\alpha_{ij}]P_\rho| = |[\alpha_{ij}]| \, \text{sgn } \rho$$

$$= |[\alpha_{ij}]| \, |P_\rho|$$

Proof: Since $|P_\rho| = \text{sgn } \rho$, it suffices to show that

$$|[\alpha_{ij}]P_\rho| = |[\alpha_{ij}]| (\text{sgn } \rho)$$

Since $[\alpha_{ij}]P_\rho = [\alpha_{i,j\rho^{-1}}]$

$$|[\alpha_{ij}]P_\rho| = |[\alpha_{i,j\rho^{-1}}]|$$

$$= \sum_{\sigma \in S_n} (\text{sgn } \sigma)(\alpha_{1,1\rho^{-1}\sigma}\alpha_{2,2\rho^{-1}\sigma} \cdots \alpha_{n,n\rho^{-1}\sigma})$$

$$= \sum_{\sigma \in S_n} (\text{sgn } \rho)(\text{sgn } \rho^{-1})(\text{sgn } \sigma)(\alpha_{1,1\rho^{-1}\sigma} \cdots \alpha_{n,n\rho^{-1}\sigma})$$

Hence,

$$|[\alpha_{ij}]P_\rho| = (\text{sgn } \rho) \sum_{\sigma \in S_n} (\text{sgn } \rho^{-1}\sigma)(\alpha_{1,1\rho^{-1}\sigma} \cdots \alpha_{n,n\rho^{-1}\sigma})$$

But as σ ranges over S_n, $\rho^{-1}\sigma$ does also and so we have

$$|[\alpha_{ij}]P_\rho| = (\text{sgn } \rho) \sum_{\rho^{-1}\sigma \in S_n} (\text{sgn } \rho^{-1}\sigma)(\alpha_{1,1\rho^{-1}\sigma} \cdots \alpha_{n,n\rho^{-1}\sigma})$$

$$= (\text{sgn } \rho)|[\alpha_{ij}]|$$

$$= |[\alpha_{ij}]| \, |P_\rho| \quad \square$$

Proposition 11

If $[\alpha_{ij}]$ is an $n \times n$ matrix and $E_{pq}{}^1$ is a linear transformation as defined in §2, then $|[\alpha_{ij}]E_{pq}{}^1| = |[\alpha_{ij}]|$; that is, adding one column of a matrix to another does not change the determinant of the matrix.

Proof: The concepts in the following argument are not overly difficult but the notation does get a bit messy. The main idea is that the resulting determinant is the original determinant plus a certain sum. In this sum, each term that appears with a plus sign also appears with a minus sign, and therefore the sum is 0.

If $[\alpha_{ij}]E_{pq}{}^1 = [\beta_{ij}]$, then $\beta_{ij} = \alpha_{ij} + \delta_{qj}\alpha_{ip}$, because the columns of $[\beta_{ij}]$ are the same as those of $[\alpha_{ij}]$ except for the qth column, which is the sum of the qth column of $[\alpha_{ij}]$ and the pth column of $[\alpha_{ij}]$. Then

$$|[\beta_{ij}]| = \sum_{\sigma \in S_n} (\text{sgn } \sigma)\left(\prod_{m=1}^{n} \beta_{m,m\sigma}\right)$$

For a particular permutation σ, examine the product

$$\beta_{1,1\sigma}\beta_{2,2\sigma}\cdots\beta_{n,n\sigma}$$

If $k\sigma \neq q$, then $\beta_{k,k\sigma} = \alpha_{k,k\sigma}$.
If $k\sigma = q$, then $\beta_{k,k\sigma} = \alpha_{k,k\sigma} + \alpha_{k,p}$.

Therefore,

$$\prod_{m=1}^{n} \beta_{m,m\sigma} = \left(\prod_{\substack{m=1 \\ m \neq k}}^{n} \alpha_{m,m\sigma}\right)(\alpha_{k,k\sigma} + \alpha_{k,p})$$

$$= \left(\prod_{\substack{m=1 \\ m \neq k}}^{n} \alpha_{m,m\sigma}\right)(\alpha_{k,k\sigma}) + \left(\prod_{\substack{m=1 \\ m \neq k}}^{n} \alpha_{m,m\sigma}\right)(\alpha_{k,p})$$

and therefore,

$$\prod_{m=1}^{n} \beta_{m,m\sigma} = \left(\prod_{m=1}^{n} \alpha_{m,m\sigma}\right) + \left(\prod_{\substack{m=1 \\ m \neq k}}^{n} \alpha_{m,m\sigma}\right)(\alpha_{k,p})$$

Since $k\sigma = q$, $k = q\sigma^{-1}$ and we can rewrite this as

$$\prod_{m=1}^{n} \beta_{m,m\sigma} = \left(\prod_{m=1}^{n} \alpha_{m,m\sigma}\right) + \left(\prod_{\substack{m=1 \\ m \neq q\sigma^{-1}}}^{n} \alpha_{m,m\sigma}\right)(\alpha_{q\sigma^{-1},p})$$

Thus,

$$\sum_{\sigma \in S_n} (\text{sgn } \sigma) \prod_{m=1}^{n} \beta_{m,m\sigma}$$

$$= \sum_{\sigma \in S_n} (\text{sgn } \sigma) \left[\prod_{m=1}^{n} \alpha_{m,m\sigma} + \left(\prod_{\substack{m=1 \\ m \neq q\sigma}}^{n}{}_{-1} \alpha_{m,m\sigma} \right) (\alpha_{q\sigma^{-1},p}) \right]$$

$$= \sum_{\sigma \in S_n} (\text{sgn } \sigma) \prod_{m=1}^{n} \alpha_{m,m\sigma} + \sum_{\sigma \in S_n} (\text{sgn } \sigma) \left(\prod_{\substack{m=1 \\ m \neq q\sigma}}^{n}{}_{-1} \alpha_{m,m\sigma} \right) (\alpha_{q\sigma^{-1},p})$$

$$= |[\alpha_{ij}]| + \sum_{\sigma \in S_n} (\text{sgn } \sigma) \left(\prod_{\substack{m=1 \\ m \neq q\sigma}}^{n}{}_{-1} \alpha_{m,m\sigma} \right) (\alpha_{q\sigma^{-1},p})$$

Therefore,

$$|[\alpha_{ij}]E_{pq}{}^1| = |[\beta_{ij}]| = |[\alpha_{ij}]| + \sum_{\sigma \in S_n} (\text{sgn } \sigma) \left(\prod_{\substack{m=1 \\ m \neq q\sigma}}^{n}{}_{-1} \alpha_{m,m\sigma} \right) (\alpha_{q\sigma^{-1},p})$$

The proof will be finished if we can show that the sum is 0.
Notice first that

$$\left(\prod_{\substack{m=1 \\ m \neq q\sigma}}^{n}{}_{-1} \alpha_{m,m\sigma} \right) (\alpha_{q\sigma^{-1},p}) = \left(\prod_{\substack{m=1 \\ m \neq q\sigma_{-1} \\ m \neq p\sigma}}^{n}{}_{-1} \alpha_{m,m\sigma} \right) (\alpha_{q\sigma^{-1},p})(\alpha_{p\sigma^{-1},p})$$

If we let $\lambda = \sigma(p \quad q)$, then $m \neq q\sigma^{-1}$ implies $m\sigma \neq q$, which implies $m\sigma(p \quad q) \neq p$, that is, $m\lambda \neq p$ and $m \neq p\lambda^{-1}$. Similarly, if $m \neq p\sigma^{-1}$, $m \neq q\lambda^{-1}$.
Also, if $m\sigma \neq p$ and $m\sigma \neq q$, then $m\sigma = m\lambda$.
Furthermore, $q\sigma^{-1} = p\lambda^{-1}$ and $p\sigma^{-1} = q\lambda^{-1}$.
Then

$$\left(\prod_{\substack{m=1 \\ m \neq q\lambda^{-1} \\ m \neq p\lambda^{-1}}}^{n} \alpha_{m,m\lambda} \right) (\alpha_{p\lambda^{-1},p})(\alpha_{q\lambda^{-1},p}) = \left(\prod_{\substack{m=1 \\ m \neq q\sigma_{-1} \\ m \neq p\sigma}}^{n}{}_{-1} \alpha_{m,m\sigma} \right) (\alpha_{q\sigma^{-1},p})(\alpha_{p\sigma^{-1},p})$$

Thus, the products are equal for $\lambda = \sigma(p \quad q)$, and therefore in the sum the product appears with the sign of σ and the sign of λ. Since $\lambda = \sigma(p \quad q)$, λ and σ have opposite signs, and therefore the two products cancel each other in the summation process. Since each product in the summation is cancelled by its negative, the summation is zero. $\square$

The following corollary now follows from Propositions 9, 10, 11 and induction.

Corollary 12

If $[\alpha_{ij}]$ is an $n \times n$ matrix and $T_1, \ldots, T_m$ are elementary matrices of the types $(b)_p$, P_p, and E_{pq}^1, then

$$| [\alpha_{ij}] T_1 \cdots T_m | = | [\alpha_{ij}] | \, | T_1 | \, | T_2 | \cdots | T_m |$$

If we let $[\alpha_{ij}] = I$, this corollary states that the determinant of the product of elementary matrices is equal to the product of the corresponding determinants. Notice that since $E_{pq}^b = (b)_p E_{pq}^1 (b^{-1})_p$,

$$| E_{pq}^b | = | (b)_p | \, | E_{pq}^1 | \, | (b^{-1})_p |$$
$$= b \cdot 1 \cdot b^{-1}$$
$$= 1$$

that is,

$$| E_{pq}^b | = 1$$

Using this result and Corollary 12 we see that adding b times the pth column to the qth column of a matrix does not change the value of the determinant. Furthermore, if two columns of $[\alpha_{ij}]$, say the pth and qth columns, are the same, then since $[\alpha_{ij}] E_{pq}^{-1}$ has the same determinant as $[\alpha_{ij}]$ and $[\alpha_{ij}] E_{pq}^{-1}$ has a column of zeros, $| [\alpha_{ij}] | = 0$. This can be generalized somewhat by the following proposition; we leave the proof as an exercise.

Proposition 13

If $[\alpha_{ij}]$ is an $n \times n$ matrix, p and q are integers, and β and γ are field elements, not both zero, such that

$$\beta \alpha_{ip} = \gamma \alpha_{iq} \qquad \text{for each } i, 1 \le i \le n$$

or

$$\beta \alpha_{pj} = \gamma \alpha_{qj} \qquad \text{for each } j, 1 \le j \le n$$

then

$$| [\alpha_{ij}] | = 0$$

Proof: Exercise. (Hint: Use E_{pq}^b for a suitable b to obtain a matrix with a column or row of zeros.)

Proposition 14

If $[\alpha_{ij}]$ is an $n \times n$ matrix, then $[\alpha_{ij}]$ is invertible iff $| [\alpha_{ij}] | \ne 0$.

Proof: If $[\alpha_{ij}]$ is invertible, then our technique for computing inverses implies that

$$[\alpha_{ij}] T_1 \cdots T_m = [\delta_{ij}] \qquad \text{for some elementary matrices } T_1, \ldots, T_m$$

Then $|[\alpha_{ij}]T_1 \cdots T_m| = 1$, which implies

$$|[\alpha_{ij}]| \, |T_1| \cdots |T_n| = 1 \qquad \text{and therefore} \quad |[\alpha_{ij}]| \neq 0$$

If $[\alpha_{ij}]$ is an $n \times n$ matrix that is not invertible, then our technique for computing inverses fails and there are elementary matrices $T_1, \ldots, T_m$ such that $[\alpha_{ij}]T_1 \cdots T_m$ is a matrix in which at least one column is all zeros. Such a matrix has determinant zero, but since $0 = |[\alpha_{ij}]T_1 \cdots T_n| = |[\alpha_{ij}]| \, |T_1| \cdots |T_n|$ and each $|T_i| \neq 0$, it follows that $|[\alpha_{ij}]| = 0$. □

An $n \times n$ matrix $[\alpha_{ij}]$ is said to be *nonsingular* if $|[\alpha_{ij}]| \neq 0$. Proposition 14 could be restated as follows:

An $n \times n$ matrix is nonsingular iff it is invertible.

Theorem 15

If $[\alpha_{ij}]$ and $[\beta_{ij}]$ are $n \times n$ matrices, then $|[\alpha_{ij}][\beta_{ij}]| = |[\alpha_{ij}]| \, |[\beta_{ij}]|$.

Proof: First assume both $[\alpha_{ij}]$ and $[\beta_{ij}]$ are invertible. Since every invertible matrix is the product of elementary matrices,

$$[\alpha_{ij}] = (T_1 \cdots T_m) \quad \text{and} \quad [\beta_{ij}] = (S_1 \cdots S_q)$$

where each T_i and S_j is an elementary matrix.
By Corollary 12, then

$$\begin{aligned}
|[\alpha_{ij}][\beta_{ij}]| &= |(T_1 \cdots T_m)(S_1 \cdots S_q)| \\
&= |T_1| \, |T_2| \cdots |T_m| \, |S_1| \cdots |S_q| \\
&= |T_1 \cdots T_m| \, |S_1 \cdots S_q| \\
&= |[\alpha_{ij}]| \, |[\beta_{ij}]|
\end{aligned}$$

If $[\alpha_{ij}]$ is not invertible, then $[\alpha_{ij}][\beta_{ij}]$ is not invertible and

$$|[\alpha_{ij}][\beta_{ij}]| = 0 = 0 \, |[\beta_{ij}]| = |[\alpha_{ij}]| \, |[\beta_{ij}]|$$

A similar argument holds if $[\beta_{ij}]$ is not invertible. □

We can now use these results to simplify our computation of determinants. If we start with $[\alpha_{ij}]$, we can multiply on the right or left by suitable matrices to obtain a product that is triangular and, in fact, if we follow portions of our technique for computing inverses, the resulting matrix will have only 1's or 0's on the diagonal. That is, $|[\alpha_{ij}]T_1 \cdots T_m| = |[\beta_{ij}]|$ and $[\beta_{ij}]$ is triangular.
Then

$$|[\alpha_{ij}]| \, |T_1| \cdots |T_m| = \beta_{11}\beta_{22} \cdots \beta_{nn}$$

and

$$|[\alpha_{ij}]| = (\beta_{11}\beta_{22} \cdots \beta_{nn})(|T_1| \cdots |T_m|)^{-1}$$

We can either use enough elementary matrices to assure that each $\beta_{ii} = 0$ or $\beta_{ii} = 1$ or else be content as long as $[\beta_{ij}]$ is triangular. Since we know $|T_i|$ for each T_i, $|[\alpha_{ij}]|$ is then easily computed.

Examples

$$\text{(i)} \quad \begin{vmatrix} 1 & 2 & 3 \\ 2 & 3 & 5 \\ 1 & 4 & 6 \end{vmatrix} = \begin{vmatrix} 1 & 0 & 3 \\ 2 & -1 & 5 \\ 1 & 2 & 6 \end{vmatrix}$$

Adding (-2) times the first column to the second column does not change the determinant.

$$= \begin{vmatrix} 1 & 0 & 0 \\ 2 & -1 & -1 \\ 1 & 2 & 3 \end{vmatrix}$$

Adding (-3) times the first column to the third column does not change the determinant.

$$= \begin{vmatrix} 1 & 0 & 0 \\ 2 & -1 & 0 \\ 1 & 2 & 1 \end{vmatrix}$$

Adding (-1) times the second column to the third column does not change the determinant.

$$= -1$$

The determinant of a triangular matrix is the product of its diagonal elements.

$$\text{(ii)} \quad \begin{vmatrix} 1 & 2 & 2 & 5 \\ 2 & 7 & 4 & -3 \\ 5 & 8 & 10 & 2 \\ 3 & 4 & 6 & 9 \end{vmatrix} = \begin{vmatrix} 1 & 2 & 0 & 5 \\ 2 & 7 & 0 & -3 \\ 5 & 8 & 0 & 2 \\ 3 & 4 & 0 & 9 \end{vmatrix}$$

Adding (-2) times the first column to the third does not change the determinant.

$$= 0$$

because the matrix has a column of zeros.

$$\text{(iii)} \quad \begin{vmatrix} 1 & 0 & 0 \\ 2 & 0 & 4 \\ 3 & 2 & 3 \end{vmatrix} = (-1)\begin{vmatrix} 1 & 0 & 0 \\ 2 & 4 & 0 \\ 3 & 3 & 2 \end{vmatrix}$$

Because interchanging the second and third columns is equivalent to multiplying by $P_{(2\ 3)}$ and $|P_{(2\ 3)}| = (-1)$.

$$= (-1)8 = -8$$

because the matrix is triangular.

(iv) $\begin{vmatrix} 1 & 3 & 5 & 4 \\ 2 & -2 & 3 & 2 \\ 3 & 1 & 7 & -5 \\ 4 & 5 & 1 & 6 \end{vmatrix}$

$= \begin{vmatrix} 1 & 0 & 0 & 0 \\ 2 & -8 & -7 & -6 \\ 3 & -8 & -2 & -17 \\ 4 & -7 & -11 & -10 \end{vmatrix}$

Adding multiples of the first column to others does not change the value of the determinant.

$= (-8)\begin{vmatrix} 1 & 0 & 0 & 0 \\ 2 & 1 & -7 & -6 \\ 3 & 1 & -2 & -17 \\ 4 & \frac{7}{8} & -11 & -10 \end{vmatrix}$

Multiplying by $(-\frac{1}{8})_2$ multiplies the determinant by $(-\frac{1}{8})$.

$= (-8)\begin{vmatrix} 1 & 0 & 0 & 0 \\ 2 & 1 & 0 & 0 \\ 3 & 1 & 5 & -11 \\ 4 & \frac{7}{8} & -\frac{39}{8} & -\frac{19}{4} \end{vmatrix}$

Adding multiples of the second column to the third and fourth does not change the value of the determinant.

$= (-8)(5)\begin{vmatrix} 1 & 0 & 0 & 0 \\ 2 & 1 & 0 & 0 \\ 3 & 1 & 1 & -11 \\ 4 & \frac{7}{8} & -\frac{39}{40} & -\frac{19}{4} \end{vmatrix}$

Multiplying the third column by $\frac{1}{5}$ multiplies the determinant by $\frac{1}{5}$.

$= (-8)(5)\begin{vmatrix} 1 & 0 & 0 & 0 \\ 2 & 1 & 0 & 0 \\ 3 & 1 & 1 & 0 \\ 4 & \frac{7}{8} & -\frac{39}{40} & \frac{239}{40} \end{vmatrix}$

Adding -11 times the third column to the fourth.

$= (-40)(239/40) = (-239)$

All of our examples used techniques involving columns, and it is left to the reader to describe the corresponding row techniques. After all, multiplying on the left by elementary matrices just operates with the rows instead of the columns, and thus the processes are essentially the same. In fact, one can use both row and column operations in computing determinants. Care should be exercised, however, not to get into this habit when computing inverses, because when computing inverses either one solves $X[\alpha_{ij}] = I$ or $[\alpha_{ij}]X = I$, and all multiplication should be done only on one side.

EXERCISE SET C

1. Allowing 0.001 second for each multiplication or addition, approximately how long would it take to use the definition and compute the determinant of a 10×10 matrix?
2. Prove Corollary 3.
3. Prove Proposition 11 directly for the cases $n = 3$ and $n = 4$.
4. Prove Proposition 13.
5. Compute the determinants of the following matrices.

(a) $\begin{bmatrix} 1 & 9 & 1 \\ 4 & 3 & 1 \\ 7 & 2 & 7 \end{bmatrix}$ (b) $\begin{bmatrix} 2 & 5 & 7 \\ 3 & 2 & 5 \\ 4 & 1 & 5 \end{bmatrix}$ (c) $\begin{bmatrix} 1 & 4 & 3 & 2 \\ 2 & 3 & 1 & 7 \\ -7 & 4 & -1 & 3 \\ -8 & 2 & 1 & 3 \end{bmatrix}$

(d) $\begin{bmatrix} 4 & 3 & 1 & 0 \\ 3 & 1 & 0 & 1 \\ 1 & 2 & 1 & 1 \\ 7 & -2 & 0 & 1 \end{bmatrix}$ (e) $\begin{bmatrix} 1 & 3 & 1 & 0 \\ 1 & 4 & 1 & 1 \\ 1 & 1 & 2 & 1 \\ 1 & 2 & 3 & 4 \end{bmatrix}$ (f) $\begin{bmatrix} 1 & 0 & 1 & 2 & 0 \\ 2 & 0 & 3 & 2 & 0 \\ 4 & 1 & 4 & 5 & 0 \\ 7 & 2 & 5 & 3 & 4 \\ 5 & 3 & 1 & 1 & 2 \end{bmatrix}$

6. Solve the following for x.

(a) $\begin{vmatrix} 2-x & 4 \\ 3 & 3-x \end{vmatrix} = 0$ (b) $\begin{vmatrix} 1-x & 0 & 0 \\ 2 & 2-x & 0 \\ 3 & 4 & 4-x \end{vmatrix} = 0$

(c) $\begin{vmatrix} 1-x & 1 & 1 \\ 10 & 3-x & 0 \\ 0 & 1 & 4-x \end{vmatrix} = 0$

7. Prove: If $[\alpha_{ij}]$ is invertible, then $|[\alpha_{ij}]|^{-1} = |[\alpha_{ij}]^{-1}|$.
8. Prove: If $[\alpha_{ij}]$ is invertible, then $|[\alpha_{ij}][\beta_{ij}][\alpha_{ij}]^{-1}| = |[\beta_{ij}]|$.
9. Prove: If $[\alpha_{ij}]$ and $[\beta_{ij}]$ are invertible, then

$$|[\alpha_{ij}][\beta_{ij}][\alpha_{ij}]^{-1}[\beta_{ij}]^{-1}| = 1$$

10. Prove: If $[\alpha_{ij}]$ is an $n \times n$ matrix, the equation $(x_1, \ldots, x_n)[\alpha_{ij}] = (c_1, \ldots, c_n)$ has a unique solution if $|[\alpha_{ij}]| \neq 0$.

**11. Prove: If $[\alpha_{ij}]$ is an $n \times n$ matrix and $|[\alpha_{ij}]| = 0$, the equation $(x_1, \ldots, x_n)[\alpha_{ij}] = (c_1, \ldots, c_n)$ either has no solution or at least as many as there are elements in F. What do this exercise and Exercise 10 say about systems of equations?

**12. Prove: If $[\alpha_{ij}]$ is an $n \times n$ matrix, then there exist elementary matrices $T_1, \ldots, T_n$ such that $[\beta_{ij}] = [\alpha_{ij}]T_1 \cdots T_n$ is a matrix satisfying the following conditions.

    (i) $[\beta_{ij}]$ is lower triangular.

    (ii) If a column in $[\beta_{ij}]$ is not all zeros, then its first nonzero entry is 1.

    (iii) If $\beta_{pq} = 1$ is the first nonzero entry in the qth column, then $\beta_{pj} = 0$ for all $j \neq q$.

    (iv) The rank of $[\alpha_{ij}]$ is the number of nonzero columns of $[\beta_{ij}]$.

    (v) The nullity of $[\alpha_{ij}]$ is the number of zero columns of $[\beta_{ij}]$.

13. Find a matrix $[\beta_{ij}]$ as described in Exercise 12, for each of the following matrices

$$\text{(a)} \begin{bmatrix} 1 & 3 \\ 2 & 6 \end{bmatrix} \quad \text{(b)} \begin{bmatrix} 1 & 5 & 6 \\ 2 & 7 & 9 \\ 3 & 8 & 11 \end{bmatrix} \quad \text{(c)} \begin{bmatrix} 1 & 2 & 4 \\ 2 & 3 & 7 \\ 4 & 6 & 14 \end{bmatrix}$$

14. Give a basis for the kernel of each of the matrices in Exercise 13.

**15. If $[\alpha_{ij}]$ is an $n \times n$ matrix, define the A_{pq}th *minor* of $[\alpha_{ij}]$ to be the $(n-1) \times (n-1)$ matrix $[\beta_{ij}]$, where

$$\beta_{ij} = \alpha_{ij} \qquad \text{if } i < p \quad \text{and} \quad j < q$$

$$\beta_{ij} = \alpha_{i,j+1} \qquad \text{if } i < p \quad \text{and} \quad j \geq q$$

$$\beta_{ij} = \alpha_{i+1,j} \qquad \text{if } i \geq p \quad \text{and} \quad j < q$$

$$\beta_{ij} = \alpha_{i+1,j+1} \qquad \text{if } i \geq p \quad \text{and} \quad j \geq q$$

Prove that

$$|[\alpha_{ij}]| = \sum_{q=1}^{n} (-1)^{p+q} \alpha_{pq} |A_{pq}| \qquad \text{for each } p, 1 \leq p \leq n$$

$$= \sum_{p=1}^{n} (-1)^{p+q} \alpha_{pq} |A_{pq}| \qquad \text{for each } q, 1 \leq q \leq n$$

§4. Equivalence and Similarity

If V and W are finite-dimensional F-spaces of dimension n and m, respectively, and T is a linear transformation from V into W, the matrix $[\alpha_{ij}]$ is said to *represent T with respect to the basis* $\{v_1, \ldots, v_n\}$ *for V and the basis* $\{y_1, \ldots, y_m\}$ *for W* if $[\alpha_{ij}]$ is the matrix for $\phi^{-1}T\lambda$, where ϕ is the isomorphism from V to F^n given by $v_i\phi = e_i$ and λ is the isomorphism from W to F^m given by $y_j\lambda = e_j$.

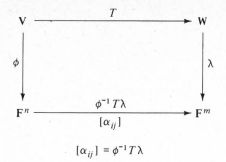

$$[\alpha_{ij}] = \phi^{-1}T\lambda$$

The matrix $[\beta_{ij}]$ represents T with respect to some bases for V and W if there exists an isomorphism μ from V onto F^n and an isomorphism θ from W onto F^m such that $[\beta_{ij}] = \mu^{-1}T\theta$. For if we let $x_i = e_i\mu^{-1}$ and $z_j = e_j\theta^{-1}$, $\{x_1, \ldots x_n\}$ and $\{z_1, \ldots, z_m\}$ are bases for V and W respectively, and $[\beta_{ij}]$ represents T with respect to these bases. If $[\alpha_{ij}]$ and $[\beta_{ij}]$ are two matrices representing T with respect to some bases, then there are isomorphisms ϕ and μ from V onto F^n and λ and θ from W onto F^m such that $[\alpha_{ij}] = \phi^{-1}T\lambda$ and $[\beta_{ij}] = \mu^{-1}T\theta$. Then $\phi[\alpha_{ij}]\lambda^{-1} = T$, and therefore

$$[\beta_{ij}] = \mu^{-1}(\phi[\alpha_{ij}]\lambda^{-1})\theta$$

or

$$[\beta_{ij}] = (\mu^{-1}\phi)[\alpha_{ij}](\lambda^{-1}\theta)$$

Since $\mu^{-1}\phi$ is the composition of two isomorphisms, $\mu^{-1}\phi$ is an automorphism of F^n and, similarly, $\lambda^{-1}\theta$ is an automorphism of F^m. Thus, if $[\alpha_{ij}]$ and $[\beta_{ij}]$ are two matrices representing T with respect to some bases for V and W, then $[\beta_{ij}]$ is the product of an automorphism of F^n, the matrix $[\alpha_{ij}]$, and an automorphism of F^m. Thus, $[\beta_{ij}]$ is the product of an invertible $n \times n$ matrix, the matrix $[\alpha_{ij}]$, and an invertible $m \times m$ matrix; that is,

$$[\beta_{ij}] = [p_{ij}][\alpha_{ij}][q_{ij}] \qquad \text{where } [p_{ij}] \text{ and } [q_{ij}] \text{ are invertible}$$

Conversely, suppose $[\beta_{ij}]$ and $[\alpha_{ij}]$ satisfy the above conditions; that is, suppose $[\beta_{ij}] = [p_{ij}][\alpha_{ij}][q_{ij}]$, where $[p_{ij}]$ is an invertible $n \times n$ matrix and $[q_{ij}]$ is an invertible $m \times m$ matrix.

Let $\{v_1, \ldots, v_n\}$ and $\{y_1, \ldots, y_m\}$ be bases for V and W, respectively, and let ϕ and λ be the isomorphisms from V to F^n and W to F^m defined by $v_i\phi = e_i$ and $y_j\lambda = e_j$. Then $T = \phi[\beta_{ij}]\lambda^{-1}$ is a linear transformation from V into W and $[\beta_{ij}]$ represents T, since $\phi^{-1}T\lambda = [\beta_{ij}]$. Furthermore, $[\alpha_{ij}]$ also represents T because:

(i) Since $[\alpha_{ij}] = [p_{ij}]^{-1}[\beta_{ij}][q_{ij}]^{-1}$,

$$[\alpha_{ij}] = [p_{ij}]^{-1}\phi^{-1}T\lambda[q_{ij}]^{-1}$$

and therefore

$$[\alpha_{ij}] = (\phi[p_{ij}])^{-1}T(\lambda[q_{ij}]^{-1})$$

(ii) Since $[p_{ij}]$ is an automorphism of F^n, $\phi[p_{ij}]$ is an isomorphism from V onto F^n.

(iii) Since $[q_{ij}]$ and $[q_{ij}]^{-1}$ are automorphisms of F^m, $\lambda[q_{ij}]^{-1}$ is an isomorphism from W onto F^m.

The above discussion establishes the following assertion.

Proposition 16

If $[\alpha_{ij}]$ and $[\beta_{ij}]$ are $n \times m$ matrices with entries from the field F, then $[\alpha_{ij}]$ and $[\beta_{ij}]$ represent a common linear transformation iff $[\alpha_{ij}] = [p_{ij}][\beta_{ij}][q_{ij}]$ for some invertible $n \times n$ matrix $[p_{ij}]$ and some invertible $m \times m$ matrix $[q_{ij}]$.

In a similar manner we can establish the following related assertion.

Proposition 17

If V and W are n- and m-dimensional F-spaces, respectively, and T and Q are linear transformations from V into W, then T and Q are represented by a common $n \times m$ matrix iff $T = PQS$ for some automorphism P of V and some automorphism S of W.

Proof: Suppose T and W are linear transformations from V into W, both of which are represented by the $n \times m$ matrix $[\alpha_{ij}]$. Then there are isomorphisms ϕ and μ from V onto F^n and λ and θ from W onto F^m such that

$$[\alpha_{ij}] = \phi^{-1}T\lambda \quad \text{and} \quad [\alpha_{ij}] = \mu^{-1}Q\theta$$

Hence, $\phi^{-1}T\lambda = \mu^{-1}Q\theta$, which implies

$$T = (\phi\mu^{-1})Q(\theta\lambda^{-1})$$

where $\phi\mu^{-1}$ is an automorphism of V, and $\theta\lambda^{-1}$ is an automorphism of W.

Now suppose T and Q are linear transformations from V to W such that $T = PQS$, where P is an automorphism of V and S is an automorphism of W. If $[\alpha_{ij}]$ represents T, then since $[\alpha_{ij}] = \phi^{-1}T\lambda$ for some isomorphisms ϕ from V onto F^n and λ from W onto F^m,

$$[\alpha_{ij}] = \phi^{-1}T\lambda$$
$$= \phi^{-1}(PQS)\lambda$$
$$= (\phi^{-1}P)Q(S\lambda)$$
$$= (P^{-1}\phi)^{-1}Q(S\lambda)$$

Since P is an automorphism of V, P^{-1} is also and $P^{-1}\phi$ is an isomorphism from V onto F^n. Similarly, $S\lambda$ is an isomorphism from W onto F^m, and therefore $[\alpha_{ij}]$ represents Q. $\square$

If V and W are n- and m-dimensional F-spaces, two linear transformations T and Q from V into W are said to be *equivalent* if $T = PQS$ for some automorphisms P of V and S of W. Two $n \times m$ matrices $[\alpha_{ij}]$ and $[\beta_{ij}]$ are said to be *equivalent* iff $[\alpha_{ij}] = [p_{ij}][\beta_{ij}][q_{ij}]$, where $[p_{ij}]$ and $[q_{ij}]$ are invertible $n \times n$ and $m \times m$ matrices.

Proposition 18

If V is an n-dimensional F-space and W is an m-dimensional F-space, the relation "equivalent" is an equivalence relation on $\mathcal{L}(V, W)$.

Proof: Exercise.

By Propositions 16 and 17 we see that two linear transformations are equivalent iff they are represented by a common matrix and two matrices are equivalent iff they represent a common linear transformation.

If T and Q are linear transformations from V into W, where V and W are n- and m-dimensional F-spaces, how can we tell if they are equivalent? How can we tell if two matrices are equivalent? Suppose $[\alpha_{ij}]$ represents T and $[\beta_{ij}]$ represents Q; if we multiply each of them on the right by suitable elementary matrices, the resulting matrices are lower triangular. Furthermore, by multiplying on the left by elementary matrices, the resulting matrices become upper triangular and the diagonal entries are all 1's and 0's. Thus, using suitable invertible $n \times n$ matrices P_1 and P_2 and suitable $m \times m$ matrices S_1 and S_2 we obtain

$$P_1[\alpha_{ij}]S_1 \quad \text{and} \quad P_2[\beta_{ij}]S_2$$

which are matrices of the form

$$\begin{bmatrix} 1 & 0 & \cdots & & 0 \\ 0 & 1 & & & \\ \vdots & & 1 & & \vdots \\ \vdots & & & \ddots & \\ 0 & \cdots & & & \end{bmatrix}$$

If $P_1[\alpha_{ij}]S_1 = P_2[\beta_{ij}]S_2$, then

$$[\alpha_{ij}] = P_1^{-1}P_2[\beta_{ij}]S_2S_1^{-1}$$

and $[\alpha_{ij}]$ represents both T and Q.

If $P_1[\alpha_{ij}]S_1 \neq P_2[\beta_{ij}]S_2$, the number of nonzero rows of one is more than the number of nonzero rows of the other, and therefore the kernel of one has dimension greater than the kernel of the other. Since dim ker $P_1[\alpha_{ij}]S_1 = $ dim ker $[\alpha_{ij}] = $ dim ker T and dim ker $P_2[\beta_{ij}]S_2 = $ dim ker $[\beta_{ij}] = $ dim ker Q, T and Q are represented by a common matrix iff dim ker $T = $ dim ker Q (Chapter 7, Exercise 10, Set E).

By using matrix techniques we have established the following assertion.

Proposition 19

If V and W are n- and m-dimensional F-spaces, respectively, two linear transformations T and Q from V into W are equivalent iff rank T = rank Q iff nullity T = nullity Q.

A more interesting but more complicated situation occurs when T is a linear transformation from V into V and V is an n-dimensional F-space. We say that $[\alpha_{ij}]$ represents T if there is an isomorphism ϕ from V onto F^n such that $[\alpha_{ij}] = \phi^{-1}T\phi$ or equivalently, $\phi[\alpha_{ij}]\phi^{-1} = T$.

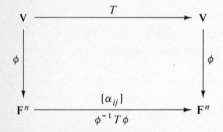

As before, if $[\alpha_{ij}]$ and $[\beta_{ij}]$ both represent T, then for isomorphisms ϕ and λ, $[\alpha_{ij}] = \phi^{-1}T\phi$ and $[\beta_{ij}] = \lambda^{-1}T\lambda$. Then

$$\phi[\alpha_{ij}]\phi^{-1} = T = \lambda[\beta_{ij}]\lambda^{-1}$$
$$[\alpha_{ij}] = \phi^{-1}(\lambda[\beta_{ij}]\lambda^{-1})\phi$$
$$= (\phi^{-1}\lambda)[\beta_{ij}](\lambda^{-1}\phi)$$

and

$$[\alpha_{ij}] = (\phi^{-1}\lambda)[\beta_{ij}](\phi^{-1}\lambda)^{-1}$$

But $\phi^{-1}\lambda$ is an automorphism of F^n and therefore can be represented by an $n \times n$ invertible matrix $[p_{ij}]$; therefore,

$$[\alpha_{ij}] = [p_{ij}][\beta_{ij}][p_{ij}]^{-1}$$

Conversely, if $[\alpha_{ij}] = [p_{ij}][\beta_{ij}][p_{ij}]^{-1}$, where all are $n \times n$ matrices and $[p_{ij}]$ is invertible, then $[\alpha_{ij}]$ and $[\beta_{ij}]$ represent a common linear transformation. We can argue in a manner similar to the proof given for equivalence, or by letting $V = F^n$ and letting $[p_{ij}]$ be the desired isomorphism. In this case the common transformation is just given by the matrix $[\alpha_{ij}]$.

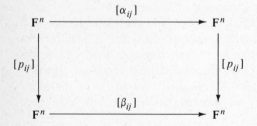

The preceding discussion contains the proof of the following assertion.

Proposition 20

If V is an n-dimensional F-space, two $n \times n$ matrices $[\alpha_{ij}]$ and $[\beta_{ij}]$ represent a common linear transformation from V into V iff $[\alpha_{ij}] = [p_{ij}][\beta_{ij}][p_{ij}]^{-1}$ for some invertible $n \times n$ matrix $[p_{ij}]$.

We also can prove the following assertion that is analogous to Proposition 17.

Proposition 21

If V is an n-dimensional F-space and T and Q are linear transformations of V, then T and Q are represented by a common matrix iff $T = P^{-1}QP$ for some automorphism P of V.

Proof: Suppose T and Q are linear transformations of the n-dimensional F-space V such that $T = P^{-1}QP$ for some automorphism P of V. If $[\alpha_{ij}]$ represents T, then $[\alpha_{ij}]$ also represents Q for if

$$[\alpha_{ij}] = \phi^{-1}T\phi$$

then

$$[\alpha_{ij}] = \phi^{-1}P^{-1}QP\phi$$

that is,

$$[\alpha_{ij}] = (P\phi)^{-1}Q(P\phi)$$

and $P\phi$ is an isomorphism from V to F^n.

Conversely, if $[\alpha_{ij}]$ represents T and Q, then

$$[\alpha_{ij}] = \phi^{-1}T\phi \quad \text{and} \quad [\alpha_{ij}] = \lambda^{-1}Q\lambda$$

for some isomorphisms ϕ and λ from V onto F^n. Then

$$\phi^{-1}T\phi = \lambda^{-1}Q\lambda$$
$$T = \phi\lambda^{-1}Q\lambda\phi^{-1}$$

and

$$T = (\lambda\phi^{-1})^{-1}Q(\lambda\phi^{-1})$$

where $\lambda\phi^{-1}$ is an automorphism of V. $\square$

If V is an n-dimensional F-space and T and Q are linear transformations from V into V, we say that T is *similar* to Q if $T = P^{-1}QP$ for some automorphism P of V. By Proposition 20, two matrices are similar if they represent a common linear transformation from V into V. By Proposition 21, T is similar to Q iff T and Q are represented by a common $n \times n$ matrix.

Proposition 22

If V is a finite-dimensional F-space, the relation of similarity is an equivalence relation on $\mathcal{L}(V)$.

Proof: Exercise.

Proposition 23

Similar linear transformations are equivalent.

Proof: Immediate from the definitions.

Proposition 24

If $[\alpha_{ij}]$ and $[\beta_{ij}]$ are similar, then $|[\alpha_{ij}]| = |[\beta_{ij}]|$.

Proof: Exercise. (See Exercise 8, Set C.)

If T is a linear transformation from V into V, a nonzero vector u is said to be a *characteristic vector* (or *eigenvector*) of T if $uT = cu$ for some scalar c in F. If c is a scalar in F such that $cu = uT$ for some nonzero vector u, c is said to be a *characteristic value* (or *eigenvalue*) of T.

Examples. $(1, 0)$ and $(0, 1)$ are characteristic vectors of

$$\begin{bmatrix} 2 & 0 \\ 0 & 3 \end{bmatrix}$$

and 2 and 3 are characteristic values because

$$(1, 0)\begin{bmatrix} 2 & 0 \\ 0 & 3 \end{bmatrix} = (2, 0) = 2(1, 0)$$

and

$$(0, 1)\begin{bmatrix} 2 & 0 \\ 0 & 3 \end{bmatrix} = (0, 3) = 3(0, 1)$$

The vectors $(-1, 1)$ and $(2, -1)$ are characteristic vectors with corresponding characteristic values 2 and 3 for the matrix

$$\begin{bmatrix} 4 & -1 \\ 2 & 1 \end{bmatrix}$$

since

$$(-1, 1)\begin{bmatrix} 4 & -1 \\ 2 & 1 \end{bmatrix} = (-2, 2) = 2(-1, 1)$$

and

$$(2, -1)\begin{bmatrix} 4 & -1 \\ 2 & 1 \end{bmatrix} = (6, -3) = 3(2, -1)$$

The linear transformation

$$\begin{bmatrix} 2 & 0 \\ 0 & 3 \end{bmatrix}$$

distorts the plane along the X and Y axes by doubling X coordinates and tripling Y coordinates. The transformation

$$\begin{bmatrix} 4 & -1 \\ 2 & 1 \end{bmatrix}$$

distorts the plane by doubling distances along the line $x = -y$ and tripling distances along the line $x = -2y$. In some geometric sense they are alike in that the distortions are somewhat alike. The transformations are, in fact, similar because

$$\begin{bmatrix} -1 & 1 \\ 2 & -1 \end{bmatrix} = \begin{bmatrix} 1 & 1 \\ 2 & 1 \end{bmatrix}^{-1}$$

and

$$\begin{bmatrix} -1 & 1 \\ 2 & -1 \end{bmatrix}\begin{bmatrix} 4 & -1 \\ 2 & 1 \end{bmatrix}\begin{bmatrix} 1 & 1 \\ 2 & 1 \end{bmatrix} = \begin{bmatrix} 2 & 0 \\ 0 & 3 \end{bmatrix}$$

The intuitive notion of distorting in the same manner is exemplified well if T is a linear transformation of an n-dimensional F-space V that has n characteristic values $c_1, \ldots, c_n$ and there are n linearly independent vectors $u_1, \ldots, u_n$ in V such that

$$c_1 u_1 = u_1 T, \quad c_2 u_2 = u_2 T, \ldots, c_n u_n = u_n T$$

Since $\{u_1, \ldots, u_n\}$ is a basis for V, if we let ϕ be the isomorphism from V to F^n defined by $u_i \phi = e_i$, the matrix representing T will be

$$\begin{bmatrix} c_1 & 0 & \cdots & 0 \\ 0 & c_2 & & \vdots \\ \vdots & & \ddots & 0 \\ 0 & & \cdots & c_n \end{bmatrix}$$

This is the best possible situation we could expect, and we now see that this is in fact the case if $c_1, \ldots, c_n$ are distinct nonzero characteristic values.

Proposition 25

If V is an n-dimensional F-space, $T \in \mathcal{L}(V)$, $c_1, \ldots, c_m$ are m distinct characteristic values of T, and $u_1, \ldots, u_m$ are characteristic vectors corresponding to each c_i, then $\{u_1, \ldots, u_m\}$ is linearly independent.

Proof: By induction on m.

If $m = 1$, $u_1 \neq 0$ and $\{u_1\}$ is linearly independent. Now assume that the assertion holds for any collection of fewer than m values and vectors.

Suppose $\{u_1, \ldots, u_m\}$ is not linearly independent, then some u_i can be expressed as a linear combination of the others. Without loss of generality let it be u_m, that is,

$$u_m = \alpha_1 u_1 + \cdots + \alpha_{m-1} u_{m-1}$$

Then

$$u_m T = \alpha_1 u_1 T + \cdots + \alpha_{m-1} u_{m-1} T$$

and therefore

$$c_m u_m = \alpha_1 c_1 u_1 + \cdots + \alpha_{m-1} c_{m-1} u_{m-1}$$

then

$$c_m(\alpha_1 u_1 + \cdots + \alpha_{m-1} u_{m-1}) = \alpha_1 c_1 u_1 + \cdots + \alpha_{m-1} c_{m-1} u_{m-1}$$

and

$$(\alpha_1 c_1 - c_m \alpha_1)u_1 + (\alpha_2 c_2 - c_m \alpha_2)u_2$$
$$+ \cdots + (\alpha_{m-1} c_{m-1} - c_m \alpha_{m-1})u_{m-1} = 0$$

that is,

$$\alpha_1(c_1 - c_m)u_1 + \alpha_2(c_2 - c_m)u_2 + \cdots + \alpha_{m-1}(c_{m-1} - c_m)u_{m-1} = 0$$

By the induction hypothesis $\{u_1, \ldots, u_{m-1}\}$ is linearly independent and each coefficient $\alpha_i(c_i - c_m) = 0$. But $c_i \neq c_m$, therefore each $\alpha_i = 0$, and $u_m = 0$, which is impossible. Therefore, $\{u_1, \ldots, u_m\}$ is linearly independent. $\square$

The following assertion now follows immediately.

Proposition 26

If V is an n-dimensional F-space, T is a linear transformation of V, and $c_1, \ldots, c_n$ are distinct characteristic values of T, then T can be represented by a diagonal matrix with diagonal elements $c_1, \ldots, c_n$.

Proposition 27

If V is an n-dimensional F-space, T and Q are linear transformations on V, and $c_1, \ldots, c_n$ are n distinct elements of F that are characteristic values of both T and Q, then T and Q are similar.

Proof: Both T and Q are represented by the diagonal matrix

$$\begin{bmatrix} c_1 & 0 & \cdots & 0 \\ 0 & c_2 & & \vdots \\ \vdots & & \ddots & \vdots \\ 0 & & \cdots & c_n \end{bmatrix}$$

By Proposition 21, T is similar to Q. □

Proposition 28

If T is a linear transformation on V, and c is a characteristic value of T, then c is a characteristic value of any transformation similar to T.

Proof: Let P be an automorphism of V, let u be a characteristic vector of T corresponding to c, and let $v = uP$.

Then

$$\begin{aligned} vP^{-1}TP &= (uP)P^{-1}TP \\ &= uTP \\ &= (cu)P \\ &= c(uP) \\ &= cv \end{aligned}$$

Thus, c is a characteristic value for $P^{-1}TP$. Since P was an arbitrary automorphism of V, c is a characteristic value of any transformation similar to T.

Proposition 29

If T is a linear transformation on V, and u is a characteristic vector of T, then c is a characteristic value corresponding to u iff $u \in ker\ (T - cI)$, where I denotes the identity transformation.

Proof: Exercise.

Proposition 30

If $[\alpha_{ij}]$ is an $n \times n$ matrix, then c is a characteristic value of $[\alpha_{ij}]$ iff $|\ [\alpha_{ij}] - cI\ | = 0$.

Proof: Exercise.

By Proposition 30 all characteristic values of a matrix $[\alpha_{ij}]$ can be determined by solving the equation $|\ [\alpha_{ij}] - xI\ | = 0$, that is, by setting $|\ [\alpha_{ij}] - x[\delta_{ij}]\ | = 0$ and solving for x. Notice that $[\alpha_{ij}] - x[\delta_{ij}]$ is just the matrix

$$\begin{bmatrix} \alpha_{11} - x & \alpha_{12} & \cdots & \alpha_{1n} \\ \alpha_{21} & \alpha_{22} - x & \cdots & \alpha_{2n} \\ \vdots & & & \\ \alpha_{n1} & \cdots & & \alpha_{nn} - x \end{bmatrix}$$

Example. Find the characteristic roots of the matrix

$$\begin{bmatrix} 2 & 1 \\ 3 & 4 \end{bmatrix}$$

Setting

$$\left| \begin{bmatrix} 2 & 1 \\ 3 & 4 \end{bmatrix} - x \begin{bmatrix} 1 & 0 \\ 0 & 1 \end{bmatrix} \right| = 0$$

we have

$$\begin{vmatrix} 2 - x & 1 \\ 3 & 4 - x \end{vmatrix} = 0$$

$$(2 - x)(4 - x) - 3 = 0$$
$$x^2 - 6x + 8 - 3 = 0$$
$$x^2 - 6x + 5 = 0$$
$$(x - 5)(x - 1) = 0$$

and the characteristic values are 1 and 5.

Characteristic vectors corresponding to 1 are in

$$\ker \begin{bmatrix} 2 - 1 & 1 \\ 3 & 4 - 1 \end{bmatrix}$$

that is, in

$$\ker \begin{bmatrix} 1 & 1 \\ 3 & 3 \end{bmatrix}$$

The vector $(3, -1)$ clearly is in the kernel of this matrix and is, therefore, a characteristic vector corresponding to 1. Characteristic vectors corresponding to 5 are in

$$\ker \begin{bmatrix} 2-5 & 1 \\ 3 & 4-5 \end{bmatrix} = \ker \begin{bmatrix} -3 & 1 \\ 3 & -1 \end{bmatrix}$$

The vector $(1, 1)$ is in the kernel of this matrix and hence is a characteristic vector corresponding to 5.

If $[\alpha_{ij}]$ is an $n \times n$ matrix, the equation $|[\alpha_{ij}] - xI| = 0$ is called the *characteristic equation of the matrix* $[\alpha_{ij}]$. Remember that $|[\alpha_{ij}] - xI| = 0$ is, in fact, a polynomial equation in one variable, x. The polynomial $|[\alpha_{ij}] - xI|$ is called the *characteristic polynomial* of $[\alpha_{ij}]$. If $[p_{ij}]$ is invertible, then

$$
\begin{aligned}
|[\alpha_{ij}] - xI| \cdot 1 &= |[\alpha_{ij}] - xI| \, |[p_{ij}]| \, |[p_{ij}]|^{-1} \\
&= |[p_{ij}]| \, |[\alpha_{ij}] - xI| \, |[p_{ij}]^{-1}| \\
&= |[p_{ij}]([\alpha_{ij}] - xI)[p_{ij}]^{-1}| \\
&= |[p_{ij}][\alpha_{ij}][p_{ij}]^{-1} - x[p_{ij}]I[p_{ij}]^{-1}| \\
&= |[p_{ij}][\alpha_{ij}][p_{ij}]^{-1} - xI|
\end{aligned}
$$

Thus, the polynomials $|[\alpha_{ij}] - xI|$ and $|[p_{ij}][\alpha_{ij}][p_{ij}]^{-1} - xI|$ are exactly the same, and therefore, if $T \in \mathcal{L}(V)$, all matrices representing T will have the same characteristic polynomial and the same characteristic equation. We can, therefore, determine all characteristic values of T if we determine all roots of $|[\alpha_{ij}] - xI| = 0$ for some matrix $[\alpha_{ij}]$ representing T. The polynomial $|[\alpha_{ij}] - xI|$ is called the *characteristic polynomial* of T.

The general problem of determining whether or not two matrices or linear transformations are similar is not as easy as the corresponding problem for equivalence, but there are several tests one can try that may give negative answers:

(i) Different ranks imply not similar.
(ii) Different determinants imply not similar.
(iii) Different characteristic values imply not similar.

The problem of determining a complete set of invariants for similarity is quite complicated, and we shall not pursue the topic any further. Remember that we use matrices to describe linear transformations and the simpler form we can get for a matrix representing a particular transformation, the more information we have at our fingertips. One problem in similarity theory is to find a relatively simple matrix that represents a given transformation. We would like to have the most simple form possible and, under certain hypotheses on the field F, certain *canonical* forms can be described. We suggest that the interested student consult one of the reference books in linear algebra.

EXERCISE SET D

1. Which of the following matrices are equivalent?

(a) $\begin{bmatrix} 1 & 5 & 2 & 1 \\ 2 & 6 & 4 & 2 \\ 3 & 7 & 7 & -1 \\ 4 & 8 & 3 & -1 \end{bmatrix}$ (b) $\begin{bmatrix} 1 & 1 & 3 & 2 \\ 4 & 1 & 6 & 1 \\ 3 & 4 & 11 & 9 \\ 9 & 7 & 23 & 4 \end{bmatrix}$

(c) $\begin{bmatrix} 1 & 2 & 4 & 1 \\ 2 & 3 & 7 & 2 \\ 3 & 5 & 11 & 4 \\ 4 & 6 & 14 & 3 \end{bmatrix}$ (d) $\begin{bmatrix} 1 & 2 & 1 & 3 \\ 5 & 2 & 2 & 7 \\ 7 & 1 & 1 & 8 \\ 9 & 2 & 2 & 11 \end{bmatrix}$

2. Give an example of two equivalent matrices that are not similar.
3. Prove Proposition 18.
4. Prove Proposition 22.
5. Prove Proposition 24.
6. Prove Proposition 29.
7. Prove Proposition 30.
8. Show that

$$\begin{bmatrix} 2 & 5 \\ 4 & 3 \end{bmatrix}$$

is similar to

$$\begin{bmatrix} 2 & 4 \\ 5 & 3 \end{bmatrix}$$

Find characteristic vectors for each.

9. Show that

$$\begin{bmatrix} 1 & 1 \\ 5 & 4 \end{bmatrix}$$

is not similar to

$$\begin{bmatrix} 1 & 1 \\ 4 & 5 \end{bmatrix}$$

10. Show that

$$\begin{bmatrix} 1 & -1 \\ 1 & 1 \end{bmatrix}$$

is not similar to a diagonal matrix with real entries.

11. Show that

$$\begin{bmatrix} 2 & 1 & 0 \\ 1 & 2 & 0 \\ 0 & 3 & 6 \end{bmatrix}$$

is similar to a diagonal matrix. Give the matrix.

12. Prove: If T can be represented by a triangular $n \times n$ matrix with n distinct elements on the diagonal, then T can be represented by a diagonal matrix with these same diagonal elements.

13. Show that the hypothesis of distinct diagonal elements cannot be eliminated from Exercise 12 by showing that

$$\begin{bmatrix} 1 & 0 \\ 0 & 1 \end{bmatrix} \quad \text{and} \quad \begin{bmatrix} 1 & 1 \\ 0 & 1 \end{bmatrix}$$

are not similar.

14. By means of an example show that two matrices that are not similar can have the same characteristic polynomial.

**15. Let V be an F-space, and let $T \in \mathcal{L}(V)$. Define ϕ_T from $F[X]$ to $\mathcal{L}(V)$ by $p(X)\phi_T = p(T)$, where $T^0 = 1$.

(a) Prove that ϕ_T is a ring homomorphism.

(b) Show that V can be viewed as an $F[X]$ module.

(c) Prove that if V is finite dimensional, then ker ϕ_T is not trivial. [Hint: $\{1, T, T^2, \ldots, T^{n^2}\}$ is linearly dependent in $\mathcal{L}(V)$.]

(d) Show that there is a nonzero monic polynomial $g(X)$ of smallest degree such that $g(T) = 0$. This polynomial is called the *minimal polynomial* of T.

(e) Show that if f is a polynomial such that $f(T) = 0$, then $g(X) \mid f(X)$.

(f) Let $p(X)$ be an irreducible polynomial. Show that $p(X)$ divides $g(X)$ the minimum polynomial iff $up(T) = 0$ for some nonzero vector u.

(g) Show that if c is a characteristic value of T, then $x - c$ divides $g(X)$.

(h) Show that if $g(X)$ has n distinct nonzero roots, then T can be represented by a diagonal matrix.

§5. Systems of Linear Equations

Systems of linear equations such as

$$\begin{aligned} 2x + 3y &= 5 \\ x - y &= 7 \end{aligned} \quad \text{or} \quad \begin{aligned} 2x + y - z &= 9 \\ x - y + 2z &= 3 \\ 3x + y - z &= 6 \end{aligned}$$

can be viewed as matrix equations if we write them in-the forms

$$\begin{bmatrix} 2 & 3 \\ 1 & -1 \end{bmatrix} \begin{bmatrix} x \\ y \end{bmatrix} = \begin{bmatrix} 5 \\ 7 \end{bmatrix} \quad \text{and} \quad \begin{bmatrix} 2 & 1 & -1 \\ 1 & -1 & 2 \\ 3 & 1 & -1 \end{bmatrix} \begin{bmatrix} x \\ y \\ z \end{bmatrix} = \begin{bmatrix} 9 \\ 3 \\ 6 \end{bmatrix}$$

or in the forms

$$(x, y) \begin{bmatrix} 2 & 1 \\ 3 & -1 \end{bmatrix} = (5, 7) \quad \text{and} \quad (x, y, z) \begin{bmatrix} 2 & 1 & 3 \\ 1 & -1 & 1 \\ -1 & 2 & -1 \end{bmatrix} = (9, 3, 6)$$

In the second form the set of solutions is the inverse image of a vector under a linear transformation. For example, the set of solutions of

$$(x, y) \begin{bmatrix} 2 & 1 \\ 3 & -1 \end{bmatrix} = (5, 7)$$

is the inverse image of $(5, 7)$ under the linear transformation

$$\begin{bmatrix} 2 & 1 \\ 3 & -1 \end{bmatrix}$$

In a case such as this, when the linear transformation is invertible, there is just one solution and it is given by

$$(5, 7) \begin{bmatrix} 2 & 1 \\ 3 & -1 \end{bmatrix}^{-1} = (5, 7) \begin{bmatrix} 0.2 & 0.2 \\ 0.6 & -0.4 \end{bmatrix} = (5.2, -1.8)$$

Similarly, if we view the second system of equations as a matrix equation, we have

$$(x, y, z) \begin{bmatrix} 2 & 1 & 3 \\ 1 & -1 & 1 \\ -1 & 2 & -1 \end{bmatrix} = (9, 3, 6) \quad \text{and if} \quad \begin{bmatrix} 2 & 1 & 3 \\ 1 & -1 & 1 \\ -1 & 2 & -1 \end{bmatrix}$$

is invertible, we can just multiply on the right by the inverse and obtain

$$(x, y, z) = (9, 3, 6) \begin{bmatrix} 2 & 1 & 3 \\ 1 & -1 & 1 \\ -1 & 2 & -1 \end{bmatrix}^{-1}$$

which will be the only solution to the system. Theoretically, this seems very nice and if we had, at hand, a table of inverses of matrices, then such computations would be very simple. The fact is, however, that we have to *compute* such inverses and the process of doing so will itself produce the solution.

Recall that our technique for computing inverses was to multiply the matrix on the right by suitable elementary matrices and to multiply the identity matrix by the same matrices; that is, to solve the matrix equation $[x_{ij}][\alpha_{ij}] = [\delta_{ij}]$ we multiplied by suitable matrices $T_1, \ldots, T_n$ and from $[x_{ij}][\alpha_{ij}]T_1 \cdots T_n = [\delta_{ij}]T_1 \cdots T_n$ we obtained $[x_{ij}] = T_1 \cdots T_n$. If we consider the matrix equation

$$(x, y, z)\begin{bmatrix} 2 & 1 & 3 \\ 1 & -1 & 1 \\ -1 & 2 & -1 \end{bmatrix} = (9, 3, 6)$$

and apply the same technique, that is, multiply by appropriate elementary matrices, then if the matrix

$$\begin{bmatrix} 2 & 1 & 3 \\ 1 & -1 & 1 \\ -1 & 2 & -1 \end{bmatrix}$$

has an inverse we shall obtain

$$(x, y, z)[\delta_{ij}] = (9, 3, 6)T_1 T_2 \cdots T_n$$

that is,

$$(x, y, z) = (9, 3, 6)T_1 T_2 \cdots T_n$$

Thus, the solution can be obtained by multiplying (9, 3, 6) by suitable elementary matrices.

Example

If

$$(x, y, z)\begin{bmatrix} 2 & 1 & 3 \\ 1 & -1 & 1 \\ -1 & 2 & -1 \end{bmatrix} = (9, 3, 6)$$

then

$$(x, y, z)\begin{bmatrix} 1 & 2 & 3 \\ -1 & 1 & 1 \\ 2 & -1 & -1 \end{bmatrix} = (3, 9, 6)$$

(interchanging first and second columns)

$$(x, y, z)\begin{bmatrix} 1 & 0 & 3 \\ -1 & 3 & 1 \\ 2 & -5 & -1 \end{bmatrix} = (3, 3, 6)$$

(adding (-2) times first column to second column)

$$(x, y, z) \begin{bmatrix} 1 & 0 & 0 \\ -1 & 3 & 4 \\ 2 & -5 & -7 \end{bmatrix} = (3, 3, -3) \qquad \text{(adding } (-3) \text{ times first column to third column)}$$

$$(x, y, z) \begin{bmatrix} 1 & 0 & 0 \\ -1 & 3 & 0 \\ 2 & -5 & -\frac{1}{3} \end{bmatrix} = (3, 3, -7) \qquad \text{(adding } (-\frac{4}{3}) \text{ times second column to third)}$$

$$(x, y, z) \begin{bmatrix} 1 & 0 & 0 \\ -1 & 3 & 0 \\ 2 & -5 & 1 \end{bmatrix} = (3, 3, 21) \qquad \text{(multiplying third column by } (-3))$$

$$(x, y, z) \begin{bmatrix} 1 & 0 & 0 \\ -1 & 3 & 0 \\ 2 & 0 & 1 \end{bmatrix} = (3, 108, 21) \qquad \text{(adding } (5) \text{ times third column to second)}$$

$$(x, y, z) \begin{bmatrix} 1 & 0 & 0 \\ -1 & 1 & 0 \\ 2 & 0 & 1 \end{bmatrix} = (3, 36, 21) \qquad \text{(multiplying second column by } (\frac{1}{3}))$$

$$(x, y, z) \begin{bmatrix} 1 & 0 & 0 \\ -1 & 1 & 0 \\ 0 & 0 & 1 \end{bmatrix} = (-39, 36, 21) \qquad \text{(adding } (-2) \text{ times third column to first column)}$$

$$(x, y, z) \begin{bmatrix} 1 & 0 & 0 \\ 0 & 1 & 0 \\ 0 & 0 & 1 \end{bmatrix} = (-3, 36, 21) \qquad \text{(adding second column to the first)}$$

Hence,

$$(x, y, z) = (-3, 36, 21)$$

Once we obtained the equation

$$(x, y, z) \begin{bmatrix} 1 & 0 & 0 \\ -1 & 3 & 0 \\ 2 & -5 & -\frac{1}{3} \end{bmatrix} = (3, 3, -7)$$

we were assured a unique solution, because the determinant of the matrix is nonzero and therefore an inverse exists. In equation form this is

$$x - y + 2z = 3$$
$$3y - 5z = 3$$
$$-\tfrac{1}{3}z = -7$$

At this point we can solve for z, then substitute in the second equation, solve for y, and then solve for x by substituting in the first equation. This technique is probably more common than completing the process as in the example.

Now that we have studied a specific example, what is the general case? A system of n equations in m unknowns

$$a_{11}x_1 + a_{21}x_2 + \cdots + a_{m1}x_m = b_1$$
$$a_{12}x_1 + a_{22}x_2 + \cdots + a_{m2}x_m = b_2$$
$$\vdots$$
$$a_{1n}x_1 + a_{2n}x_2 + \cdots + a_{mn}x_m = b_n$$

where each a_{ij} and b_i are elements from a field F, can be viewed as a matrix equation,

$$(x_1, \ldots, x_m) \begin{bmatrix} a_{11} & a_{12} & \cdots & a_{1n} \\ a_{21} & a_{22} & \cdots & a_{2n} \\ \vdots & \vdots & & \vdots \\ a_{m1} & a_{m2} & \cdots & a_{mn} \end{bmatrix} = (b_1, \ldots, b_n)$$

with

$$\begin{bmatrix} a_{11} & a_{12} & \cdots & a_{1n} \\ \vdots & \vdots & & \vdots \\ a_{m1} & a_{m2} & \cdots & a_{mn} \end{bmatrix}$$

viewed as a linear transformation from F^m to F^n. Since multiplication on the right by elementary matrices is reversible, any equation so obtained from $(x_1, \ldots, x_m)[a_{ij}] = (b_1, \ldots, b_n)$ is equivalent to the original equation and will have the same set of solutions. What are the possible sets of solutions to $(x_1, \ldots, x_m)[a_{ij}] = (b_1, \ldots, b_n)$? We are looking for the inverse image of $(b_1, \ldots, b_n)$ under the linear transformation $[a_{ij}]$, and therefore the system has a solution iff $(b_1, \ldots, b_n)$ is in the image of F^m under the linear transformation $[a_{ij}]$. Since the image of F^m under $[a_{ij}]$ is precisely the subspace spanned by the rows of $[a_{ij}]$, a solution exists iff $(b_1, \ldots, b_n)$ can be expressed as a linear combination of the rows of $[a_{ij}]$ and, in fact, the coefficients used will be a solution.

What is the form of the inverse image of a vector under a linear transformation? If the set is not empty then, from earlier studies, we know that it is a coset of the kernel of the transformation; that is, if ϕ is a linear transformation from V into W with $v \in V$ and $w \in W$ such that $v\phi = w$, then $w\phi^{-1} = K + v$, where K is the kernel of ϕ. Thus, if we know the kernel of $[a_{ij}]$ and one vector in the inverse image of $(b_1, \ldots, b_n)$, we know all such vectors. For this reason, the first step in trying to solve such a system is to determine the kernel of $[a_{ij}]$.

A system of equations

$$a_{11}x_1 + a_{21}x_2 + \cdots + a_{n1}x_n = 0$$
$$\vdots$$
$$a_{1m}x_1 + a_{2m}x_2 + \cdots + a_{nm}x_n = 0$$

or

$$(x_1, \ldots, x_n)[a_{ij}] = (0, \ldots, 0)$$

is called a *homogeneous system* of linear equations. Solving such a system means finding the kernel of $[a_{ij}]$, that is, finding $(0, \ldots, 0)[a_{ij}]^{-1}$. Since the zero vector is always in the kernel, a homogeneous system will always have a solution. If $n > m$, then $[a_{ij}]$ maps F^n into F^m, and since the dimension of the image must be less than n, the kernel must be non-trivial. Thus, we have the following assertion.

Proposition 31

If $n > m$, then a system of m homogeneous equations in n unknowns, over the field F, has a nontrivial solution.

In general, an $n \times m$ matrix $[a_{ij}]$ has a nontrivial kernel iff its rank is less than n. Thus, a nontrivial solution may exist even though $m \geq n$, but we are assured of a nontrivial solution if $m < n$.

If $[a_{ij}]$ is one-to-one, then there is only one solution. If we apply our technique to the system $(x_1, \ldots, x_n)[a_{ij}] = (0, \ldots, 0)$, we know that we will eventually obtain a matrix that has 0's above the diagonal, that is,

$$(x_1, \ldots, x_n)[a_{ij}] = (0, \ldots, 0)$$

is equivalent to an equation

$$(x_1, \ldots, x_n)[c_{ij}] = (0, \ldots, 0) \qquad \text{where } i < j \text{ implies } c_{ij} = 0$$

In general, when we apply our technique of multiplying on the right by elementary matrices, we obtain a matrix in which

(i) All columns of zeros are on the right.
(ii) The first nonzero entry in each other column is a 1.
(iii) If the number of nonzero columns is p, then for each i, $i = 1, \ldots, p$, e_i appears as one of the rows.

(iv) All other rows are linear combinations of $e_1, \ldots, e_p$.

(v) Each e_i is the image, under the matrix, of some e_j with $e_i \leq e_j$.

(vi) The kernel is spanned by $n - p$ vectors.

Consider now the following example.

From

$$(x_1, x_2, x_3, x_4) \begin{bmatrix} 1 & 1 & 4 & 3 \\ 2 & -1 & 7 & 2 \\ 4 & 1 & 15 & 8 \\ 5 & -1 & 18 & 7 \end{bmatrix} = (0, 0, 0, 0)$$

it follows that

$$(x_1, x_2, x_3, x_4) \begin{bmatrix} 1 & 0 & 0 & 0 \\ 2 & -3 & -1 & -4 \\ 4 & -3 & -1 & -4 \\ 5 & -6 & -2 & -8 \end{bmatrix} = (0, 0, 0, 0)$$

and hence

$$(x_1, x_2, x_3, x_4) \begin{bmatrix} 1 & 0 & 0 & 0 \\ 2 & 1 & 1 & 4 \\ 4 & 1 & 1 & 4 \\ 5 & 2 & 2 & 8 \end{bmatrix} = (0, 0, 0, 0)$$

Thus,

$$(x_1, x_2, x_3, x_4) \begin{bmatrix} 1 & 0 & 0 & 0 \\ 2 & 1 & 0 & 0 \\ 4 & 1 & 0 & 0 \\ 5 & 2 & 0 & 0 \end{bmatrix} = (0, 0, 0, 0)$$

and

$$(x_1, x_2, x_3, x_4) \begin{bmatrix} 1 & 0 & 0 & 0 \\ 0 & 1 & 0 & 0 \\ 2 & 1 & 0 & 0 \\ 1 & 2 & 0 & 0 \end{bmatrix} = (0, 0, 0, 0)$$

The image of

$$\begin{bmatrix} 1 & 0 & 0 & 0 \\ 0 & 1 & 0 & 0 \\ 2 & 1 & 0 & 0 \\ 1 & 2 & 0 & 0 \end{bmatrix}$$

is spanned by its rows. The image of e_1 is e_1, and the image of e_2 is e_2. The image of e_3 is $(2, 1, 0, 0)$, which is the sum of $2e_1 + e_2$. Hence, $2e_1 + e_2 - e_3$ is in the kernel. Similarly, since $(1, 2, 0, 0)$ is the image of e_4 and is the sum of $e_1 + 2e_2$, $e_1 + 2e_2 - e_4$ is in the kernel. Thus, we have $(2, 1, -1, 0)$ and $(1, 2, 0, -1)$ in the kernel. We know the kernel is two dimensional because the image is two dimensional. (dim V = dim $V\phi$ + dim ker ϕ.) Clearly, $(2, 1, -1, 0)$ and $(1, 2, 0, -1)$ are linearly independent and hence span the kernel.

Our technique also works to find at least one solution to a system if such a solution exists.

Consider the system

$$(x_1, x_2, x_3, x_4) \begin{bmatrix} 1 & 1 & 4 & 3 \\ 2 & -1 & 7 & 2 \\ 4 & 1 & 15 & 8 \\ 5 & -1 & 18 & 7 \end{bmatrix} = (6, 3, 23, 14)$$

If we follow the same technique as before we get

$$(x_1, x_2, x_3, x_4) \begin{bmatrix} 1 & 0 & 0 & 0 \\ 0 & 1 & 0 & 0 \\ 2 & 1 & 0 & 0 \\ 1 & 2 & 0 & 0 \end{bmatrix} = (4, 1, 0, 0)$$

and clearly $(4, 1, 0, 0)$ is a solution to this system. Now from the fact that $(2, 1, -1, 0)$ and $(1, 2, 0, -1)$ span the kernel, it follows that the set of all solutions is $\langle(2, 1, -1, 0), (1, 2, 0, -1)\rangle + (4, 1, 0, 0)$ or the set of all vectors of the form

$$\alpha(2, 1, -1, 0) + \beta(1, 2, 0, -1) + (4, 1, 0, 0) \qquad \text{for } \alpha \text{ and } \beta \text{ in } F$$

which is the same as all vectors of the form

$$(2\alpha + \beta + 4, \alpha + 2\beta + 1, -\alpha, -\beta) \qquad \text{for } \alpha \text{ and } \beta \text{ in } F$$

The technique can also be used to show that no solution exists. Consider the system

$$(x_1, x_2, x_3, x_4) \begin{bmatrix} 1 & 1 & 4 & 3 \\ 2 & -1 & 7 & 2 \\ 4 & 1 & 15 & 8 \\ 5 & -1 & 18 & 7 \end{bmatrix} = (2, 5, 9, 11)$$

If we follow the same technique as before we get

$$(x_1, x_2, x_3, x_4) \begin{bmatrix} 1 & 0 & 0 & 0 \\ 0 & 1 & 0 & 0 \\ 2 & 1 & 0 & 0 \\ 1 & 2 & 0 & 0 \end{bmatrix} = (-1, 1, -6, 12)$$

This system has no solution, since the image of any vector under

$$\begin{bmatrix} 1 & 0 & 0 & 0 \\ 0 & 1 & 0 & 0 \\ 2 & 1 & 0 & 0 \\ 1 & 2 & 0 & 0 \end{bmatrix}$$

must have 0's as the last two components.

In general, the system $(x_1, x_2, \ldots, x_n)[a_{ij}] = (b_1, b_2, \ldots, b_m)$ is equivalent to a matrix equation $(x_1, \ldots, x_n)[c_{ij}] = (d_1, \ldots, d_m)$ obtained by multiplying on the right by elementary matrices and where $[c_{ij}]$ is lower triangular of rank $k \leq m$, and the vectors $e_1, \ldots, e_k$ each appear just once as rows of $[c_{ij}]$. The system will have a solution iff $d_i = 0$ for $i > k$. One of the solutions will be given by

$$d_1 e_{j_1} + d_2 e_{j_2} + d_3 e_{j_3} + \cdots + d_k e_{j_k}$$

where $e_{j_i}[c_{ij}] = e_i$.

The rest of the solutions are obtained by adding vectors in the kernel of $[c_{ij}]$ to this vector.

For example, the solutions of

$$(x_1, x_2, x_3, x_4, x_5) \begin{bmatrix} 1 & 0 & 0 & 0 & 0 & 0 \\ 2 & 0 & 0 & 0 & 0 & 0 \\ 0 & 1 & 0 & 0 & 0 & 0 \\ 5 & 3 & 0 & 0 & 0 & 0 \\ 0 & 0 & 1 & 0 & 0 & 0 \end{bmatrix} = (4, 7, -3, 0, 0, 0)$$

are given by $K + (4, 0, 7, 0, -3)$, where K is the kernel of the matrix. The rank of the matrix is 3, the kernel is of dimension 2, and is spanned by $(2, -1, 0, 0, 0)$ and $(5, 0, 3, -1, 0)$. Thus, the set of all solutions is the set of all vectors of the form

$$\alpha(2, -1, 0, 0, 0) + \beta(5, 0, 3, -1, 0) + (4, 0, 7, 0, -3)$$

or

$$(2\alpha + 5\beta + 4, -\alpha, 3\beta + 7, -\beta, -3) \quad \text{for } \alpha \text{ and } \beta \text{ in } F$$

The solutions of

$$(x_1, x_2, x_3, x_4, x_5, x_6, x_7) \begin{bmatrix} 1 & 0 & 0 & 0 & 0 & 0 & 0 & 0 \\ 0 & 1 & 0 & 0 & 0 & 0 & 0 & 0 \\ 2 & 3 & 0 & 0 & 0 & 0 & 0 & 0 \\ 0 & 0 & 1 & 0 & 0 & 0 & 0 & 0 \\ 4 & 5 & 3 & 0 & 0 & 0 & 0 & 0 \\ -1 & 2 & -4 & 0 & 0 & 0 & 0 & 0 \\ 0 & 0 & 0 & 1 & 0 & 0 & 0 & 0 \end{bmatrix}$$

$$= (4, 2, -3, 6, 0, 0, 0, 0)$$

are given by $K + (4, 2, 0, -3, 0, 0, 6)$, where K is the kernel of the matrix, is of dimension 3, and is spanned by $(2, 3, -1, 0, 0, 0, 0)$, $(4, 5, 0, 3, -1, 0, 0)$, and $(-1, 2, 0, -4, 0, -1, 0)$.

The system of equations $(x_1, \ldots x_n)[a_{ij}] = (b_1, \ldots, b_m)$ is equivalent to the matrix equation

$$(x_1, \ldots, x_n, -1) \begin{bmatrix} a_{11} & a_{12} & \cdots & a_{1m} \\ a_{21} & a_{22} & \cdots & a_{2m} \\ \vdots & & & \\ a_{n1} & a_{n2} & \cdots & a_{nm} \\ b_1 & b_2 & \cdots & b_m \end{bmatrix} = (0, 0, \ldots, 0)$$

The matrix $[a_{ij}]$ is called the *coefficient* matrix for the system and

$$\begin{bmatrix} a_{11} & \cdots & a_{1m} \\ \vdots & & \\ a_{n1} & \cdots & a_{nm} \\ b_1 & \cdots & b_m \end{bmatrix}$$

is called the *augmented* matrix. By applying our technique we get the matrix equation

$$(x_1, \ldots, x_n) \begin{bmatrix} c_{11} & 0 & \cdots & & 0 \\ c_{21} & c_{22} & 0 & \cdots & 0 \\ \vdots & & & \ddots & \vdots \\ & & & & 0 \\ c_{n1} & c_{n2} & & \cdots & c_{nm} \end{bmatrix} = (d_1, \ldots, d_m)$$

where the d_i's are obtained from the b_i's by applying the same transformations that we apply to $[a_{ij}]$. This system is equivalent to

$$(x_1, \ldots, x_n, -1) \begin{bmatrix} c_{11} & 0 & \cdots & & 0 \\ c_{21} & c_{22} & 0 & \cdots & 0 \\ \vdots & & & \ddots & \vdots \\ & & & & 0 \\ c_{n1} & c_{n2} & & \cdots & c_{nm} \\ d_1 & \cdots & & & d_m \end{bmatrix} = (0, \ldots, 0)$$

From our previous discussion we know that the system

$$(x_1, \ldots, x_n)[c_{ij}] = (d_1, \ldots, d_m)$$

has a solution iff the number of nonzero columns of $[c_{ij}]$ is the same as the number of nonzero columns of the augmented matrix and one solution is obtained easily from $(d_1, \ldots, d_m)$; that is, if the rank of $[c_{ij}]$ is equal to the rank of the augmented matrix. Thus, for simplicity in computation, we frequently work only with the augmented matrix and thus perform the column operations on $[a_{ij}]$ and $(b_1, \ldots, b_m)$ simultaneously. Let us look at some examples of this technique.

Example 1. Consider the equation previously solved.

$$(x_1, x_2, x_3, x_4) \begin{bmatrix} 1 & 1 & 4 & 3 \\ 2 & -1 & 7 & 2 \\ 4 & 1 & 15 & 8 \\ 5 & -1 & 18 & 7 \end{bmatrix} = (6, 3, 23, 14)$$

is equivalent to

$$(x_1, x_2, x_3, x_4, -1) \begin{bmatrix} 1 & 1 & 4 & 3 \\ 2 & -1 & 7 & 2 \\ 4 & 1 & 15 & 8 \\ 5 & -1 & 18 & 7 \\ 6 & 3 & 23 & 14 \end{bmatrix} = (0, 0, 0, 0)$$

Applying our technique to this matrix yields

$$(x_1, x_2, x_3, x_4, -1) \begin{bmatrix} 1 & 0 & 0 & 0 \\ 0 & 1 & 0 & 0 \\ 2 & 1 & 0 & 0 \\ 1 & 2 & 0 & 0 \\ 4 & 1 & 0 & 0 \end{bmatrix} = (0, 0, 0, 0)$$

which is easily seen to have $(4, 1, 0, 0, -1)$ as a solution, and therefore $(4, 1, 0, 0)$ is a solution to the original system.

Example 2

$$(x, y, z) \begin{bmatrix} 1 & 3 & 2 \\ 4 & 6 & 1 \\ 3 & 1 & 2 \end{bmatrix} = (1, 2, -1)$$

The augmented matrix is

$$\begin{bmatrix} 1 & 3 & 2 \\ 4 & 6 & 1 \\ 3 & 1 & 2 \\ 1 & 2 & -1 \end{bmatrix}$$

and the equivalent system is

$$(x, y, z, -1) \begin{bmatrix} 1 & 3 & 2 \\ 4 & 6 & 1 \\ 3 & 1 & 2 \\ 1 & 2 & -1 \end{bmatrix} = (0, 0, 0)$$

Applying our technique to the matrix yields the following sequence of equivalent matrices:

$$\begin{bmatrix} 1 & 3 & 2 \\ 4 & 6 & 1 \\ 3 & 1 & 2 \\ 1 & 2 & -1 \end{bmatrix}, \quad \begin{bmatrix} 1 & 0 & 0 \\ 4 & -6 & -7 \\ 3 & -8 & -4 \\ 1 & -1 & -3 \end{bmatrix}, \quad \begin{bmatrix} 1 & 0 & 0 \\ 4 & 6 & 7 \\ 3 & 8 & 4 \\ 1 & 1 & 3 \end{bmatrix},$$

$$\begin{bmatrix} 1 & 0 & 0 \\ 4 & 6 & 1 \\ 3 & 8 & -4 \\ 1 & 1 & 2 \end{bmatrix}, \quad \begin{bmatrix} 1 & 0 & 0 \\ 4 & 0 & 1 \\ 3 & 32 & -4 \\ 1 & -11 & 2 \end{bmatrix}, \quad \begin{bmatrix} 1 & 0 & 0 \\ 4 & 1 & 0 \\ 3 & -4 & 32 \\ 1 & 2 & -11 \end{bmatrix}$$

At this point we know that the rank of the coefficient matrix is equal to the rank of the augmented matrix, since both are of rank 3. We then continue to find a solution.

We obtain

$$\begin{bmatrix} 1 & 0 & 0 \\ 0 & 1 & 0 \\ 19 & -4 & 32 \\ -7 & 2 & -11 \end{bmatrix}, \quad \begin{bmatrix} 1 & 0 & 0 \\ 0 & 1 & 0 \\ 19 & -4 & 1 \\ -7 & 2 & -\frac{11}{32} \end{bmatrix},$$

and finally

$$\begin{bmatrix} 1 & 0 & 0 \\ 0 & 1 & 0 \\ 0 & 0 & 1 \\ -\frac{1}{32} & \frac{5}{8} & -\frac{11}{32} \end{bmatrix}$$

Thus, $\left(-\frac{1}{32}, \frac{5}{8}, -\frac{11}{32}, -1\right)$ is a solution to

$$(x, y, z, -1) \begin{bmatrix} 1 & 3 & 2 \\ 4 & 6 & 1 \\ 3 & 1 & 2 \\ 1 & 2 & -1 \end{bmatrix} = (0, 0, 0)$$

and therefore $\left(-\frac{1}{32}, \frac{5}{8}, -\frac{11}{32}\right)$ is a solution to

$$(x, y, z) \begin{bmatrix} 1 & 3 & 2 \\ 4 & 6 & 1 \\ 3 & 1 & 2 \end{bmatrix} = (1, 2, -1)$$

Since the rank of the coefficient matrix is 3, it is invertible and $\left(-\frac{1}{32}, \frac{5}{8}, -\frac{11}{32}\right)$ is the only solution.

Example 3

$$(x, y, z, w) \begin{bmatrix} 1 & 2 & 3 & 4 & 5 \\ 2 & 3 & 1 & 5 & 4 \\ 1 & 2 & 1 & 3 & 4 \\ -1 & 1 & -1 & 1 & -1 \end{bmatrix} = (3, -2, 5, 7, 6)$$

If we form the augmented matrix we have

$$\begin{bmatrix} 1 & 2 & 3 & 4 & 5 \\ 2 & 3 & 1 & 5 & 4 \\ 1 & 2 & 1 & 3 & 4 \\ -1 & 1 & -1 & 1 & -1 \\ 3 & -2 & 5 & 7 & 6 \end{bmatrix}$$

which is equivalent to

$$\begin{bmatrix} 1 & 0 & 0 & 0 & 0 \\ 2 & -1 & -5 & -3 & -6 \\ 1 & 0 & -2 & -1 & -1 \\ -1 & 3 & 2 & 5 & 4 \\ 3 & -8 & -4 & -5 & -9 \end{bmatrix}$$

Continuing we get the sequence

$$\begin{bmatrix} 1 & 0 & 0 & 0 & 0 \\ 2 & 1 & 5 & 3 & 6 \\ 1 & 0 & 2 & 1 & 1 \\ -1 & -3 & -2 & -5 & -4 \\ 3 & 8 & 4 & 5 & 9 \end{bmatrix}, \quad \begin{bmatrix} 1 & 0 & 0 & 0 & 0 \\ 2 & 1 & 0 & 0 & 0 \\ 1 & 0 & 2 & 1 & 1 \\ -1 & -3 & 13 & 4 & 14 \\ 3 & 8 & -36 & -19 & -39 \end{bmatrix},$$

$$\begin{bmatrix} 1 & 0 & 0 & 0 & 0 \\ 2 & 1 & 0 & 0 & 0 \\ 1 & 0 & 1 & 2 & 1 \\ -1 & -3 & 4 & 13 & 14 \\ 3 & 8 & -19 & -36 & -39 \end{bmatrix}, \quad \begin{bmatrix} 1 & 0 & 0 & 0 & 0 \\ 2 & 1 & 0 & 0 & 0 \\ 1 & 0 & 1 & 0 & 0 \\ -1 & -3 & 4 & 5 & 10 \\ 3 & 8 & -19 & 2 & -20 \end{bmatrix},$$

$$\begin{bmatrix} 1 & 0 & 0 & 0 & 0 \\ 2 & 1 & 0 & 0 & 0 \\ 1 & 0 & 1 & 0 & 0 \\ -1 & -3 & 4 & 5 & 0 \\ 3 & 8 & -19 & 2 & -24 \end{bmatrix}$$

At this point we know there is no solution, since the coefficient matrix is of rank 4, whereas the augmented matrix is of rank 5.

EXERCISE SET E

1. Determine the kernels of the following matrices.

(a) $\begin{bmatrix} 3 & 1 \\ -6 & -2 \end{bmatrix}$ (b) $\begin{bmatrix} 4 & 7 \\ 3 & 1 \end{bmatrix}$ (c) $\begin{bmatrix} 1 & 4 & 7 \\ 2 & 9 & 3 \\ 0 & 1 & -1 \end{bmatrix}$ (d) $\begin{bmatrix} 1 & 2 & 4 & 3 \\ 6 & 1 & 3 & 9 \\ 2 & 1 & 2 & 1 \\ 3 & -2 & -3 & 5 \end{bmatrix}$

2. Solve the following systems of equations if such solutions exist.

   (a) $3x - 6y = 15$
       $x - 2y = 5$

   (b) $4x + 3y = 2$
       $7x + y = 1$

   (c) $x + 2y = 1$
       $4x + 9y + z = 10$
       $7x + 3y - z = 7$

   (d) $x + 6y + 2z + 3w = 12$
       $2x + y + z - 2w = 0$
       $4x + 3y + 2z - 3w = 3$
       $3x + 9y + z + 5w = 4$

3. Solve the following systems of equations if such solutions exist.

   (a) $(x, y) \begin{bmatrix} 2 & 3 \\ 4 & 6 \end{bmatrix} = (1, -1)$

   (b) $(x, y, z) \begin{bmatrix} 1 & 2 & 3 \\ 3 & 4 & 1 \\ 1 & 0 & -5 \end{bmatrix} = (2, 6, 14)$

   (c) $(x, y, z) \begin{bmatrix} 1 & 4 & 5 \\ 6 & -1 & 3 \\ 7 & 1 & 2 \end{bmatrix} = (8, 5, 7)$

(d) $(x, y, z, w) \begin{bmatrix} 1 & 1 & 2 & -3 \\ 7 & 1 & 3 & 2 \\ 4 & -3 & 1 & 5 \\ 6 & 1 & 3 & 2 \end{bmatrix} = (5, 6, 1, -12)$

(e) $(x, y, z, w) \begin{bmatrix} 1 & 3 & 4 & 2 \\ 2 & 1 & -3 & -8 \\ 4 & 5 & 6 & -2 \\ 6 & -2 & 1 & -11 \end{bmatrix} = (1, 2, 3, 2)$

(f) $(x, y, z, w) \begin{bmatrix} 1 & 0 & 4 & 6 \\ 0 & 1 & 5 & 2 \\ 2 & 3 & 0 & 7 \\ 6 & 1 & 3 & 0 \end{bmatrix} = (5, -2, 7, -1)$

APPENDIX 1
THE ARITHMETIC OF 2×2
AND 3×3 MATRICES

In many colleges and universities a brief introduction to vector spaces and matrices is included in the sequence of calculus courses. We include this appendix for those students who do not have such a background or who may have forgotten some of the elementary ideas. We shall consider only 2×2 and 3×3 matrices over $\mathbf{R}$ and shall, in fact, prove very few of our assertions but shall leave most verifications to the reader.

A 2×2 matrix is a square array of four real numbers such as

$$\begin{bmatrix} 2 & 4 \\ 3 & -1 \end{bmatrix}$$

and a 3×3 matrix is a square array of nine real numbers such as

$$\begin{bmatrix} 1 & 0.2 & 3 \\ \frac{1}{2} & \frac{2}{7} & -4 \\ -3.1 & 5 & \frac{6}{11} \end{bmatrix}$$

The first *row* of

$$\begin{bmatrix} 2 & 4 \\ 3 & -1 \end{bmatrix}$$

is $[2 \quad 4]$ and the first *column* of

$$\begin{bmatrix} 2 & 4 \\ 3 & -1 \end{bmatrix}$$

is

$$\begin{bmatrix} 2 \\ 3 \end{bmatrix}$$

Columns go up and down, rows go across. To discuss matrices in general we usually use a subscript notation such as

$$\begin{bmatrix} a_{11} & a_{12} \\ a_{21} & a_{22} \end{bmatrix} \quad \begin{bmatrix} b_{11} & b_{12} & b_{13} \\ b_{21} & b_{22} & b_{23} \\ b_{31} & b_{32} & b_{33} \end{bmatrix}$$

Notice that the first subscript indicates the row and the second indicates the column; for example, b_{23} is in the *second* row in the *third* column.

We add two 2×2 matrices or two 3×3 matrices by adding the corresponding entries or components. Thus,

$$\begin{bmatrix} 2 & 3 \\ 4 & 1 \end{bmatrix} + \begin{bmatrix} 1 & -2 \\ 3 & -4 \end{bmatrix} = \begin{bmatrix} 3 & 1 \\ 7 & -3 \end{bmatrix}$$

and in general we have

$$\begin{bmatrix} a_{11} & a_{12} \\ a_{21} & a_{22} \end{bmatrix} + \begin{bmatrix} b_{11} & b_{12} \\ b_{21} & b_{22} \end{bmatrix} = \begin{bmatrix} a_{11} + b_{11} & a_{12} + b_{12} \\ a_{21} + b_{21} & a_{22} + b_{22} \end{bmatrix}$$

and

$$\begin{bmatrix} a_{11} & a_{12} & a_{13} \\ a_{21} & a_{22} & a_{23} \\ a_{31} & a_{32} & a_{33} \end{bmatrix} + \begin{bmatrix} b_{11} & b_{12} & b_{13} \\ b_{21} & b_{22} & b_{23} \\ b_{31} & b_{32} & b_{33} \end{bmatrix}$$

$$= \begin{bmatrix} a_{11} + b_{11} & a_{12} + b_{12} & a_{13} + b_{13} \\ a_{21} + b_{21} & a_{22} + b_{22} & a_{23} + b_{23} \\ a_{31} + b_{31} & a_{32} + b_{32} & a_{33} + b_{33} \end{bmatrix}$$

Notice that it makes no sense to talk about adding a 2×2 and a 3×3 matrix.

It is easy to show that matrix addition is associative and commutative. For convenience, let us from now on denote the 2×2 matrices by M_2 and the 3×3 matrices by M_3. The 2×2 matrix with all 0 entries is an additive identity in M_2, and the 3×3 matrix with all 0 entries is an additive identity in M_3. The matrix

$$\begin{bmatrix} -a_{11} & -a_{12} \\ -a_{21} & -a_{22} \end{bmatrix}$$

is easily seen to be an additive inverse for

$$\begin{bmatrix} a_{11} & a_{12} \\ a_{21} & a_{22} \end{bmatrix}$$

and an analogous statement holds for 3×3 matrices. Thus, M_2 and M_3 are both abelian groups with respect to matrix addition.

The 2×2 matrices

$$\begin{bmatrix} a_{11} & a_{12} \\ a_{21} & a_{22} \end{bmatrix} \quad \text{and} \quad \begin{bmatrix} b_{11} & b_{12} \\ b_{21} & b_{22} \end{bmatrix}$$

are multiplied by the "turn, multiply, and add" or "row by column" rule,

$$\begin{bmatrix} a_{11} & a_{12} \\ a_{21} & a_{22} \end{bmatrix}\begin{bmatrix} b_{11} & b_{12} \\ b_{21} & b_{22} \end{bmatrix} = \begin{bmatrix} a_{11}b_{11} + a_{12}b_{21} & a_{11}b_{12} + a_{12}b_{22} \\ a_{21}b_{11} + a_{22}b_{21} & a_{21}b_{12} + a_{22}b_{22} \end{bmatrix}$$

and 3×3 matrices are multiplied similarly:

$$\begin{bmatrix} a_{11} & a_{12} & a_{13} \\ a_{21} & a_{22} & a_{23} \\ a_{31} & a_{32} & a_{33} \end{bmatrix}\begin{bmatrix} b_{11} & b_{12} & b_{13} \\ b_{21} & b_{22} & b_{23} \\ b_{31} & b_{32} & b_{33} \end{bmatrix} =$$

$$\begin{bmatrix} a_{11}b_{11}+a_{12}b_{21}+a_{13}b_{31} & a_{11}b_{12}+a_{12}b_{22}+a_{13}b_{32} & a_{11}b_{13}+a_{12}b_{23}a_{13}b_{33} \\ a_{21}b_{11}+a_{22}b_{21}+a_{23}b_{31} & a_{21}b_{12}+a_{22}b_{22}+a_{23}b_{32} & a_{21}b_{13}+a_{22}b_{23}+a_{23}b_{33} \\ a_{31}b_{11}+a_{32}b_{21}+a_{33}b_{31} & a_{31}b_{12}+a_{32}b_{22}+a_{33}b_{32} & a_{31}b_{13}+a_{32}b_{23}+a_{33}b_{33} \end{bmatrix}$$

It can be shown that matrix multiplication is associative and satisfies the distributive law, but is, in general, not commutative. Thus, both M_2 and M_3 are examples of rings (see Chapter 5). The matrices

$$\begin{bmatrix} 1 & 0 \\ 0 & 1 \end{bmatrix} \quad \text{and} \quad \begin{bmatrix} 1 & 0 & 0 \\ 0 & 1 & 0 \\ 0 & 0 & 1 \end{bmatrix}$$

are easily seen to be multiplicative identities. A matrix is said to be *invertible* if it has an inverse, that is, if there is a matrix whose product with the given matrix is the corresponding identity matrix. For example,

$$\begin{bmatrix} 2 & 1 \\ 1 & 1 \end{bmatrix} \quad \text{is invertible because} \quad \begin{bmatrix} 2 & 1 \\ 1 & 1 \end{bmatrix}\begin{bmatrix} 1 & -1 \\ -1 & 2 \end{bmatrix} = \begin{bmatrix} 1 & 0 \\ 0 & 1 \end{bmatrix}$$

If A and B are invertible matrices, then AB is also invertible because if $AC = I$ and $BD = I$, where I denotes the identity matrix, then

$$(AB)(DC) = A(BD)C = AIC = AC = I$$

Furthermore, it is not too hard to show that if $AB = I$, then $BA = I$ for either 2×2 or 3×3 matrices. Thus, the invertible 2×2 matrices form a group and the invertible 3×3 matrices form a group.

The *determinant* of the matrix

$$\begin{bmatrix} a_{11} & a_{12} \\ a_{21} & a_{22} \end{bmatrix}$$

is denoted by

$$\begin{vmatrix} a_{11} & a_{12} \\ a_{21} & a_{22} \end{vmatrix}$$

and defined by

$$\begin{vmatrix} a_{11} & a_{12} \\ a_{21} & a_{22} \end{vmatrix} = a_{11}a_{22} - a_{21}a_{12}$$

The determinant of

$$\begin{bmatrix} a_{11} & a_{12} & a_{13} \\ a_{21} & a_{22} & a_{23} \\ a_{31} & a_{32} & a_{33} \end{bmatrix}$$

is similarly denoted and defined by

$$\begin{vmatrix} a_{11} & a_{12} & a_{13} \\ a_{21} & a_{22} & a_{23} \\ a_{31} & a_{32} & a_{33} \end{vmatrix} = a_{11}a_{22}a_{33} - a_{11}a_{23}a_{32} + a_{12}a_{23}a_{31} \\ - a_{12}a_{21}a_{33} + a_{13}a_{21}a_{32} - a_{13}a_{22}a_{31}$$

By direct but somewhat messy computation it can be shown that the determinant of a product of two matrices is the product of their determinants. Since the determinant of I is 1, invertible matrices must have nonzero determinants. It can further be shown that matrices with nonzero determinants are invertible. In the case of 2×2 matrices this can be done by solving the equation

$$\begin{bmatrix} a & b \\ c & d \end{bmatrix} \begin{bmatrix} x & y \\ z & w \end{bmatrix} = \begin{bmatrix} 1 & 0 \\ 0 & 1 \end{bmatrix}$$

for x, y, z, and w. The fact that $ad - bc \neq 0$ will assure that the desired solutions exist. A similar but more complicated technique works for 3×3 matrices.

APPENDIX 2
POLYNOMIALS

In Chapter 8 we formally develop a theory of polynomials, but previous experience with polynomials over the integers, rational numbers, and real numbers make them valuable as illustrations and examples of many types of structures that are studied earlier in the text. In this appendix we provide some of the terminology and notation frequently used in earlier courses that is sometimes forgotten as the student progresses in mathematics.

A polynomial in the one variable x over the integers, rational numbers, or real numbers is a formal sum of the form

$$a_0 + a_1 x + a_2 x^2 + \cdots + a_n x^n$$

where each of the a_i's are, respectively, integers, rational numbers, or real numbers. The sum is formal in the sense that we can, for example, only indicate the sum of x and x^2 by $x + x^2$; we have no simpler form for this element. If we have two polynomials

$$a_0 + a_1 x + a_2 x^2 + \cdots + a_n x^n \quad \text{and} \quad b_0 + b_1 x + b_2 x^2 + \cdots + b_n x^n$$

we add them by adding the corresponding coefficients and obtain

$$(a_0 + b_0) + (a_1 + b_1)x + (a_2 + b_2)x^2 + \cdots + (a_n + b_n)x^n$$

Multiplying polynomials is not as simple in that the coefficients of the product are obtained by adding together several products of coefficients; that is, the coefficient of x^k in the product is obtained by adding

$$a_0 b_k + a_1 b_{k-1} + a_2 b_{k-2} + \cdots + a_{k-1} b_1 + a_k b_0$$

since when we perform the multiplication in the usual manner

$$(a_0)(b_k x^k) = (a_0 b_k)x^k, \quad (a_1 x)(b_{k-1} x^{k-1}) = a_1 b_{k-1} x^k, \text{ etc.}$$

Summation notation is very convenient for polynomials if we adopt the convention that $x^0 = 1$. Then we can denote

307

$$a_0 + a_1 x + a_2 x^2 + \cdots + a_n x^n \quad \text{by} \quad \sum_{i=0}^{n} a_i x^i$$

The rule for addition becomes

$$\sum_{i=0}^{n} a_i x^i + \sum_{i=0}^{n} b_i x^i = \sum_{i=0}^{n} (a_i + b_i) x^i$$

and the rule for multiplication becomes

$$\left(\sum_{i=0}^{n} a_i x^i \right) \left(\sum_{j=0}^{m} b_j x^j \right) = \sum_{k=0}^{n+m} c_k x^k$$

where $c_k = \Sigma_{k=i+j} a_i b_j$.

Two polynomials $\Sigma_{i=0}^{n} a_i x^i$ and $\Sigma_{i=0}^{n} b_i x^i$ are equal if $a_i = b_i$ for each $i, i = \{0, 1, \ldots, n\}$. If we assume commutativity and associativity of the formal addition, we can then prove that the set of all polynomials over the integers, rational numbers, or real numbers satisfies the first eight axioms given for **Z** in Chapter 2.

A polynomial $p(x)$ is said to have degree n if n is a nonnegative integer such that $p(x) = \Sigma_{i=0}^{n} a_i x^i$ and $a_n \neq 0$.

The degree of a polynomial p is frequently denoted by δp. The zero polynomial can be considered to degree $-\infty$, where $-\infty$ obeys the rules $(-\infty)n = -\infty, (-\infty) + m = -\infty$ for any positive integers n and m and also $(-\infty) + (-\infty) = -\infty$. We shall also consider $-\infty$ to be less than any nonnegative integer. This may seem like an artificial definition, but it enables us to state many propositions more simply by giving the zero polynomial a degree. For example, the polynomial $\Sigma_{i=0}^{n} a_i x^i$ is assured of having degree less than or equal to n.

We identify constant polynomials with numbers and, in fact, just view numbers as polynomials of degree 0 or $-\infty$.

By applying the rules for computation we can show that the degree of the sum of two polynomials never exceeds the maximum of the degrees of the summands and that for polynomials over the integers, rational numbers, or real numbers the degree of the product of two polynomials is the sum of their degrees.

From calculus recall that the derivative of a polynomial is given by

$$D\left(\sum_{i=0}^{n} a_i x^i \right) = \sum_{i=1}^{n} i a_i x^{i-1}$$

and that

$$\int \sum_{i=0}^{n} a_i x^i = \sum_{i=0}^{n} \frac{1}{i+1} a_i x^{i+1} + C$$

where C denotes the set of all constant polynomials.

If we denote $\sum_{i=0}^{n} a_i x^i$ by $p(x)$, and b is a number, by $p(b)$ we mean $\sum_{i=0}^{n} a_i b^i$. (As with x, $0^0 = 1$ in this case.) The number b is a *root* or *zero* of $p(x)$ if $p(b) = 0$.

The set of all polynomials over the integers is frequently denoted by $\mathbf{Z}[x]$, the set of polynomials over the rational numbers by $\mathbf{Q}[x]$, and the set of polynomials over the real numbers by $\mathbf{R}[x]$.

BIBLIOGRAPHY

The following list of books is intended only as a starting point for the reader who wishes to go further into some of the topics developed in the text. It is, by no means, intended as a complete bibliography.

[1] Barnes, Wilfred, *Introduction to Abstract Algebra*, Boston: Heath, 1963.

[2] Birkhoff, Garrett, and MacLane, Saunders, *A Survey of Modern Algebra* (3rd ed.), New York: Macmillan, 1965.

[3] Hall, Marshall, *The Theory of Groups*, New York: Macmillan, 1959.

[4] Halmos, P. R., *Finite Dimensional Vector Spaces* (2nd ed.), Princeton: Van Nostrand Reinhold, 1958.

[5] Halmos, P. R., *Naive Set Theory*, Princeton: Van Nostrand Reinhold, 1960.

[6] Herstein, I. N., *Topics in Algebra*, New York: Blaisdell, 1964.

[7] Krause, Eugene F., *Introduction to Linear Algebra*, New York: Holt, Rinehart & Winston, Inc., 1970.

[8] Kurosh, A. G., *The Theory of Groups* (2nd English ed.), New York: Chelsea House, 1960.

[9] Ledermann, Walter, *Introduction to the Theory of Finite Groups* (5th ed.), New York: Interscience, 1964.

[10] LeVeque, William J., *Elementary Theory of Numbers*, Reading, Mass.: Addison-Wesley, 1962.

[11] McCoy, N. H., *Introduction to Modern Algebra*, Boston: Allyn & Bacon, 1962.

[12] McCoy, N. H., *The Theory of Rings*, New York: Macmillan, 1964.

[13] Paley, Hiram, and Weichsel, Paul M., *A First Course in Abstract Algebra*, New York: Holt, Rinehart & Winston, 1966.

[14] Thrall, Robert M., and Tornheim, Leonard, *Vector Spaces and Matrices*, New York: Wiley, 1957.

[15] Wilder, R. L., *Introduction to the Foundations of Mathematics*, New York: Wiley, 1965.

ANSWERS TO SELECTED EXERCISES

Chapter 1

Exercise Set A

1. (e) True: Since $A \subseteq A \cup B$ and $A \cap B \subseteq B$, $A \cup B \subseteq A \cap B$ implies $A \subseteq B$. Similarly, since $B \subseteq A \cup B$ and $A \cap B \subseteq A$, $B \subseteq A$. Therefore $A = B$.

   (g) False: For example, let $A = \{1, 2, 3\}$, $B = \{4, 5\}$, and $C = \{3, 4, 5\}$.

   (m) False: For example, let $A = \{1, 2\}$ and $B = \phi$.

Exercise Set B

1. (d) No, because $(1, 1)$ and $(1, -1)$ are both in the set.

   (e) No, because $\tan \pi/2$ is undefined.

2. (c) $[0, 35]$

   (d) $\{-2\}$

3. (d) ϕ

   (f) $\mathbf{R}$

4. (b) $0T^{-1}$ is a plane through the origin, perpendicular to $(2, 3, 4)$.

   (c) $4T^{-1}$ is a plane through $(0, 0, 1)$, perpendicular to $(2, 3, 4)$.

7. Hint: Strictly increasing functions are one-to-one.

8. $x(f \circ g) = 2x^2 - 1$.

Exercise Set C

1. (e) is the only equivalence relation.

2. (b) $\neq$

   (f) The relation $\{(a, a), (b, b)\}$ on the set $\{a, b, c\}$.

6. (a) $\bar{2} = \{2, -2\}$.

   (b) $\bar{2} = \{2, -4\}$, $\bar{3} = \{3, -5\}$.

   (c) $\bar{b} = \{b, -b\}$.

8. $4\overline{\tfrac{1}{2}} = [4, 5)$.

9. The classes for R are X, V, and W.

Exercise Set D

1. $(a, b) \sim (c, d)$ iff $a^2 + b^2 = c^2 + d^2$.
2. $(a, b) \sim (c, d)$ iff $b - d = 2(a - c)$.
3. $f \sim g$ iff f and g have the same constant term.
4. (a) $2x$
7. $f \sim g$ iff $f(5) = g(5)$.
   $\{p : p \sim 0\}$ is the set of all polynomials with 5 as a root.

Exercise Set E

1. (h), (i), and (j) are all closed.
2. No

Chapter 2

Exercise Set A

5. All but Axiom 4.
7. All but Axiom 7.

Exercise Set B

4. If $b \leq 0$, then Proposition 11 can be used to show that $ab \leq 0$.
7. If $a < 0$, then $(1 - b)a \geq 0$ since $(1 - b) \leq 0$.

Exercise Set C

3. (a) 205
   (d) 14
5. (a) 5
11. Let $T = \{z \in N : m \mid z \text{ and } n \mid z\}$ and show that the smallest element in T is the LCM of m and n.

Exercise Set D

3. $xg^2 = x^4 + 2x^2 + 2$.
5. How many 2's appear in the prime factorization of x^2?

Exercise Set E

4. With respect to $\tilde{6}$, $\bar{2} = \{\ldots, -4, 2, 8, 14, \ldots\}$.
7. (i) $\bar{3}$
   (iv) $\{\bar{2}, \bar{8}\}$
   (vi) No solutions
10. No. For example, in $\mathbf{Z}_6$, $\bar{2} \cdot \bar{3} = \bar{0}$.

Chapter 3

Exercise Set A

1. $x(f \circ g) = (2x + 1)^3 + 3$.
   $xf^{-1} = \frac{1}{2}x - \frac{1}{2}$, $xg^{-1} = \sqrt[3]{x - 3}$,
   $x(f \circ g)^{-1} = x(g^{-1} \circ f^{-1}) = \frac{1}{2}\sqrt[3]{x - 3} - \frac{1}{2}$.

2. $xf = x$ for $x \leq 0$, $xf = x + 1$ for $x > 0$

$xg = x$ for $x \leq 0$, $xg = 0$ for $0 < x \leq 1$, $xg = x - 1$ for $x > 1$

3. $xp = x$ if x is rational and $xp = -x$ if x is irrational. The inverse of p is p.

5. (a) $\begin{pmatrix} 1 & 2 & 3 \\ 3 & 2 & 1 \end{pmatrix}$

(c) $\begin{pmatrix} 1 & 2 & 3 \\ 2 & 3 & 1 \end{pmatrix}$

6. (a) $\begin{pmatrix} 1 & 2 & 3 & 4 \\ 4 & 1 & 2 & 3 \end{pmatrix}$

7. (a) $\begin{pmatrix} 1 & 2 & 3 & 4 \\ 1 & 3 & 4 & 2 \end{pmatrix}$

Exercise Set B

4. (a) 2

(b) 3

6. (a) $\{1, 3\}, \{2, 4\}$

(b) $\{1, 2, 3\}, \{4\}$

Exercise Set C

2. (a) $(1\ 2\ 5\ 7\ 4\ 6)$

(d) $(1\ 8\ 5\ 6)(2\ 3\ 4)$

3. (a) $(1\ 2)(1\ 5)(1\ 7)(1\ 4)(1\ 6)$, odd

(d) $(1\ 8)(1\ 5)(1\ 6)(2\ 3)(2\ 4)$, odd

4. $(1) = (1\ 2)(1\ 2)$.

10. $A_3 = \{I, (1\ 2\ 3), (1\ 3\ 2)\}$.

13. The order of $(1\ 2)(3\ 4\ 5)$ is 6.

Chapter 4

Exercise Set A

4. (c), (h), (i), and (j) are groups.

8. ϕ with the operation ϕ.

Exercise Set B

4.

	I	R	T	U	V	W
I	I	R	T	U	V	W
R	R	T	I	V	W	U
T	T	I	R	W	U	V
U	U	W	V	I	T	R
V	V	U	W	R	I	T
W	W	V	U	T	R	I

7. (a) $(1\ 2)(3\ 4)$
8. (b) $(1\ 3)(2\ 4\ 5)$
11. $(1\ 3\ 5\ 4\ 2)^{-1} = (1\ 2)(1\ 4)(1\ 5)(1\ 3)$.

Exercise Set C

2. (a) $\{\bar{0}, \bar{2}, \bar{4}, \bar{6}, \bar{8}, \overline{10}\}$
  (c) $\{(1\ 2)(3\ 4\ 5),\ (3\ 5\ 4),\ (1\ 2),\ (3\ 4\ 5),\ (1\ 2)(3\ 5\ 4),\ I\}$
7. See Chapter 3, §2.
11. First prove the assertion for positive integers p by means of induction.

Exercise Set D

2. (b), (e), (g), and (h) are subgroups.
3. (a) and (c) are subgroups.
4. (a), (c), and (e) are subgroups.
6. $(x - 5)(x - 7)$ is in $F \cap S$.
7. $\cos \pi x$ is in $\cap Aq$.
9. $\cap \mathcal{W}$ is the set of all integral multiples of 6.

Exercise Set E

1. (a) $\{I,\ (1\ 2)\}$
  (c) S_3
2. $\langle \bar{3} \rangle = \{\bar{0}, \bar{3}, \bar{6}, \bar{9}, \overline{12}, \overline{15}\}$
9. See Exercise 6, Set E, Chapter 1.

Exercise Set F

1. $S_3(1\ 2) = S_3$.
  $S_3(1\ 4) = \{(1\ 4),\ (1\ 2\ 4),\ (1\ 3\ 4),\ (2\ 3)(1\ 4),\ (1\ 2\ 3\ 4),\ (1\ 3\ 2\ 4)\}$.
  etc.
5. (a) No
  (b) Yes
  (c) Yes
6. (a) Yes
  (b) No
  (c) Yes
  (d) Yes
  (e) No
  (f) Yes
7. $3x^2 - 4 \in W$.

Exercise Set G

1. (a)
7. $\langle \bar{3} \rangle = \{\bar{0}, \bar{3}, \bar{6}, \bar{9}\}$.
  $\langle \bar{3} \rangle + \bar{1} = \{\bar{1}, \bar{4}, \bar{7}, \overline{10}\}$.
  $\langle \bar{3} \rangle + \bar{2} = \{\bar{2}, \bar{5}, \bar{8}, \overline{11}\}$.

	$\langle \bar{3} \rangle$	$\langle \bar{3} \rangle + \bar{1}$	$\langle \bar{3} \rangle + \bar{2}$
$\langle \bar{3} \rangle$	$\langle \bar{3} \rangle$	$\langle \bar{3} \rangle + \bar{1}$	$\langle \bar{3} \rangle + \bar{2}$
$\langle \bar{3} \rangle + \bar{1}$	$\langle \bar{3} \rangle + \bar{1}$	$\langle \bar{3} \rangle + \bar{2}$	$\langle \bar{3} \rangle$
$\langle \bar{3} \rangle + \bar{2}$	$\langle \bar{3} \rangle + \bar{2}$	$\langle \bar{3} \rangle$	$\langle \bar{3} \rangle + \bar{1}$

Exercise Set H

1. $\mathbf{Z}_2 \oplus \mathbf{Z}_3 = \langle (\bar{1}, \bar{1}) \rangle$.
3. $6(\bar{a}, \bar{b}) = (\bar{0}, \bar{0})$ for all $(\bar{a}, \bar{b})$ in $\mathbf{Z}_2 \oplus \mathbf{Z}_6$ and, therefore, $\mathbf{Z}_2 \oplus \mathbf{Z}_6$ contains no element of order 12.
6. $\langle (\bar{2}, \bar{3}) \rangle = \{(\bar{2}, \bar{3}), (\bar{0}, \bar{6}), (\bar{2}, \bar{0}), (\bar{0}, \bar{3}), (\bar{2}, \bar{6}), (\bar{0}, \bar{0})\}$. The table is a 6×6 table.

Chapter 5

Exercise Set A

1. (c), (d), (e), (f), (g), (i), (j)
2. No
3. No
4. $\bar{2}$ has no multiplicative inverse.
14. The crucial point is to show that $1/(a + b\sqrt[3]{2} + c\sqrt[3]{4}) = p + q\sqrt[3]{2} + t\sqrt[3]{4}$ for some rational numbers p, q, and t if one of a, b, or c is different from 0.
15. Let $xg = 0$ for $x \leq 0$ and $xg = x^2$ for $x > 0$.
    Let $xf = x^2$ for $x \leq 0$ and $xf = 0$ for $x > 0$.
    Both f and g are differentiable, but $f \cdot g = 0$.
18. $b + b = (b + b)(b + b) = b(b + b) + b(b + b) = bb + bb + bb + bb = b + b + b + b$.

Exercise Set B

3. The set of all units in $\mathbf{Z}_6$ is $\{\bar{1}, \bar{5}\}$.
   The set of all units in $\mathbf{Z}_8$ is $\{\bar{1}, \bar{3}, \bar{5}, \bar{7}\}$.
7. (a) m and n are relatively prime iff $mx + ny = 1$ for some integers x and y
      iff $ny \equiv 1 \pmod{m}$ for some integer y,
      iff $\bar{n}\bar{y} = \bar{1}$ for some integer y.

Exercise Set C

1. (a), (b), (d), and (f) are subrings.
2. The subrings of $\mathbf{Z}_9$ are $\{\bar{0}\}$, $\mathbf{Z}_9$, and $\{\bar{0}, \bar{3}, \bar{6}\}$.
3. $\langle \bar{8} \rangle = \{\bar{0}, \bar{8}, \bar{4}\}$.
6. $\{a\sqrt{3}{:}a \in \mathbf{Z}\}$.
7. $\{a + b\sqrt{3}{:}b \in \mathbf{Z}, a \in \mathbf{Z}, \text{ and } 3 \mid a\}$.
11. $\mathbf{R}(x^2)$
14. $\{a + b\sqrt{5} + c\sqrt{7} + d\sqrt{35}{:}a, b, c, \text{ and } d \text{ are rational}\}$.
15. No

16. (b) and (d) are ideals.
17. All of them.

Exercise Set D

1. $\{\bar{0}, \bar{2}, \bar{4}, \bar{6}, \bar{8}, \overline{10}\}$, $\{\bar{0}, \bar{3}, \bar{6}, \bar{9}\}$, $\{\bar{0}, \bar{4}, \bar{8}\}$, $\{\bar{0}, \bar{6}\}$, $\{\bar{0}\}$, $\mathbf{Z}_{12}$.
3. $\frac{1}{2} \cdot 2$ is not even.
4. The constant function 1 is differentiable. Since $|x|$ is continuous and not differentiable, $1 \cdot |x|$ is not differentiable.

Exercise Set E

4. The crucial point in the proof is noticing that $(r_1, 0) \cdot (r_2, s_2) = (r_1 r_2, 0)$.
12. The "1" of $R \oplus S$ will be $(1, 1)$.
14. If $a \neq 0$ and $a \in R$, then aR is an ideal in R and $aR \neq \{0\}$. Hence $aR = R$ and, since $1 \in R$, $ab = 1$ for some b in R.
16. $0\phi^{-1} = A$.
19. $(c - b)a^{-1}$ is a solution.

Chapter 6

Exercise Set A

2. Let $\bar{3}\phi = \bar{1}$ and define the other images as needed, e.g., $\bar{2}\phi = (\bar{3} \cdot \bar{3})\phi = \bar{3}\phi + \bar{3}\phi = \bar{1} + \bar{1} = \bar{2}$.
7. Adapt the techniques from Exercise 6, that is, $\bar{1}\phi = (\bar{1}, \bar{1})$, and extend as needed.
15. See Proposition 8 of this chapter.
17. See Proposition 10 of this chapter.
22. If $a \in S$, then $b\phi = a$ for some b in R, because ϕ is onto. Then $a(1\phi) = (b\phi)(1\phi) = (b \cdot 1)\phi = b\phi = a$ and, by similar reasoning, $(1\phi)a = a$.

Exercise Set B

1. If $|G| = 5$, then every nontrivial element of G has order 5. Thus, for some element b in G, $G = \langle b \rangle = \{e, b, b^2, b^3, b^4\}$. One needs to verify that ϕ, defined by $e\phi = \bar{0}$, $b\phi = \bar{1}$, $b^2\phi = \bar{2}$, $b^3\phi = \bar{3}$, and $b^4\phi = \bar{4}$ is an isomorphism.

4.

	e	x	x^2	x^3	y	xy	x^2y	x^3y
e	e	x	x^2	x^3	y	xy	x^2y	x^3y
x	x	x^2	x^3	e	xy	x^2y	x^3y	y
x^2	x^2	x^3	e	x	x^2y	x^3y	y	xy
x^3	x^3	e	x	x^2	x^3y	y	xy	x^2y
y	y	x^3y	x^2y	xy	x^2	x	e	x^3
xy	xy	y	x^3y	x^2y	x^3	x^2	x	e
x^2y	x^2y	xy	y	x^3y	e	x^3	x^2	x
x^3y	x^3y	x^2y	xy	y	x	e	x^3	x^2

Exercise Set C

2. Yes to all these questions; $(xy)g = (xy)^3 = x^3y^3 = (xg)(yg)$.
6. $\bar{0}\phi^{-1} = A_5$.
7. $\{(x, y):(x, y)T = 0\}$ is a line through the origin and $(3, -2)$.
8. $0\gamma^{-1}$ is the set of all functions with zero derivative at 5; it is a subgroup.
14. See Exercise 22, Set A.
17. $e \in K\phi^{-1}$, because $e\phi = e$ and $e \in K$. If $x, y \in K\phi^{-1}$, then $x\phi \in K$ and $y\phi \in K$. Hence $(xy^{-1})\phi = (x\phi)(y\phi)^{-1} \in K$ and $xy^{-1} \in K\phi^{-1}$.

Exercise Set D

1. $P_R = (I\,R\,S\,T)(A\,M\,B\,N)$.
   $P_A = (I\,A)(R\,N)(S\,B)(T\,M)$.
4. We know from exercises in Chapter 3 that a permutation of a finite set can be expressed as a product of disjoint cycles in which the support of each cycle is an orbit of the permutation. If $y \in G$ and $a \neq e$, $ya \neq y$ and hence the support of p_a is all of G. If $x \in G$, then the orbit containing x is $\{x, xp_a, xp_a^2, \ldots, xp_a^{n-1}\}$, which is just $\{x, xa, xa^2, \ldots, xa^{n-1}\}$. Since the order of p_a is n, each orbit has n elements and the cycles are of the form $(x\,xa\,xa^2\,\cdots\,xa^{n-1})$.

Exercise Set E

1. $\ker D^3$ is the set of all polynomials of degree 3 or less.
5. Argue that $|\ker \phi| = 17$ or $|\ker \phi| = 1$.
6. What are the possibilities for $\ker \lambda$?
9. (a) Determinant
10. $\ker \phi$ is the line $x = y$; $3\phi^{-1}$ is the line with equation $y = x - 3$.
12. $\ker \phi$ is A_4.
14. (a) $x^7/210 + Q$, where Q is all polynomials of degree 2 or less
16. $(p\phi)\phi^{-1} = K + p$

Exercise Set F

1. Define ϕ by $(\bar{x})\phi = \overline{2x}$.
3. Define λ by $(f)\lambda = f(5)$.
5. $\mathbf{Z}_7$ and a group with just one element.
6. S_3, a group with one element, and $\mathbf{Z}_2$ because the only normal subgroups of S_3 are $\{I\}$, S_3, and A_3.

Exercise Set G

1. Multiplying each element of $\mathbf{Z}_4$ by $\bar{3}$ preserves addition, maps $\mathbf{Z}_4$ onto $\mathbf{Z}_4$, maps $\bar{1}$ to $\bar{3}$, and $\bar{3}$ to $\bar{1}$. Thus the function ϕ defined by $\bar{x}\phi = \overline{3x}$ is an automorphism. An automorphism of $\mathbf{Z}_4$ must either map $\bar{1}$ to $\bar{1}$ or $\bar{1}$ to $\bar{3}$, because 1 must be mapped to an element of order 4. If $\bar{1}$ is mapped to $\bar{1}$, then every element is mapped to itself. Thus the only automorphisms of $\mathbf{Z}_4$ are the identity and the map ϕ above and therefore $A(\mathbf{Z}_4) \cong \mathbf{Z}_2$. Notice that $A(\mathbf{Z}_4) \cong \{\bar{1}, \bar{3}\}$ under multiplication modulo 4.
2. $A(\mathbf{Z}_6) \cong \{\bar{1}, \bar{5}\}$ under multiplication modulo 6.

Exercise Set H

3. If D is a field, then $a/b = ab^{-1}/1$ and the embedding maps D onto $Q(D)$.
6. Show that a/a is a multiplicative identity for any $a \neq 0$. Fix $a \neq 0$ and define ϕ from D into $Q(D)$ by $x\phi = xa/a$.

Exercise Set I

1. The characteristic of $\mathbf{Z}_2 \oplus \mathbf{Z}_3$ is 6, of $\mathbf{Z}_2 \oplus \mathbf{Z}_4$ is 4, of $\mathbf{Z}_3 \oplus \mathbf{Z}_5$ is 15, and of $\mathbf{Z}_6 \oplus \mathbf{Z}_9$ is 36.
4. They have the same characteristic.
7. 0

Exercise Set J

1. $E(\mathbf{Z}_4) \cong \mathbf{Z}_4$.
3. Once the image of 1 is determined, the endomorphism is determined. Define T from $E(\mathbf{Z})$ to $\mathbf{Z}$ by $\phi T = (1)\phi$. The crucial point is that $1\phi = 1 \cdot (1\phi)$.
7. (c) the functions in which $ad - bc = \pm 1$.

Chapter 7

Exercise Set A

1. (a), (b), (c), (f), and (g) are vector spaces.
5. Use Proposition 1.
7. Use Exercise 6 and Proposition 9.

Exercise Set B

1. (a), (b), (d), (e), (g), and (j) are subspaces.
2. (b), (d), (e), and (f) are subspaces.
5. (a) The line through $(0, 0, 0)$ and $(1, 2, 3)$
   (b) The XY plane
   (c) The XY plane
6. From Chapter 3, we know that $S + T$ is a subgroup. Show that it is closed under scalar multiplication.

Exercise Set C

2. W is the line through $(0, 0, 0)$ and $(1, 1, 1)$. $(\mathbf{R} \oplus \mathbf{R} \oplus \mathbf{R})/W$ is the set of all lines parallel to W.
10. This is analogous to Exercise 12, Set G, Chapter 4 and to Exercise 15, Set E, Chapter 5.

Exercise Set D

2. (a), (b), (c), and (f) are linear transformations.

Exercise Set E

1. (b), (c), (d), (f), and (g) are linearly independent.
2. (a) $\{(1, 1, -1), (2, 4, 3), (4, 6, 5)\}$

3. (a) $(4, 1, 3) = (-2)(1, 1, 0) + 3(1, 1, 1) + 3(1, 0, 0)$
5. (a) $\{(2, 3, 1), (1, 4, 7), (3, 7, -1)\}$
6. The central idea is that if dim $V = n$, then $V \cong F^n$ and therefore $|V| = |F^n| = |F|^n$.
9. If $T \in \mathcal{L}(V)$ and T is invertible, then one can show that T is a unit. If T is not invertible, then ker $T \neq \{0\}$. If $\{y_1, \ldots, y_m\}$ is a basis for ker T, let $\{y_1, \ldots, y_m, y_{m+1}, \ldots, y_n\}$ be a basis for V, and define the linear transformation P by $(\Sigma_{i=1}^n a_i y_i)P = \Sigma_{i=1}^m a_i y_i$. Then $PT = 0$.
10. Show that ker $TQ = $ ker T.

Exercise Set F

1. If M is an R-module and K is a subset of M, K is a submodule of M if K is an R-module with the same addition and scalar multiplication as defined for M. (See the definition of subspace.) Definitions of linear transformation and isomorphism are given analogously.
2. Let ϕ be the embedding into $E(V)$ such that $1\phi = I$. For each vector v and scalar α, define $v\alpha$ to be $(v)(\alpha\phi)$. The verification of vector space axioms is straightforward using the fact that ϕ preserves operations and maps F into $E(V)$. For example:

$$v(\alpha + \beta) = (v)[(\alpha + \beta)\phi] \quad \text{by definition of scalar multiplication}$$
$$= (v)[\alpha\phi + \beta\phi] \quad \text{because } \phi \text{ preserves addition}$$
$$= (v)(\alpha\phi) + (v)(\beta\phi) \quad \text{by the definition of addition of functions}$$
$$= v\alpha + v\beta \quad \text{by definition of scalar multiplication}$$

3. Your proofs in Exercise 2 should translate for this problem. You do not need to use inverses or commutativity of multiplication in your arguments.

Chapter 8

Exercise Set A

3. ker E_5 is the set of all polynomials in $\mathbf{R}[x]$ that have 5 for a root. This set is just the set of all polynomials of the form $(x - 5)g(x)$.
4. The set of all polynomials of the form $(x^2 - 2)g(x)$, where $g(x) \in \mathbf{Q}[x]$.

Exercise Set B

6. If p is irreducible and $p \mid fg$ and $p \nmid f$, then, by Proposition 18, p and f are relatively prime. Thus for some polynomials A and B, $pA + fB = 1$. Then $(pA + fB)g = 1 \cdot g$ and $pAg + fgB = g$. Since $p \mid pAg$ and $p \mid fgB$, $p \mid (pAg + fgB)$ and therefore $p \mid g$.

Exercise Set C

1. Let f and g be polynomials in $F[x]$. By Theorem 21, the ideal generated by f and g is generated by one element, that is, $\langle f, g \rangle = \langle d \rangle$ for some d in $F[x]$ and we can choose d to be monic. Since $\langle d \rangle = dF[x]$ and $f \in \langle d \rangle$ and $g \in \langle d \rangle$, $d \mid f$ and $d \mid g$. If $c \mid f$ and $c \mid g$, then $f \in \langle c \rangle$ and $g \in \langle c \rangle$. Therefore $\langle f, g \rangle \subseteq \langle c \rangle$. Since $\langle f, g \rangle = \langle d \rangle$, $\langle d \rangle \subseteq \langle c \rangle$ and therefore $d \in cF[x]$. Thus $d = cq$ for some polynomial q and $c \mid d$.

Exercise Set D

2. Let dim $K = n$ and let $\beta \in K$. If for distinct nonnegative integers p and q, $\beta^p = \beta^q$, then β is a root of $X^p - X^q$ and β is algebraic. Otherwise $1, \beta, \beta^2, \ldots, \beta^n$ are all distinct and therefore $\{1, \beta, \ldots, \beta^n\}$ is dependent in K. Thus there are elements $a_0, a_1, \ldots, a_n$ in F, not all zero such that $a_0 1 + a_1\beta + a_2\beta^2 + \cdots + a_b\beta^n = 0$. Thus β is a root of $a_0 + a_1 X + a_2 X^2 + \cdots + a_n X^n$ and is algebraic.

Chapter 9

Exercise Set A

1. $\begin{bmatrix} 8 & 35 & 12 \\ -2 & -5 & 1 \end{bmatrix}$

3. $\begin{bmatrix} \alpha & 0 & \cdots & 0 \\ 0 & \alpha & \cdots & 0 \\ \vdots & & & \vdots \\ 0 & \cdots & 0 & \alpha \end{bmatrix}$

4. (b) $\begin{bmatrix} 0 & 1 & 0 & 0 \\ 0 & 0 & 1 & 0 \\ 0 & 0 & 0 & 1 \\ 1 & 0 & 0 & 0 \end{bmatrix}$

6. $E_{23}^1 = \begin{bmatrix} 1 & 0 & 0 \\ 0 & 1 & 1 \\ 0 & 0 & 1 \end{bmatrix}$

12. (a) $\begin{bmatrix} 2 & -3 \\ -1 & 2 \end{bmatrix}$

Exercise Set B

1. (a) $\begin{bmatrix} -2 & 1 \\ \frac{3}{2} & -\frac{1}{2} \end{bmatrix}$ Check the others by multiplying.

2. $(-2, 6, -9)$

9. $(E_{pq}^b)^t = E_{qp}^b$.

Exercise Set C

1. (a) 225
   (f) -14
6. (a) $x = 6, -1$.
   (b) $x = 1, 2,$ or 4.

13. (b) $\begin{bmatrix} 1 & 0 & 0 \\ 0 & 1 & 0 \\ -\frac{5}{3} & \frac{7}{3} & 0 \end{bmatrix}$

14. The kernel of the matrix (b) is $\langle(-\frac{5}{3}, \frac{7}{3}, -1)\rangle$.

Exercise Set D

2. See Exercise 13.
8. Both matrices have characteristic values 7 and -2 and, therefore, are similar by Proposition 27.
10. The characteristic polynomial is $(1 - x)^2 + 1$, which has no real roots.
11. Hint: Show that the matrix has three distinct characteristic values.
14. See Exercise 13.

Exercise Set E

1. (a) $\langle(2, 1)\rangle$
   (d) $\{(0, 0, 0, 0)\}$
2. (c) $(\frac{11}{5}, -\frac{3}{5}, \frac{33}{5})$
3. (b) $\langle(-2, 1, -1)\rangle + (5, -1, 0)$

LIST OF SPECIAL SYMBOLS

$\emptyset$ the empty set, 3

$y \in A$, 1

$A \subseteq B$, 1

$A \cup B$ and $A \cap B$, 3

$\cup \mathcal{a}$ and $\cap \mathcal{a}$, 3

(m, n) ordered pair, 4

$M \times N$ cartesian product, 4

$f|_M$ function restricted to M, 9

$R|_M$ relation restricted to M, 15

$f \circ g$ f composed with g, 12

$\bar{a}$ equivalence class for a, 16

$\mathbf{R}$ the real numbers, 10

$\mathbf{R}[x]$ the polynomials in x over $\mathbf{R}$, 30, 309

$\mathbf{Z}$ the integers, 26

$\mathbf{Z}[x]$ the polynomials in x over $\mathbf{Z}$, 309

$\mathbf{Q}$ the rational numbers, 85

$\mathbf{Q}[x]$ the polynomials in x over $\mathbf{Q}$, 309

$R[X]$ the polynomials in X over the ring R, 220

$\mathbf{N}$ the natural numbers, 26

$x \mid y$ x divides y, 32

$p(x) \mid q(x)$ $p(x)$ divides $q(x)$, 223

GCD greatest common divisor, 35, 225

LCM least common multiple, 39, 229

$\equiv$ congruent modulo an integer, 42; modulo a subgroup, 92; modulo a subring, 122; modulo a subspace, 192

$\mathbf{Z}_m$ the integers modulo m, 44

iff   if and only if

Π product notation, 63

$\sum$ summation notation, 216, 307

f^{-1} the inverse of the function f, 8, 49

$P(S)$ the set of permutations of the set S, 52

S_n the permutations of $\mathbf{Z}$ whose support is contained in $\{1, \ldots, n\}$, 61

A_n the set of even permutations in S_n, 62

Δ, 63

$|B|$ order of the group B, 90

$|A|$ the determinant of the matrix A, 262

$[\alpha_{ij}]$ the matrix with α_{ij} in the ith row and jth column, 239

$|\alpha_{ij}|$ the determinant of the matrix $[\alpha_{ij}]$, 262

$[G:H]$ the index of H in G, 96

G/H G modulo H, 102

$A \oplus B$ the direct sum or product of A and B, 107

$\prod_{i \in I} A_i$, 109

$\bigoplus_{i \in I} A_i$, 109

$\langle S \rangle$ subgroup, subring, or subspace generated by the set S, 89, 119, 189

$\langle a \rangle$ subgroup, subring, or subspace generated by the element a, 89, 119, 189

$N \triangleleft G$ N is normal in G, 99

$C(a)$ centralizer of a, 92

$N(H)$ normalizer of H, 97

$Z(G)$ center of G, 170

$A(G)$ automorphisms of G, 169

$I(G)$ inner automorphisms of G, 170

η natural function from G to G/H, 148

Ha right coset of H in G determined by a, 93

aH left coset of H in G determined by a, 93

$\cong$ is isomorphic to, 136

$\bar{\phi}$ the isomorphism associated with the homomorphism ϕ, 166

ϕ_a the inner automorphism determined by a, 170

$\mathfrak{L}(V, W)$ the set of linear transformations from V to W, 197

$\mathfrak{L}(V)$ the set of linear transformations from V to V, 197

F^n $F \oplus \cdots \oplus F$ with n summands, 210

$E(M)$ the set of endomorphisms of M, 180

A^t the transpose of the matrix A, 261

δ_{ij} the Kronecker delta, 249

$(b)_p$, 250

E_{pq}^1 and E_{pq}^b, 250

P_p, 255

E_b the evaluation map, 222

δp the degree of p, 220, 308

INDEX

72 73 74 7 6 5 4 3 2 1